Ecology and Management of Giant Hogweed
(*Heracleum mantegazzianum*)

ECOLOGY AND MANAGEMENT OF GIANT HOGWEED

(*Heracleum mantegazzianum*)

Edited by

P. Pyšek

Academy of Sciences of the Czech Republic
Institute of Botany, Průhonice, Czech Republic

M.J.W. Cock

CABI Switzerland Centre
Delémont, Switzerland

W. Nentwig

Community Ecology, University of Bern
Bern, Switzerland

H.P. Ravn

Forest and Landscape, The Royal Veterinary and Agricultural
University, Hørsholm, Denmark

www.cabi.org

CABI is a trading name of CAB International

CAB International Head Office
Nosworthy Way
Wallingford
Oxfordshire OX10 8DE
UK

Tel: +44 (0)1491 832111
Fax: +44 (0)1491 833508
E-mail: cabi@cabi.org
Website: www.cabi.org

CABI North American Office
875 Massachusetts Avenue
7th Floor
Cambridge, MA 02139
USA

Tel: +1 617 395 4056
Fax: +1 617 354 6875
E-mail: cabi-nao@cabi.org

A catalogue record for this book is available from the British Library, London, UK.

A catalogue record for this book is available from the Library of Congress, Washington, DC.

ISBN-13: 978 1 84593 206 0

Typeset by MRM Graphics Ltd, Winslow, Bucks.
Printed and bound in the UK by Athenaeum Press, Gateshead.

Contents

Contributors

Rita Merete Buttenschøn
Danish Centre for Forest, Landscape and Planning, The Royal Veterinary and Agricultural University, Kvak Møllevej 31, DK-7100 Vejle, Denmark; e-mail: rmb@kvl.dk, fax: +45-35281512

Matthew J.W. Cock
CABI Switzerland Centre, Rue des Grillons 1, CH-2800 Delémont, Switzerland; e-mail: m.cock@cabi.org, fax: +41-32-4214871

R. Lutz Eckstein
Institute of Landscape Ecology and Resource Management, Division of Landscape Ecology and Landscape Planning, University of Giessen, Heinrich-Buff-Ring 26-32, D-35392 Giessen, Germany; e-mail: Lutz.Eckstein@agrar.uni-giessen.de, fax: +49-641-9937169

Harry C. Evans
CABI UK Centre (Ascot), Silwood Park, Buckhurst Road, Ascot, Berkshire SL5 7TA, UK; e-mail: h.evans@cabi.org, fax: +44-1491-829123

Lars Fröberg
Botanical Museum, Lund University, Östra Vallgatan 18, SE-223 61 Lund, Sweden; e-mail: lars.froberg@botmus.lu.se

Dmitry Geltman
Komarov Botanical Institute of the Russian Academy of Science, Prof. Popov Street, 2, St Petersburg 197376, Russia; e-mail: geltman@binran.ru

Zigmantas Gudžinskas
Institute of Botany, Laboratory of Flora and Geobotany, Žaliųjų Ežerų Str. 49, LT-08406 Vilnius, Lithuania; e-mail: zigmantas.g@botanika.lt

Steen O. Hansen
Community Ecology, Zoological Institute, University of Bern, Baltzerstrasse 6, CH-3012 Bern, Switzerland; fax: +41-31-631-4888

Jan Hattendorf
Community Ecology, Zoological Institute, University of Bern, Baltzerstrasse 6, CH-3012 Bern, Switzerland; fax: +41-31-631-4888

Jörg Hüls
Department of Landscape Ecology and Resource Management, Interdisciplinary Research Centre, Justus-Liebig-University Giessen, Heinrich-Buff-Ring 26-32, DE-35392 Giessen, Germany; e-mail: Joerg.Huels@agrar.uni-giessen.de

Šárka Jahodová
Department of Ecology, Faculty of Science, Charles University Prague, Viničná 7, CZ-128 01 Praha 2, Czech Republic; e-mail: jahodova@natur.cuni.cz
(Institute of Botany, Academy of Sciences of the Czech Republic, CZ-252 43 Průhonice, Czech Republic)

Vojtěch Jarošík
Department of Ecology, Faculty of Science, Charles University, Viničná 7, CZ-128 01 Praha 2, Czech Republic; e-mail: jarosik@cesnet.cz
(Institute of Botany, Academy of Sciences of the Czech Republic, Z-252 43 Průhonice, Czech Republic)

Angela Karp
Plant and Invertebrate Ecology Division, Rothamsted Research, Harpenden, AL5 2JQ, UK; e-mail: angela.karp@bbsrc.ac.uk

Lukáš Krinke
Museum of Local History, Zádušní 1841, CZ-27280, Kladno, Czech Republic

Lenka Moravcová
Institute of Botany, Academy of Sciences of the Czech Republic, CZ-252 43 Průhonice, Czech Republic; e-mail: moravcova@ibot.cas.cz, fax: +420-267750031

Jana Müllerová
Institute of Botany, Academy of Sciences of the Czech Republic, CZ-252 43 Průhonice, Czech Republic; e-mail: mullerova@ibot.cas.cz, fax: +420-267750031

Nana Nehrbass
UFZ Centre for Environmental Research Leipzig-Halle, Department of Ecological Modelling, Leipzig, Germany; e-mail: nana.nehrbass@ufz.de, fax: +49-341-235-3500

Wolfgang Nentwig
Community Ecology, Zoological Institute, University of Bern, Baltzerstrasse 6, CH-3012 Bern, Switzerland; e-mail: wolfgang.nentwig@zos.unibe.ch, fax: +41-31-631-4888

Charlotte Nielsen
Danish Centre for Forest, Landscape and Planning, The Royal Veterinary and Agricultural University, Hørsholm Kongevej 11, DK-2970 Hørsholm, Denmark; e-mail: chn@kvl.dk, fax: +45-3528-1517

Annette Otte

Institute of Landscape Ecology and Resource Management, Division of Landscape Ecology and Landscape Planning, University of Giessen, Heinrich-Buff-Ring 26-32, D-35392 Giessen, Germany; e-mail: Annette.Otte@agrar.uni-giessen.de, fax: +49-641-9937169

Jan Pergl

Institute of Botany, Academy of Sciences of the Czech Republic, CZ-252 43 Průhonice, Czech Republic; e-mail: pergl@ibot.cas.cz, fax: +420-267750031

Irena Perglová

Institute of Botany, Academy of Sciences of the Czech Republic, CZ-252 43 Průhonice, Czech Republic; e-mail: perglova@ibot.cas.cz, fax: +420-267750031

Ilze Priekule

Latvian Plant Protection Research Centre, Lielvārdes iela 36/38, Riga LV-1006, Latvia; e-mail: ilze.priekule@laapc.lv, fax: +371-7551-226

Petr Pyšek

Institute of Botany, Academy of Sciences of the Czech Republic, CZ-252 43 Průhonice, Czech Republic; e-mail: pysek@ibot.cas.cz, fax: +420-267750031 (Department of Ecology, Faculty of Science, Charles University, Viničná 7, CZ-128 01 Praha 2, Czech Republic)

Hans Peter Ravn

Danish Centre for Forest, Landscape and Planning, The Royal Veterinary and Agricultural University, Hørsholm Kongevej 11, DK-2970 Hørsholm, Denmark; e-mail: hpr@kvl.dk, fax: +45-3528-1517

Marion K. Seier

CABI UK Centre (Ascot), Silwood Park, Buckhurst Road, Ascot, Berkshire SL5 7TA, UK; e-mail: m.seier@cabi.org, fax: +44-1491-829123

Jan Thiele

Institute of Landscape Ecology and Resource Management, Division of Landscape Ecology and Landscape Planning, University of Giessen, Heinrich-Buff-Ring 26-32, D-35392 Giessen, Germany; e-mail: Jan.Thiele@agrar.uni-giessen.de, fax: +49-641-9937169

Ken Thompson

Department of Animal and Plant Sciences, University of Sheffield, Sheffield S10 2TN, UK; e-mail: ken.thompson@sheffield.ac.uk

Olga Treikale

Latvian Plant Protection Research Centre, Lielvārdes iela 36/38, Riga LV-1006, Latvia; e-mail: olga.treikale@laapc.lv, fax: +371-7551-226

Sviatlana Trybush

Plant and Invertebrate Ecology Division, Rothamsted Research, Harpenden, AL5 2JQ, UK; e-mail: sviatlana.trybush@bbsrc.ac.uk

Ineta Vanaga

Latvian Plant Protection Research Centre, Lielvārdes iela 36/38, Riga LV-1006, Latvia; e-mail: ineta.vanaga@laapc.lv, fax: +371-7551-226

Eckart Winkler

UFZ Centre for Environmental Research Leipzig-Halle, Department of Ecological Modelling, Leipzig, Germany; e-mail: eckart.winkler@ufz.de, fax: +49-341-235-3500

Ruediger Wittenberg

CABI Bioscience Switzerland Centre, Rue des Grillons 1, CH-2800 Delémont, Switzerland, e-mail: r.wittenberg@cabi.org

Acknowledgement

RETURN OF THE GIANT HOGWEED by Mike Rutherford, Pete Gabriel, Steve Hackett and Tony Banks (pka Genesis) © 1971 Stratsong Ltd – All Rights Reserved – Lyric reproduced by kind permission of Carlin Music Corp – London NW1 8BD

Preface
All You Ever Wanted to Know About Hogweed, but Were Afraid to Ask!

Turn and run! Nothing can stop them!
(Genesis, 1971)

Significant problems with invasive alien species are a relatively recent phenomenon. In most parts of the world, the biggest impacts have been felt only in the last 50 years. There are some prominent examples of widespread invasions, presumably with impacts (although in most cases we have little or no information on this) that date back much further. For instance, Charles Darwin wrote of invasive populations of the alien milk thistle, *Silybum marianum*, and cardoon, *Cynara cardunculus*, covering square kilometres in Argentina in 1833. Countless species were added to lists of invasive aliens in the 20th century and for every region for which accurate data are available, the numbers of species at every stage of the introduction–invasion continuum are increasing. More worrisome invaders, more biodiversity affected and ever-increasing complexities for those tasked with the management of natural and semi-natural ecosystems.

Means for dealing with invasive alien species, beyond uncoordinated efforts at local scales, have had to be developed rapidly in many parts of the world to avert looming species extinctions and various ecosystem-level impacts. Such methods are needed at spatial scales ranging from landscapes to continents, and at temporal scales of days or weeks (for small-scale containment or eradication efforts) to years, decades or more (to prevent further introductions across national borders and to limit dissemination within political boundaries). The rush to develop cost-effective management strategies arose at about the same time as ecologists and biogeographers were starting to appreciate both the momentous potential of non-native species to disrupt ecosystems, and also the opportunities that invasions provided for studying fundamental issues in biogeography and ecology. Invasion ecology has arrived as a reasonably distinct, and decidedly 'hot', sub-discipline of ecology. Considerable advances have been made. We are (a little) closer to winning (some) 'science battles' (i.e. figuring out how many different factors, individually and in concert, determine which species become invasive and which systems are invaded). Unfortunately, dispatches from the many fronts suggest that the 'conservation war' is being lost (Hulme, 2003).

It is encouraging to see that many focused initiatives are underway across the globe to tackle the myriad of problems that flow from biological invasions. Such a project is the one which led to this book: 'Giant Hogweed (*Heracleum mantegazzianum*) a pernicious invasive weed: developing a sustainable strategy for alien invasive plant management in Europe (GIANT ALIEN)'. This 40-month project, conducted under the 5th Framework Programme of the European Union in 2002–2005, involved seven countries and nine institutions. Almost 40 scientists were brought together, in an integrated way, with expertise from various relevant disciplines: taxonomy, plant ecology, molecular biology, plant sociology and zoology. This programme sought to collate all available information and to collect additional data on all aspects of the giant hogweed, *Heracleum mantegazzianum*, with the aim of providing the ultimate compendium of current knowledge on this amazing invasive plant to facilitate improved management. This is a very worthwhile undertaking for many reasons. First, it is clear that management interventions against giant hogweed are often ineffective. Successes and failures in different parts of its adventive range needed to be evaluated with the aim of formulating robust guidelines for sustainable management. Second, superb data are available on the biogeography of the species. Considerable work has already been published on the invasion ecology of *H. mantegazzianum* in parts of its new range, but additional data and the multi-disciplinary nature of the team assembled for this project afford opportunities for new and expanded analyses of range expansion and population dynamics at different spatial and temporal scales. In this context, *H. mantegazzianum* provides a stunning model species for testing numerous emerging concepts in the field of invasion ecology. For instance, genetic evidence clearly shows the fundamental role of rapid and widespread dissemination of the species by humans in structuring its current range. Available evidence for hogweed in different parts of its adventive range also gives us clear insights on lag phases (the time between arrival in a region and the onset of exponential population growth). Third, superb data are available on almost every life-history trait of *H. mantegazzianum*. This paves the way for very important analyses of the role of different traits in the population dynamics of a major invasive species. These studies, which I predict will stand as model cases in plant invasion ecology, are very neatly utilized to explore options for management.

Several studies in the book address the potential of using biological control in the management of *H. mantegazzianum*. Studies undertaken as part of this project have greatly increased our knowledge of the insect fauna and other biota associated with giant hogweed. Unfortunately, prospects for bio-control using insects or pathogens are not particularly promising. Despite the additional work spurred by the GIANT ALIEN project, more work is urgently needed to achieve a more thorough inventory of biota associated with *H. mantegazzianum*. For pathogens, molecular studies are the key to progress in this field. Despite the improved guidelines for management using mechanical and chemical control that have emerged from the work reported on in this volume,

I suggest that until successful biological control agents are found, means for the sustainable management of hogweed will remain elusive.

I greatly enjoyed reading this book. The editors and authors deserve much praise. Although much work remains to be done on finding ways to manage hogweed effectively, this volume has brought together a dazzling amount of information of crucial value.

David M. Richardson
Centre for Invasion Biology
Department of Botany & Zoology
Stellenbosch University
South Africa

Reference

Hulme, P. (2003) Biological invasions: winning the science battles but losing the conservation war? *Oryx* 37, 178–193.

1

Taxonomy, Identification, Genetic Relationships and Distribution of Large *Heracleum* Species in Europe

Šárka Jahodová,[1,3] Lars Fröberg,[2] Petr Pyšek,[1,3] Dmitry Geltman,[4] Sviatlana Trybush[5] and Angela Karp[5]

[1]Charles University Prague, Praha, Czech Republic; [2]Lund University, Lund, Sweden; [3]Institute of Botany of the Academy of Sciences of the Czech Republic, Průhonice, Czech Republic; [4]Komarov Botanical Institute of the Russian Academy of Science, St Petersburg, Russia; [5]Rothamsted Research, Harpenden, UK

> He captured it and brought it home ... He came home to London and
> made a present of the Hogweed to the Royal Gardens at Kew
>
> (Genesis, 1971)

Introduction

The 65 species in the genus *Heracleum* L. (*Apiaceae*) are distributed in the temperate areas of Europe, Asia, Africa and North America (Pimenov and Leonov, 1993). Centres of the highest species diversity are China (29 species) and the Caucasus (26 species; Mandenova, 1950; Fading and Watson, 2006). *Flora Europaea* lists eight native species, the most common of which is *H. sphondylium* L. (Brummitt, 1968). Several of the large members of the genus are cultivated as ornamentals outside their native range, and some are also used as fodder crops. A group of closely related large species of the genus *Heracleum*, which were introduced into Europe, are called 'large, tall or giant' hogweeds (Nielsen *et al.*, 2005). The most distinctive characteristic of these species is their size; they can attain heights of up to 4–5 m, which ranks them among the tallest and largest herbs in Europe. Three species became invasive in Europe: *Heracleum mantegazzianum* Sommier & Levier, *H. sosnowskyi* Manden. and *H. persicum* Desf. ex Fischer.

This chapter aims to place giant hogweed, *Heracleum mantegazzianum*, the main subject of the present volume, into a wider generic context in terms of taxonomy, genetic relationships with related taxa and distribution in both its native and invaded ranges. A brief excursion into past botanical investigations, which were carried out in the Caucasus and adjacent regions of Asia and resulted in the description of the invasive *Heracleum* species in Europe, is also presented. The history of the introduction of *Heracleum* species is reviewed and the current distributions of the three major invasive representatives in Europe are compared. It was only recently recognized that the taxonomy of *Heracleum* species alien to Europe is complex and plants previously called *H. mantegazzianum* actually belong to more than one species (Holub, 1997). A comprehensive taxonomic pan-European assessment of the genus is still lacking; this chapter nevertheless provides a taxonomic treatment for Nordic countries, which constitute an area of north-eastern Europe where the distribution ranges of two major invasive species meet. In this chapter we use taxonomic names in the sense of Mandenova (1950, 1987).

History of the Taxonomic Investigations of Large *Heracleum* Species in their Native Caucasus Mountains

The native distributions of the large *Heracleum* species currently invasive in Europe are the Western Greater Caucasus for *H. mantegazzianum* (see Otte *et al.*, Chapter 2, this volume), Central and Eastern Greater Caucasus and western, central, eastern and south-western Transcaucasia as well as north-eastern Turkey for *H. sosnowskyi* and Turkey, Iran and Iraq for *H. persicum* (Fig. 1.1).

Giant hogweeds are very spectacular plants and were collected by the earliest botanical expeditions to the Caucasus Mountains. The Herbarium of the Komarov Botanical Institute (LE) contains a *Heracleum* specimen (later named *H. sosnowskyi* Manden.) collected on 19 July 1772, during the expedition of the Russian Academy of Sciences led by J.A. Güldenstädt. It is quite likely that seeds of large hogweeds were also collected by the Caucasian expeditions of P.S. Pallas (1793–1794) and Ch. Steven (1800–1805, 1807–1810).

One of the important centres of Russian botany at the beginning of the 19th century was the private botanical garden of Count A.A. Razumovsky, at Gorenki near Moscow. The catalogue of this garden for 1812 (Fischer, 1812) mentions '*Heracleum giganteum* Spr.' without any description. This name was validated later by Hornemann (1819), with the accompanying diagnosis 'foliis ternatis: foliolis oblongis lobatis, radiis umbellae et umbellulae hirsutis, hab. in Russia?' However, this name was forgotten and is no longer in taxonomic use.

In the 19th century, large *Heracleum* species from the Caucasus were known mainly as *H. pubescens* (Hoffm.) M. Bieb. This species was very briefly described as *Sphondylium pubescens* Hoffm. (Hoffmann, 1814), but was soon moved to the genus *Heracleum* by Marschall von Bieberstein (1819), who gave it a more extensive and clear description and recorded it from the Crimea and Eastern Greater Caucasus. According to catalogues and herbarium specimens, a large hogweed with this name was extensively cultivated in the Imperial Botanical Garden in St Petersburg.

In 1841, other members of the giant hogweed complex, i.e. *H. persicum* Desf. ex Fischer and *H. wilhelmsii* Fisch. et Ave-Lall., were described (Fischer

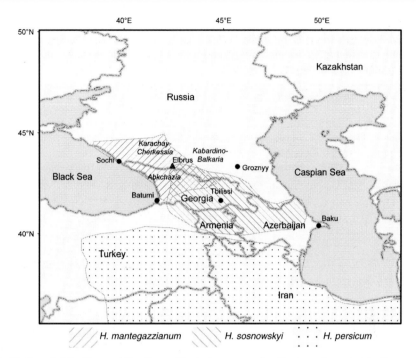

Fig. 1.1. Native distributions of the three *Heracleum* species that are invasive in Europe. Note that throughout the book, the mountain ranges of the Caucasus Mountains are referred to as the Greater and Lesser Caucasus (see Otte *et al.*, Chapter 2, this volume for details). Where geographical terms are used, the region consists of the North Caucasus and Transcaucasia (the southern part of the region).

et al., 1841) on the basis of plants grown from seeds; the former was collected by I. Szovits in Transcaucasia and Persia and the latter by Wilhelms in modern Georgia. Earlier (Fischer and Meyer, 1835), one more species, sometimes regarded as a member of giant hogweed complex, *H. trachyloma* Fisch. et C.A. Mey., was described. C.F. Ledebour (1844–1846) accepted *H. pubescens*, *H. wilhelmsii*, *H. trachyloma* and *H. caucasicum* Steven (close to *H. pubescens*) in his Russian flora. The latter taxon is dubious and not mentioned again. Boissier (1872) accepted *H. pubescens* in a broad sense, including *H. wilhelmsii* and *H. trachyloma* as varieties.

It is relevant to point out that the majority of the pre-1872 Caucasian collections come from the Transcaucasia and Central Greater Caucasus. The Western Greater Caucasus was closed to expeditions because of war. It is for this reason that large hogweeds collected at that time mainly belong to *H. sosnowskyi* (in terms of the modern nomenclature, see below), because *H. mantegazzianum* Somm. et Levier s. str., occurring in Western Greater Caucasus, was not then known to science. In addition, expeditions to high mountain areas were arduous and expensive. Therefore, it is not surprising that *H. mantegazzianum* was discovered much later, by the Caucasian expedition of S. Sommier and E. Levier in 1890 (Sommier and Levier, 1895, 1900). The

Table 1.1. Species of *Heracleum* section *Pubescentia* Manden. from Caucasus, Turkey, Iran and Middle Asia. Based on Mandenova (1950) plus two species described later (marked with *). The main discriminating features are given for the three invasive species in Europe plus remarks on other species where appropriate.

Species	Described in	Native distribution	Treatment of species by later authors	Discriminating features
H. pubescens (Hoffm.) M. Bieb	(1814) 1819	Endemic to Crimea		
H. persicum Desf. ex Fischer	(1829) 1841	Iran and Turkey		True perennial. The plants often have more than one stem, and a ± continuous violet colour on the lower part of the stem. The leaves are more elongate and have 2–3 pairs of leaflets, having broad, short lobes; the teeth on the leaf margins are short, with ± convex sides. The umbels are rather convex, and the rays have broadly triangular, whitish, ± ascending papillae, and usually no glandular hairs. The flowers have outer petals with rather broad lobes and the ovaries/fruits usually have a dense covering of more than 1 mm long, whitish hairs and slightly expanded dorsal vittae.
H. trachyloma Fischer et Meyer	1835	Southern Transcaucasia and Talysh		
H. wilhelmsii Fisch. & Lallem	1840	Endemic to Trialeti mountain ridge (Georgia)		
H. lehmannianum Bunge	1850	Pamiro-Alai mountains in Middle Asia		
H. crenatifolium Boiss.	1872	Turkey		
H. mantegazzianum Sommier and Levier	1895	Southern slope of the Western Greater Caucasus		Monocarpic perennial. The leaves are broad, with 1–2 pairs of leaflets, having elongate, biserrate lobes; the teeth on the leaf margins have concave sides. The umbels are rather flat, and the rays have narrow, translucent, patent papillae and/or glandular hairs. The flowers have outer petals with

Table 1.1. *Continued*

Species	Described in	Native distribution	Treatment of species by later authors	Discriminating features
				rather narrow lobes and the ovaries/fruits have *ca* 0.5–1 mm long glandular hairs and distinctly expanded dorsal vittae.
H. paphlagonicum Czecz.	1932	Turkey		
H. sosnowskyi I. Manden.	1944 (1950)	Eastern part of the Greater Caucasus ridge, south-western and eastern Transcaucasia	Synonym of *H. wilhelmsii* (Tamamschjan,1967)	Monocarpic perennial. Secretory vittae up to three-quarters of semifruit length on dorsal side, up to half on the commissure. Petals of marginate flowers smaller than in *H. mantegazzianum* and *H. wilhelmsii.* Flower disk tuberculate-rugose. Rays of umbels and umbellets with small hirsute indumentum. Basal leaves and lowermost stem leaves ternate or pinnate, segments broadly ovate or ovate-elongate, unequal at the base, compara-tively not deeply 3-lobate or pinnatisected into ovate lobes, apical segment almost rounded, deeply divided into 3 lobes. Shape and division of leaves is variable across the distribution area.
H. grossheimii I. Manden.	1950	Georgia: Imereti and Guria historical areas	Synonym of *H. mantegazzianum* (Satsyperova,1984; Tamamschjan, 1967; Menitsky, 1991)	
H. sommieri I. Manden.	1950	Georgia, endemic to Svaneti historical area		
*H. circassicum I. Manden.	1970	Mountains along the Black Sea coast near Tuapse	Synonym of *H. pubescens* (Satsyperova, 1984) or *H. mantegazzianum* (Menitsky, 1991)	Close to *H. mantegazzianum,* but has more densely pubescent leaves and fruits, less dissected leaves, which are similar to those of *H. sosnowskyi.*
*H. idae Kulieva	1975	Azerbaydzhan	Synonym of *H. trachyloma* (Menitsky, 1991)	

classic locality of this species is the upper part of the Kodor river basin (Georgia, Abkhasia).

Lipsky (1899) in his '*Flora of Caucasus*' (not a critical study) mentions two species of the giant hogweed group: *H. pubescens* for 'all Caucasus' at altitudes between 660 and 2700 m, and *H. mantegazzianum* for Abkhasia only.

The main period of the investigation of the Caucasian *Heracleum* taxa was associated with the activity of Ida P. Mandenova. She worked in Tbilisi, but also studied very important collections in the Herbarium of the Komarov Botanical Institute. In 1944 she described another member of the giant hogweed complex, *H. sosnowskyi*, from Georgia (Meskheti historical area; Mandenova, 1944) and verified this species later in a monograph on the Caucasian species of *Heracleum* (1950).

Mandenova (1950) mentioned, as 'characteristic and constant' features of *H. sosnowskyi*, the form of the vittae of fruits and small hirsute indumentum of rays of the umbel. She pointed out that plants from the classic locality have ternate leaves with wide ovate segments, but other plants can have also pinnate leaves with elongate and deeply dissected segments and that this latter form is more widely distributed. According to her, *H. sosnowskyi* was often collected, but wrongly recorded in 'Floras' as *H. pubescens*, which she regarded as endemic to the Crimea. She also mentions that it is quite possible that an earlier priority name for *H. sosnowskyi* exists.

Species of the giant hogweed complex were placed by Mandenova in the section *Pubescentia* I. Manden. At the time of Mandenova's work, there were six Caucasian species and five from other parts of the world in this section (Table 1.1). Mandenova described not only *H. sosnowskyi*, but also several other species of the giant hogweed complex, such as *H. circassicum* (Mandenova, 1970) from mountains along the Black Sea coast near Tuapse. This species, according to her description, is close to *H. mantegazzianum*, but has more densely pubescent leaves and fruits, and less dissected leaves, which are similar to those of *H. sosnowskyi*.

For reasons that are not clear, Mandenova did not present an account of *Heracleum* in the second edition of A. Grossheim's '*Flora of Caucasus*', which was produced by S.G. Tamamschjan (1967). Tamamschjan generally followed Mandenova's concept of species, but treated *H. sosnowskyi* as a synonym of *H. wilhelmsii* and *H. grossheimii* as a synonym of *H. mantegazzianum*.

After World War II, giant hogweeds (under the name *H. sosnowskyi*) were tested and used as forage (silage) plants in the USSR. The source of seeds was specified in one of the earliest publications on this subject (Marchenko, 1953) as Kabardino-Balkaria (North Caucasus). It is interesting that Marchenko observed a mixture of two species (*H. sosnowskyi* and *H. mantegazzianum*) but collected only *H. sosnowskyi*. Later on, seeds of large hogweeds were repeatedly collected for introduction, breeding and cultivation, mostly under the name *H. sosnowskyi*, but it is impossible to prove whether or not this determination was correct. For example, some plants in the Central Botanical Garden in Minsk (Belarus), where an extensive breeding programme on *Heracleum* took place, grown under the name *H. sosnowskyi* originated from Krasnaya Polyana (near Sochi). However, according to taxonomic studies, this species does not occur there.

The results of work on the introduction of hogweeds as forage plants are summarized by Satsyperova (1984). The book contains a chapter on *Heracleum* taxonomy in which the section *Pubescentia* is accepted with eight species: *H. trachyloma*, *H. idae* Kulieva (described from Azerbaydzhan; Kulieva, 1975), *H. lehmannianum*, *H. wilhelmsii*, *H. sosnowskyi*, *H. mantegazzianum*, *H. sommieri* and *H. pubescens*. It is clear that this author generally followed Mandenova, but treated *H. grossheimii* as a synomym of *H. mantegazzianum* and *H. circassicum* as a synonym of *H. pubescens*. As a result, *H. pubescens*, according to this concept, occurs not only in the Crimea but also in the North Caucasus (Black Sea coast).

The latest work on the subject is a brief synopsis of the Caucasian *Apiaceae* in the 'Caucasian flora conspectus' (Menitsky, 1991). For *Heracleum*, Menitsky generally followed Mandenova, but regarded *H. grossheimii* and *H. circassicum* as synonyms of *H. mantegazzianum*. He also excluded *H. pubescens* from the Caucasian flora and treated *H. idae* (accepted by Satsyperova) as a synonym of *H. trachyloma*.

It is evident that the taxonomy of the giant hogweed complex in its native area is still debatable. This is not only due to the complexity of the genus, but also the poor representation of these extremely large plants in herbaria (see Fig. 3.1) and lack of field observation. It is also necessary to take into account that Russian taxonomic tradition usually accepts a narrower species concept than western European taxonomists.

Introduction of Large *Heracleum* Species and their Distribution in Europe

The first known record of a giant hogweed in Europe is for England, 1817, when it appeared on the Kew Botanic Gardens, London seed list under the name of *H. giganteum*. It is most likely that the plant introduced under this name was *H. mantegazzianum*, since this was the start of the spread of this species in Europe. In 1828, the first naturalized population of *H. mantegazzianum* was recorded, growing wild in Cambridgeshire, England. Soon afterwards, the plant began to spread rapidly across Europe. Ranked according to the date of introduction, the UK was followed by the Netherlands, Switzerland, Germany, Ireland, Denmark and Czech Republic (Nielsen *et al.*, 2005); in the latter country (see Pyšek *et al.*, Chapter 3, this volume) the length of period between the introduction into cultivation (1862) and the first report of a spontaneous occurrence nearby (1877) is very similar to the 11 years in England. Thus, it is reasonably certain that the pattern and historical dynamics on a national scale in various countries resembles that described in Chapter 3 for the Czech Republic. In 14 of the 19 countries for which historical data are available, it was first reported before 1900, and only in Austria, Slovakia and Iceland is the first record after 1960 (Nielsen *et al.*, 2005). The main mechanism of introduction into Europe was as an ornamental curiosity. Seeds were planted in botanic gardens and the grounds of important estates. This fashion continued for most of the 19th century and only declined and eventually ceased after warnings about the dangers of the plant appeared in western European literature towards the end of the 20th century (Nielsen *et al.*, 2005).

Heracleum mantegazzianum is the most widespread of the large invasive hogweed species in Europe (Fig. 1.2A). Its distribution in Europe is clearly biased towards the central and northern part of the continent and this species is virtually absent from southern Europe. This pattern of distribution can be explained by the factors determining its distribution in the Czech Republic (Pyšek *et al.*, 1998) where January temperatures and human population density most significantly affect the abundance of this species. The spread of *H. mantegazzianum* was fast in regions with a high human population density and limited in warmer areas with a relatively high mean January temperature (Pyšek *et al.*, 1998). Assessing the relative importance of these factors reveals that the effect of population density is stronger than that of temperature.

The under-representation of *Heracleum* in regions with warm January temperatures has been interpreted in terms of the germination ecology of the species (Pyšek *et al.*, 1998). The breaking of dormancy in *H. mantegazzianum* requires chilling (Tiley *et al.*, 1996; Moravcová *et al.*, 2005), which is provided by a cold period in humid soil during winter. However, the results of recent studies (see Moravcová *et al.*, Chapter 5, this volume; Krinke *et al.*, 2005; Moravcová *et al.*, 2005) suggest that even areas with higher January temperatures are sufficiently cold for the stratification and breaking of dormancy. It seems that the negative effect of warm winters may not only operate by negatively affecting germination. Low winter temperatures could, at least in part, explain why the species is absent or very rare in south-eastern parts of Europe (Hungary, Romania, Bulgaria etc.), while thriving in more northerly and temperate areas, e.g. British Isles, Denmark, Sweden, Germany (Nielsen *et al.*, 2005). The distribution of this species is unlikely to be limited by insufficient population density in southern Europe, so climatic constraints seem the most likely. Outside Europe, *H. mantegazzianum* is naturalized in Canada (Morton, 1978) and the USA (Kartesz and Meacham, 1999).

Heracleum sosnowskyi, originally described in 1944, was introduced into Europe as an agricultural crop and was silaged to provide fodder for livestock. Because the plant is hardy and thrives in a cold climate, it has been used as a crop in north and north-west Russia since its introduction in 1947. From the 1940s onwards, it was introduced as a crop to the Baltic countries (Latvia, Lithuania and Estonia) as well as to Belarus, Ukraine and the former German Democratic Republic. The history of planting is reflected in the species' current distribution (Fig. 1.2B). This practice was eventually abandoned in the Baltic States, partly because the anise-scented plants affected the flavour of meat and milk of the animals fed this fodder and also partly because of the health risk to humans and cattle. In parts of northern Russia, in contrast, it is still cultivated (Nielsen *et al.*, 2005).

The history of invasion of *H. persicum* in Europe is the most obscure of the three species, partly because it was the first to be described (1829) and some of the subsequent records of *H. persicum* were probably of *H. mantegazzianum* or *H. sosnowskyi*. The only known wild populations of this plant in Europe are

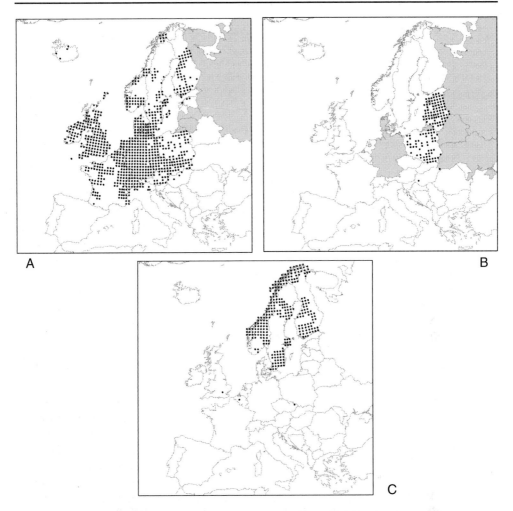

Fig. 1.2. Distributions of the three main invasive *Heracleum* species in their invaded range in Europe: *H. mantegazzianum* (A), *H. sosnowskyi* (B) and *H. persicum* (C). Taken from Nielsen *et al.* (2005), updated. Distribution data for Norway and France are based on presence or absence at the county/department level. As a result the illustrated distribution in these areas may over-represent the actual distribution. Other countries are shaded in which the species is reported to occur but the exact distribution is unknown. Note that for Poland, the distribution data do not distinguish between *H. mantegazzianum* and *H. sosnowskyi*; the plotted distributions therefore relate to both species. The occurrence of *H. persicum* in the Czech Republic is based on a record in Holub (1997). Note that records of *H. persicum* in UK and Belgium are from cultivation.

in Fennoscandia (Fig. 1.2C). The earliest record of introduction again comes from the seed list of the Kew Botanic Gardens in London, where material under the name of *H. laciniatum* was received in 1819. Seeds from London populations of a similar plant were taken by English horticulturalists and planted in northern Norway as early as 1836 (Christy, 1837; see further below). The main mechanism of spread of this species again appears to be as an ornamental.

Taxonomy and History of Invasive *Heracleum* Species in Nordic Countries

In the Nordic countries (Denmark, Finland, Iceland, Norway and Sweden) large *Heracleum* species have been known since the first third of the 19th century, but it was not appreciated until the mid 20th century that there were more than one species. A recent revision of the genus in the Nordic countries (L. Fröberg, in prep.) has shown that, besides the native *H. sphondylium* L. subsp. *sphondylium* and subsp. *sibiricum* (L.) Simonkai, two widespread introduced species occur: *H. mantegazzianum* Somm. et Lev. and *H. persicum* Desf. ex Fischer. Furthermore, several morphotypes, which currently cannot be related with certainty to any of the existing species, have also been collected. The taxonomic treatment and description of *H. mantegazzianum* and *H. persicum* is presented below; for description of *H. sosnowskyi* see Table 1.1.

Taxa recognized in Nordic countries

Plants of *H. mantegazzianum* are monocarpic, i.e. they die after flowering (Pergl *et al.*, 2006 and Chapter 6, this volume), and consist of only one vertical stem, with distinct red dots on the lower part. The leaves are broad, with 1–2 pairs of leaflets, which have elongated, biserrate lobes; the teeth on the leaf margins have concave sides. The umbels are rather flat, and the rays have narrow, translucent, patent papillae and/or glandular hairs. The flowers have

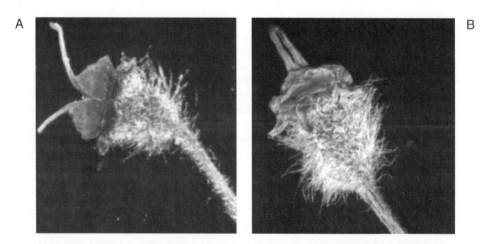

A   B

Fig. 1.3. Hairs on ovaries just after anthesis; covering in *H. mantegazzianum* (A) is of ± translucent, glandular hairs (also on pedicels), and in *H. persicum* (B) of whitish hairs, which are also somewhat longer and thinner.

outer petals with rather narrow lobes and the ovaries/fruits have ca. 0.5–1 mm long glandular hairs (Fig. 1.3A) and distinctly expanded dorsal vittae (i.e. brown-coloured oil channels on the fruit surface).

Heracleum persicum is a perennial that flowers more than once (Often and Graff, 1994). The plants often have more than one stem and a ± continuous violet coloration on the lower part of the stem. The leaves are longer and have 2–3 pairs of leaflets, having broad, non-elongated lobes; the teeth on the leaf margins are short, with ± convex sides. The umbels are rather convex, and the rays have broadly triangular, whitish, ± ascending papillae, with usually non-glandular hairs. The flowers have outer petals with rather broad lobes and the ovaries/fruits usually have a dense covering of more than 1 mm long, whitish hairs (Fig. 1.3B) and slightly expanded dorsal vittae. A comprehensive presentation of the distinguishing characters of *H. mantegazzianum* and *H. persicum* is available in Often and Graff (1994).

The identity of *H. persicum* was previously uncertain, but is now confirmed by comparison with material from Turkey. It was largely confused with *H. mantegazzianum*, and earlier recognized under the name of *H. laciniatum* Hornem. in northern Norway (Nordhagen, 1940; Lid, 1963) and Finland (Hiitonen, 1933). According to the description by Hornemann (1813), *H. laciniatum* should have entire, lobed leaves that are tomentose beneath and is probably not conspecific with *H. persicum*.

A few collections from southern Denmark and Sweden clearly differ from *H. mantegazzianum* and *H. persicum* in having leaves with distinctly rounded lobes and white-tomentose undersides. They were incorrectly referred to as *H. stevenii*, but correspond to *H. platytaenium*, which is a species native to Turkey (confirmed by comparison with material from the original distribution area).

There are other specimens from Nordic countries that do not fit with either of these species. A few specimens from Finland and northern Sweden differ from *H. persicum* in having more densely hairy lower leaf surfaces; the rays and pedicels lack hairs and have papillae that are similar to those in *H. mantegazzianum*, and the ovaries/fruits have shorter hairs than *H. persicum*. Some specimens from Denmark, southern Norway and Sweden differ in being somewhat smaller and having leaves without elongated lobes, umbels with fewer rays, and flowers with red-violet anthers and not distinctly radiating outer petals. This latter morphotype might correspond to *H. pubescens* (Hoffm.) M. Bieb., which is characterized by having shorter leaf-lobes and fewer rays than *H. mantegazzianum* (Mandenova, 1974).

Heracleum mantegazzianum and *H. persicum* are both found rather frequently in northern Europe, where several intermediate morphotypes also occur. These taxa are also probably quite variable in their original distribution area. It is possible that several genotypes were introduced into Europe over the last two centuries and may have subsequently interbred, e.g. in botanical gardens, giving rise to the large variability that can be observed in northern Europe today. However, these observations are based on herbarium material only, and several diagnostic characters are not available. For an understanding of the group, it is important to acquire a total picture of the variation that occurs in the field, including life form, size and other characters. The genetic treatment of the large *Heracleum* species, presented below, might improve our understanding of the relationships among the related taxa within this taxonomically complex group.

History of the group in Nordic countries

In the early 19th century, seeds of giant *Heracleum* species collected in southwestern Asia were sent to several European botanical gardens, under various names. Furthermore, private individuals could purchase seed material from gardens, distributing the plants even further afield (Bruun *et al.*, 2003). The first record of giant *Heracleum* species in the Nordic countries is 1836, when seeds were sent to northern Norway (Alm and Jensen, 1993). The material probably consisted of *H. persicum*, which is now established there as a troublesome weed under the vernacular name of 'Tromsø palm'. A giant *Heracleum* species is reported to have escaped from cultivation in southern Sweden in the late 19th century (Lilja, 1870). It is difficult to confirm the identity of this plant, but it may not be *H. mantegazzianum*, since this species was first verified in the Nordic countries at the beginning of the 20th century. Furthermore, a record from southern Finland, Sipoo near Helsinki, in 1876 is conspecific with *H. persicum*. The first confirmed record of *H. mantegazzianum* in Nordic countries is 1903 (Sweden, escaped near Nyköping; cultivated in the area since 1898). However, a larger proportion of later records of large *Heracleum* species in the Nordic area are for *H. mantegazzianum* and it can thus be concluded that this species has now totally outcompeted other related species, at least in the southern parts of this region.

Genetic Relationships Among Large Hogweed Taxa in Both their Invaded and Native Ranges

In central and western Europe, *H. mantegazzianum* has long been recognized as a prominent example of an invasive alien (Pyšek, 1991; Tiley *et al.*, 1996; Collingham *et al.*, 2000; Wadsworth *et al.*, 2000); however, its genetic relationship to the other large *Heracleum* taxa that are also invasive in parts of Europe (*H. sosnowskyi* and *H. persicum*) has been studied only recently (Nielsen *et al.*, 2005; Jahodová *et al.*, 2007).

Jahodová *et al.* (2007) used the molecular marker technique of Amplified Fragment Length Polymorphism (AFLP) to study genetic relationships among invasive taxa of *Heracleum* in Europe and their native ranges. The results of this study indicate that there are three genetically distinct invasive *Heracleum* species in Europe: *H. mantegazzianum*, *H. sosnowskyi* and *H. persicum* (Fig. 1.4).

Genetic relationships at the species level

AFLPs reveal a close genetic relationship among *H. mantegazzianum*, *H. sos-*

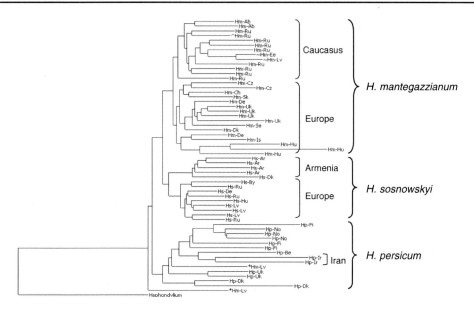

Fig. 1.4. Genetic relationships between the three *Heracleum* taxa invasive in Europe revealed by neighbour-joining analysis of AFLP data. The dendrogram includes 58 populations from both native and invaded ranges of *H. mantegazzianum* (Hm), *H. sosnowskyi* (Hs) and *H. persicum* (Hp). Dendrogram was rooted using *H. sphondylium*. Populations from 19 countries were analysed: Abkhazia (Ab), Armenia (Ar), Belgium (Be), Belarus (By), Czech Republic (Cz), Germany (De), Denmark (Dk), Estonia (Ee), Finland (Fi), Hungary (Hu), Switzerland (Ch), Iran (Ir), Iceland (Is), Latvia (Lv), Norway (No), Russia (Ru), Slovakia (Sk), Sweden (Se) and United Kingdom (Uk). * Possible hybrids; ^ originally identified as a hybrid *H. mantegazzianum × ponticum*; ~ originally identified as *H. sosnowskyi*. Based on data from Jahodová *et al.* (2007).

nowskyi and *H. persicum*, which is evident from the high Dice similarity coefficients between their samples (ranging from 0.726 to 0.921; mean ± SD, 0.843 ± 0.032) and the low among-taxa variation detected by analysis of molecular variance (with approximately 18% of the genetic variation distributed among the three taxa and 82% distributed within the taxa, Table 1.2). The AFLPs also reveal a difference in the distribution of neutral genetic variation in *H. mantegazzianum*, *H. sosnowskyi* and *H. persicum* between their native and invaded regions. In the invaded range, the within-taxon variation is higher (77.3%) than in the native range (61.0%), leaving 22.7% and 39.0% of the variation attributable to differences among taxa, respectively (Table 1.2). Furthermore, invading and native populations differ genetically in both *H. mantegazzianum* and *H. sosnowskyi*. Neighbour-joining analysis of similarity matrices produced from the AFLP data separates populations of these taxa into distinct clusters corresponding to their origins in the resultant dendrogram (Fig. 1.4) (Jahodová *et al.*, 2007).

The separation of samples from the native and invaded range, together with the high genetic variation found among samples from within the invaded range, suggests that, at the continental scale, these taxa are not affected by

Table 1.2. Analyses of molecular variance (AMOVA) of the AFLP data. All estimations are significant ($P < 0.001$). Europe stands for the invaded distribution range. Partly based on data from Jahodová *et al.* (2007).

Grouping	Number of samples	Number of populations	Number of taxon groups	Source of variation	df	Sum of squares	Variance components	% variation
H. mantegazzianum, *H. sosnowskyi* and *H. persicum* (Europe and native ranges)	146	51	3	among taxa	2	795.45	7.18	17.72
				among populations	48	2909.96	14.73	36.34
				within populations	95	1768.92	18.62	45.94
H. mantegazzianum, *H. sosnowskyi* and *H. persicum* (Europe)	92	N/A	3	among taxa	2	664.74	9.28	22.65
				within taxa	95	3010.54	31.69	77.35
H. mantegazzianum, *H. sosnowskyi* and *H. persicum* (native ranges)	45	N/A	3	among taxa	2	429.42	16.39	39.01
				within taxa	42	1076.58	25.63	60.99
H. mantegazzianum (Europe)	41	15	1	among populations	14	754.12	14.71	51.61
				within populations	26	358.58	13.79	48.39
H. mantegazzianum (native range)	28	10	1	among populations	9	324.87	5.79	22.51
				within populations	18	359.17	19.95	77.49
H. sosnowskyi (Europe)	28	10	1	among populations	9	373.96	8.29	31.02
				within populations	18	332.00	18.44	68.98
H. sosnowskyi (native range)	11	4	1	among populations	3	103.61	5.56	22.28
				within populations	7	135.67	19.38	77.72

genetic bottlenecks. Multiple introductions, intra- and inter-specific hybridizations and rapid evolution may be responsible for this pattern of variability (Sakai *et al.*, 2001).

An exception to the finding of genetic differentiation between populations from the invaded and native ranges is recorded by Jahodová *et al.* (2007) for two populations from Estonia and Latvia (determined as *H. mantegazzianum*) and one from Denmark (determined as *H. sosnowskyi*), which are genetically closer to the populations from the native than the invaded range (Fig. 1.4). Such results suggest an independent, and quite likely also a more recent, introduction at these locations.

Jahodová *et al.* (2007) noted a lack of correlation between genetic and geographical distance for all three invasive *Heracleum* taxa in Europe. This lends support to the view that, at the continental scale, they were dispersed mainly by human activities. Since all of the three taxa were introduced and spread for utilitarian purposes (i.e. as ornamentals or forage crops), their patterns of dispersal are not consistent with that expected for a gradual invasion from a single point of introduction. In the native range, however, the genetic distances between populations of *H. mantegazzianum* show a highly significant increase with increasing geographical distance, which explain 42% of the variation (Fig. 1.5). It was not possible to determine this relationship for the other two taxa in their native ranges due to insufficient numbers of sampled populations, but for *H. mantegazzianum* this relationship clearly indicates isolation by distance in the native range. As can be seen from Fig. 1.5, the increase in dissimilarity with geographic distance is a straight line on a semi-log plot which indicates that it is faster for short distances, up to 50–100 km, and decelerates for populations 100–300 km apart. This suggests that there is a threshold beyond which geographic isolation no longer promotes increase in dissimilarity between populations in *H. mantegazzianum* (Jahodová *et al.*, 2007).

Genetic relationships at the population level

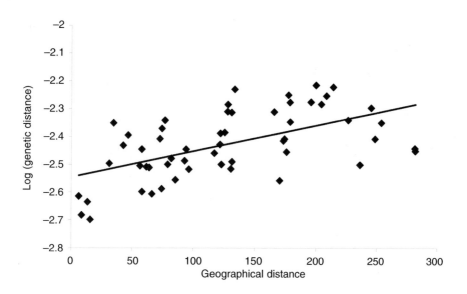

Fig. 1.5. Genetic distance (calculated as 1-Dice similarity coefficient) between *Heracleum mantegazzianum* populations in their native range in the Caucasus increases with geographical distance between populations. Eleven populations were sampled in the Caucasus and the plot shows the regression based on all possible pairwise combinations. Note the semi-logarithmic scale. Genetic distance = 0.0009 geographical distance − 2.545; F = 37.72, df = 1, 53; $P < 0.0001$. Based on data from Jahodová *et al.* (2007).

Preliminary results (based on two to four samples per population) of the genetic variation at the population level within all three *Heracleum* taxa (Š. Jahodová *et al.*, unpublished) reveal a higher within-population variation in *H. mantegazzianum* and *H. sosnowskyi* in their native ranges (both approximately 77%) than in Europe (48% and 69%, respectively; Table 1.2). This may indicate that, whilst genetic bottlenecks are not apparent at the continental scale in the invaded range, the variation within individual populations is affected by founder events and that most populations in the invaded range do not experience subsequent re-introduction and/or strong gene-flow from other established populations (Nei *et al.*, 1975; Ward, 2006).

These findings are in general agreement with those previously reported by Walker *et al.* (2003), who conducted a population-genetic study of *H. mantegazzianum* in three river catchment areas in north-east England. They also found high levels of variation within populations but did not compare their results with native populations. On the other hand, Walker *et al.* (2003) did detect inbreeding and suggested that the population genetic structure of invading populations is influenced by genetic drift. The higher between- compared to within-catchment variation is seen as indicating a large initial founder population in Britain, or multiple introductions.

Implications of the Genetic Studies for Taxonomy

The comparison of European samples with those obtained from the native ranges of *H. mantegazzianum* Sommier et Levier (native to the Western Greater Caucasus), *H. sosnowskyi* Manden. (native to Eastern and Central Greater Caucasus, Transcaucasia and north-eastern Turkey) and *H. persicum* Desf. ex Fischer (native to Turkey, Iran and Iraq) reveals close genetic relationships of samples from native and introduced areas for all three taxa (Jahodová *et al.*, 2007).

In Europe, high genetic variation is recorded at the species level (among-species variation was 23%, in contrast to 39% in the native ranges, Table 1.2), which may obscure boundaries between the taxa in the invaded range. This is especially true for *H. mantegazzianum* and *H. sosnowskyi*, which appear to be the most closely related pair of the three taxa (Dice similarity coefficient between samples of these two taxa is 0.872 ± 0.032, compared to 0.825 ± 0.020 and 0.830 ± 0.023 for *H. persicum* and *H. mantegazzianum*, resp. *H. sosnowskyi*).

Several cases of likely misidentification revealed by molecular markers are reported by Jahodová *et al.* (2007). These include a putative hybrid *H. ponticum* × *H. mantegazzianum* identified on the basis of intermediate morphological characters, which was shown by AFLP analysis to be *H. mantegazzianum*, and samples labelled as *H. sosnowskyi*, which were found to be genetically closer to *H. mantegazzianum* than to *H. sosnowskyi*. In addition, molecular analysis indicates that several samples from two city parks in Latvia, originally determined as *H. mantegazzianum*, may actually be of hybrid origin (Fig. 1.4).

Such close genetic relationships and the extremely variable nature of the morphological characters cause difficulties when determining specimens. Consequently, misidentifications are rather common.

Acknowledgements

We thank J. Kirschner and Z. Kaplan for helpful comments on the manuscript, and T. Dixon for improving our English. Logistic support from J. Pergl is greatly appreciated. We thank O. Booy, I. Dancza, D. De Meyere, A. Fehér, G. Gavrilova, S. Hansen, J. Hattendorf, J. Hüls, I. Illarionova, L. Končeková, G. Konechnaya, B. Magnusson, C. Nielsen, C.V. Nissen, A. Often, J. Pergl, I. Perglová, M.G. Pimenov, N. Portenier, I. Roze, R. Sobhian, J. Thiele, P. Uotila and R. Wittenberg for help with the collection of samples. The study was supported by the project 'Giant Hogweed (*Heracleum mantegazzianum*) a pernicious invasive weed: developing a sustainable strategy for alien invasive plant management in Europe', funded within the 'Energy, Environment and Sustainable Development Programme' (grant no. EVK2-CT-2001-00128) of the European Union 5th Framework Programme. Š.J. and P.P. were supported by institutional long-term research plans no. AV0Z60050516 of the Academy of Sciences of the Czech Republic, and no. 0021620828 of the Ministry of Education of the Czech Republic. Rothamsted Research receives grant-aided support from the Biotechnology and Biological Sciences Research Council of the UK. Flora Nordica is acknowledged for the permission to include the material to be published in the flora.

References

Alm, T. and Jensen, C. (1993) The Tromsø palm (*Heracleum laciniatum* auct. scand.) – some comments on the introduction of the species to northern Norway. *Blyttia* 51, 61–69. [In Norwegian.]

Boissier, E. (1872) *Flora Orientalis*. Vol. 2. Georg, Geneva and Basileae.

Brummitt, R.K. (1968) *Heracleum L.* In: Tutin, T.G., Heywood, V.H., Burges, N.A., Moore, D.M. and Valentine, D.H. (eds) *Flora Europaea*. Vol. 2. *Rosaceae to Umbelliferae* Cambridge University Press, Cambridge, pp. 364–366.

Bruun, H.H., Erneberg, M. and Ravn, H.P. (2003) The introduction history of giant hogweed in Denmark. *URT* 27, 43–49. [In Danish.]

Christy, W. (1837) Notes of a voyage to Alten, Hammerfest, etc. *Entomological Magazine* 4, 462–583.

Collingham, Y.C., Wadsworth, R.A., Willis, S.G., Huntley, B. and Hulme, P.E. (2000) Predicting the spatial distribution of alien riparian species: issues of spatial scale and extent. *Journal of Applied Ecology* 37 (Suppl. 1), 13–27.

Fading, P. and Watson, M.F. (2005) *Heracleum*. In: Flora of China Editorial Committee (eds) *Flora of China Vol. 14 (Apiaceae through Ericaceae)*.Science Press, Beijing and Missouri Botanical Garden Press, St Louis, p. 194.

Fischer, F.E.L. (1812) *Catalogue du Jardin des Plants de S. E. Monsieur le Comte Alexis Razoumoffsky*. Del'Imprimerie de N.S. Vsevolojsky Moscow.

Fischer, F.E.L. and Meyer, C.A. (1835) *Index seminum, quae Hortus Botanicus Imperialis*

Petropolitanus pro mutua commutacione offert. Accedunt animadversiones botanicae nonnulae, Vol. 1. Petropoli.

Fischer, F.E.L., Meyer, C.A. and Avé-Lallemant, J.L.E. (1841) *Index seminum, quae Hortus Botanicus Imperialis Petropolitanus pro mutua commutacione offert. Accedunt animadversiones botanicae nonnulae,* Vol. 7. Ex typis Academiae Caesareae Petropolitanae, Petropoli.

Hiitonen, I. (1933) *Suomen Kasvio.* Kustannusosakeyhtiö 'Otava', Helsinki.

Hoffmann, G.F. (1814) *Genera plantarum Umbelliferarum.* Typ. N.S. Vsevolozskianis, Mosqua.

Holub, J. (1997) Heracleum – hogweed. In: Slavík, B., Chrtek Jr, J. and Tomšovic, P. (eds) *Flora of the Czech Republic 5.* Academia, Praha, pp. 386–395. [In Czech.]

Hornemann, J.W. (1813) *Tyronum et Botanophilorum.* Typis E.A.H. Molleri, Hafniae.

Hornemann, J.W. (1819) *Supplementum Horti Botanici Hafniensis.* Typis Schultzii, Hafniae.

Jahodová, Š., Trybush, S., Pyšek, P., Wade, M. and Karp, A. (2007) Invasive species of Heracleum in Europe: an insight into genetic relationships and invasion history. *Diversity and Distributions* (13, 99–114).

Kartesz, J.T. and Meacham, C.A. (1999) *Synthesis of the North American Flora. Version 1.0.* North Carolina Botanical Garden, Chapel Hill, North Carolina.

Krinke, L., Moravcová, L., Pyšek, P., Jarošík, V., Pergl, J. and Perglová, I. (2005) Seed bank of an invasive alien, *Heracleum mantegazzianum*, and its seasonal dynamics. *Seed Science Research* 15, 239–248.

Kulieva, Kh.G. (1975) New species of hogweed (Heracleum L.) from Azerbaidzhan. *Novosti Systematiki Vysshykh Rastenii* 12, 246–247. [In Russian.]

Ledebour, C.F. (1844–1846) *Flora Rossica.* Vol. 2. Samtibus Librariae E. Schweizerbart, Stuttgartiae.

Lid, J. (1963) *Norwegian and Swedish flora.* Ed. 1. Det Norske Samlaget, Oslo. [In Norwegian.]

Lilja, N. (1870) *The Flora of Scania [Skåne].* Ed. 2. L.J. Hiertas Förlag, Stockholm. [In Swedish.]

Lipsky, V.I. (1899) Flora of the Caucasus. *Trudy Tiflisskogo Botanicheskogo Sada* 4, I–XV, 1–584. [In Russian.]

Mandenova, I.P. (1944) Fragments of the monograph on the Caucasian hogweeds. *Zametki po Sistematike i Geografii Rastenii* 12, 15–19. [In Russian.]

Mandenova, I.P. (1950) *Caucasian Species of the Genus Heracleum.* Georgian Academy of Sciences, Tbilisi. [In Russian.]

Mandenova, I.P. (1970) New taxa of the genus Heracleum L.. *Zametki po Sistematike i Geografii Rastenii* 28, 21–24. [In Russian.]

Mandenova, I.P. (1974) Heracleum L. In: Shishkin, B.K. (ed.) *Flora of the U.S.S.R.* [translated from Russian]. Vol. 17. Israel Program for Scientific Translations, Jerusalem, pp. 223–259.

Mandenova, I.P. (1987) Heracleum. In: Reichinger, K.H. (ed.) *Flora Iranica: Umbelliferae. Flora des iranischen Hochlandes und der umrahmenden Gebirge: Persien, Afghanistan, Teile von West-Pakistan, Nord-Iraq, Azerbaidjan, Turkmenistan.* Akademische Druck – u. Verlagsanstalt, Graz, pp. 492–502.

Marchenko, A.A. (1953) Biological characters and forage advantages of Sosnowsky hogweed (*Heracleum sosnovskyi* Manden.). Dissertation for the degree of the candidate of biological sciences. Leningrad. [In Russian, manuscript deposited in the Komarov Botanical Institute.]

Marschall von Bieberstein, F.A. (1819) *Flora Taurico-Caucasica.* Vol. 3 – *Supplementum.* Charcoviae.

Menitsky, Yu.L. (1991) Synopsis of the species of the family *Apiaceae* (*Umbelliferae*) from the

Caucasus. *Botanicheskii Zhurnal* 76, 1749–1764. [In Russian.]

Moravcová, L., Perglová, I., Pyšek, P., Jarošík, V. and Pergl, J. (2005) Effects of fruit position on fruit mass and seed germination in the alien species *Heracleum mantegazzianum* (*Apiaceae*) and the implications for its invasion. *Acta Oecologica* 28, 1–10.

Morton, J.K. (1978) Distribution of giant cow parsnip (*Heracleum mantegazzianum*) in Canada. *Canadian Field Naturalist* 92, 182–185.

Nei, M., Maruyama, T. and Chakraborty, R. (1975) The bottleneck effect and genetic variability in populations. *Evolution* 29, 1–10.

Nielsen, C., Ravn, H.P., Cock, M. and Nentwig, W. (eds) (2005) *The Giant Hogweed Best Practice Manual. Guidelines for the Management and Control of an Invasive Alien Weed in Europe*. Forest and Landscape Denmark, Hørsholm, Denmark.

Nordhagen, R. (1940) *Norwegian Flora, with Brief Comments on the Introduction of Trees, Ornamentals and Utility Plants*. H. Aschehoug & Co. (W. Nygaard), Oslo. [In Norwegian.]

Often, A. and Graff, G. (1994) Diagnostic features of *Heracleum mantegazzianum* and Tromsø palm, *H. 'laciniatum'*. *Blyttia* 52, 129–133. [In Norwegian.]

Pergl, J., Perglová, I., Pyšek, P. and Dietz, H. (2006) Population age structure and reproductive behavior of the monocarpic perennial *Heracleum mantegazzianum* (*Apiaceae*) in its native and invaded distribution ranges. *American Journal of Botany* 93, 1018–1028.

Pimenov, M.G. and Leonov, M.V. (1993) *The Genera of the Umbelliferae – A Nomenclator*. Royal Botanical Garden Kew, London and Botanical Garden, Moscow University.

Pyšek, P. (1991) *Heracleum mantegazzianum* in the Czech Republic: dynamics of spreading from the historical perspective. *Folia Geobotanica & Phytotaxonomica* 26, 439–454.

Pyšek, P., Kopecký, M., Jarošík, V. and Kotková, P. (1998) The role of human density and climate in the spread of *Heracleum mantegazzianum* in the Central European landscape. *Diversity and Distributions* 4, 9–16.

Sakai, A.K., Allendorf, F.W., Holt, J.S., Lodge, D.M., Molofsky, J., With, K.A., Baughman, S., Cabin, R.J., Cohen, J.E., Ellstrand, N.C., McCauley, D.E., O'Neil, P., Parker, I.M., Thompson, J.N. and Weller, S.G. (2001) The population biology of invasive species. *Annual Review of Ecology and Systematics* 32, 305–332.

Satsyperova, I.F. (1984) *Hogweeds in the Flora of the USSR: New Forage Plants*. Nauka, Leningrad, Russia. [In Russian.]

Sommier, S. and Levier, E. (1895) Decas Umbelliferarum novarum Caucasi. *Nuovo Giornale Botanico Italiano* 2 (2), 85–96.

Sommier, S. and Levier, E. (1900) Enumeratio plantarum anno 1890 in Caucaso lectarum. *Trudy Imperatorskago Sankt-Peterburgskago Botaniceskago Sada, Acta Horti Petropolitani*, St. Petersburg 16, I–XXIII and 1–586.

Tamamschjan, S.G. (1967) *Heracleum* L. In: Grossheim, A.A., *Flora of the Caucasus*, 2nd edn. Nauka, Leningrad, pp. 121–130. [In Russian.]

Tiley, G.E.D., Dodd, F.S. and Wade, P.M. (1996) Biological flora of the British Isles. 190. *Heracleum mantegazzianum* Sommier et Levier. *Journal of Ecology* 84, 297–319.

Wadsworth, R.A., Collingham, Y.C., Willis, S.G., Huntley, B. and Hulme, P.E. (2000) Simulating the spread and management of alien riparian weeds: are they out of control? *Journal of Applied Ecology* 37, 28–38.

Walker, N.F., Hulme, P.E. and Hoelzel, A.R. (2003) Population genetics of an invasive species, *Heracleum mantegazzianum*: implications for the role of life history, demographics and independent introductions. *Molecular Ecology* 12, 1747–1756.

Ward, S. (2006) Genetic analysis of invasive plant populations at different spatial scales. *Biological Invasions* 8, 541–552.

2 Heracleum mantegazzianum in its Primary Distribution Range of the Western Greater Caucasus

ANNETTE OTTE, R. LUTZ ECKSTEIN AND JAN THIELE

University of Giessen, Giessen, Germany

*Long ago in the Russian hills, a Victorian explorer found the regal
Hogweed by a marsh*

(Genesis, 1971)

Introduction

With the exception of the Caucasus, the distribution of *Heracleum man-tegazzianum* Sommier & Levier in Europe (see Jahodová *et al.*, Chapter 1, this volume) is caused by humans. The Caucasus Mountains stretching from the Black Sea to the Caspian Sea consist of the Greater Caucasus in the north and a smaller mountain range in the south (the Lesser Caucasus). According to Walter (1974), the Greater Caucasus can be divided into the Western Greater Caucasus, which comprises the parts west of Mt Elbrus (5633 m), the Central Greater Caucasus between Mt Elbrus and Mt Kazbek (5044 m), and the Eastern Greater Caucasus between Mt Kazbek and the Caspian Sea.

This chapter explores the character and species composition of plant communities with *H. mantegazzianum* in its native area of the Western Greater Caucasus and how the distribution and spread of the species is facilitated through human influences. According to Grossheim (1948) and Mandenova (1950), the centre of speciation of *H. mantegazzianum* lies in the Western Greater Caucasus. Nakhutsrishvili (2003a) considers the species as a relict of the Tertiary flora in Europe, whereas 'the high mountain flora of the Caucasus consists mainly of boreal and arctic-alpine elements', which are thought to have colonized during the Pleistocene. However, Feodorov (1952) and Kharadze (1960) proposed that the most typical representatives of the Caucasian high mountain flora were of autochthonous origin, based around a Tertiary nucleus and developed during the Quaternary. *Heracleum man-tegazzianum* is an endemic element of the floristic region 'Colchis' (Gagnidze,

1974) with a restricted range in the botanic–geographic districts Maykopskiy, Labinsiy, Zelenchukskiy, Abkhazskiy Kavkasioniy, Svanetsko–Rachinskiy Kavkasioniy and Cherkessko–Abkazsko–Megrelskiy (see Jahodová *et al.*, Chapter 1, this volume).

Subalpine tall-forb communities, for which Levier and Sommier have coined the term 'mammoth flora' ('Mammutflora' (Levier, 1894)), are considered to be the natural habitats of *H. mantegazzianum*. This vegetation occurs in forest clearings, i.e. humid humus-rich slopes within the montane forest belt, which 'certainly were once covered by forest' (Rikli, 1914). Enchantedly, Rikli (1914) writes:

> A great number of giant herbs of various families, in many cases head-high or even larger, cover whole depressions or wide stretches of the slopes with a flowerage beyond comparison, silhouetting delightfully with their diversified colour effects against the abounding lush-green foliage. In this dissipative glory sometimes horse and horseman disappear. (...) Candelabra-branched, enormous *Apiaceae* raise in front of us, occasionally up to three metres high, spreading their large white umbels. Especially conspicuous is *H. mantegazzianum*, which is also grown as classy ornamental plant in parks of Central Europe.[1]

Today the distribution of *H. mantegazzianum* ranges from altitudes above the timberline (concerning the timberline see Nakhutsrishvili (1999)) down to the lowlands of the north- and south-western forelands of the Greater Caucasus. Correspondingly, there is a great diversity of tall-forb communities harbouring *H. mantegazzianum*.

Habitat Description and Ecology

Methods

North and south of the main ridge of the Western Greater Caucasus on Russian territory (Krasnodar Territory: Seversk, Goryachiy Kluch, Apsheronsk, Tulski, Psebai, Sochi, Abchasia: Sukhum, Gudauta; Fig. 2.1), 94 vegetation relevés with a plot size of 5 × 5 m were recorded in stands of *H. mantegazzianum* in 2002–2004 according to the method of Braun-Blanquet (1965) using the modified 9-degree abundance scale proposed by Barkman *et al.* (1964). The relevés were sampled by D. Geltman, N. Portenier (Komarov-Institute, Academy of Science St Petersburg, Russia), A. Otte and J. Thiele (Justus Liebig-University Giessen, Germany). Subsequently D. Geltmann and N. Portenier identified taxa that could not be identified in the field and checked the nomenclature.

The relevés were analysed with respect to their structure, floristic composition, and abiotic and biotic conditions. For each relevé, we recorded region, district, locality, coordinates (measured by a hand-held global positioning system (GPS), accuracy ± 5 m), slope aspect, inclination, altitude, height (cm), coverage (%) of tree-, shrub-, herb-, moss-layer and bare soil (%), topographic

---

[1] Translation A. Otte.

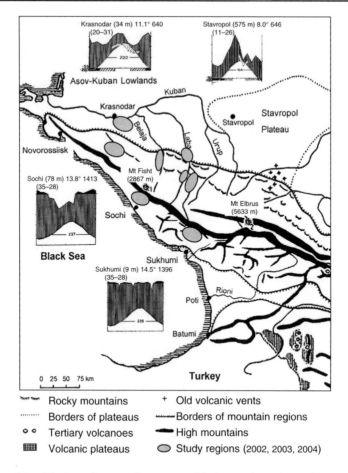

Fig. 2.1. Map of the Western Greater Caucasus with the study regions. Climate diagrams (complemented and redrawn from Walter (1974)) show that the mountain ridge divides a region with humid (perhumid season with >100 mm of precipitation in black with the scale reduced tenfold) warm temperate to subtropical climate in the south from a region with typical temperate (nemoral) climate in the north.

position of the relevé (e.g. river bank, valley bottom, slope foot, slope), position of *H. mantegazzianum* within the habitat (e.g. edge, throughout the whole habitat), degree of shading, land use (e.g. meadow, pasture, abandoned agricultural land, natural habitat), geological formation, degree of soil moisture and soil texture class (sand, silt, clay).

Major gradients in the vegetation relevés were explored by Detrended Correspondence Analysis (DCA), a method of indirect gradient analysis (Lepš and Šmilauer, 2003), using the program package CANOCO for Windows 4.5. For gradient analysis, classes of the 9-degree abundance scale were transformed to numerical ranks (values 1 to 9). Detrended Correspondence Analysis was set up with detrending by second order polynomials, downweighting of rare species, and focus on inter-sample distances. The length of the main gradient (first DCA axis) was 3.8. Therefore, Hill's scaling was applied (ter Braak and Šmilauer, 1998). In addition, response curves of selected species along the first ordination axis were calculated by Generalized Additive Models (GAM) in CANOCO.

Altitudinal distribution of vegetation, agricultural land use and habitats with *Heracleum mantegazzianum*

The ridge of the Greater Caucasus (3600 m a.s.l.), with distinctive peaks such as Mt Fisht (2852 m), Mt Elbrus, Mt Kazbek and Mt Babadagh (3637 m), is the watershed between North Caucasus and Transcaucasia (Henning, 1972) and divides the non-tropical and subtropical zonal climate (Fig. 2.1). A comparison of the distribution of precipitation (Henning, 1972) with the occurrence of *H. mantegazzianum* (as given in Grossheim, 1948) shows a strong relationship between the areas with the highest annual precipitation (> 1000 mm), which lie in the Western Greater Caucasus and the Colchis, and the original range of this species.

The habitat types where *H. mantegazzianum* is found vary considerably along the altitudinal gradient from 50 to 2200 m a.s.l. within the Western Greater Caucasus. An overview of the vegetation belts, predominant agricultural land use and habitat types of *H. mantegazzianum* at different altitudes, based on Dolukhanov (1966), Walter (1974), Nakhutsrishvili (1999) and our studies, is presented in Table 2.1.

The north-western foreland of the Caucasus (50–150 m a.s.l.), which has formerly been dominated by *Stipa*-steppe, is currently used by extensive agriculture, the dominant crops being winter wheat (*Triticum aestivum* L.), winter barley (*Hordeum vulgare* L. em. Alef.), maize (*Zea mays* L.) and sunflower (*Helianthus annuus* L.). The enormous crop fields of the area are enclosed by broad shrub belts. Only along the Kuban River and its tributaries do sparse remnants of gallery-like alluvial softwood forests occur. The foot of the mountain chain is marked by loess-covered, partly eroded ridges (original forest steppe with oak species, *Quercus petraea* Liebl., *Q. robur* L., above 150–600 m) stretching parallel with the mountain chain, which are grazed where they become too steep for crop cultivation. In the lower montane forest belt (above 600–1000 m), potentially species-rich deciduous forests with *Carpinus betulus* L. (from 600 m), *Quercus petraea* (up to 1000 m), *Fagus orientalis* Lipsky (on northern slopes from 400 m) and several species of *Acer*, *Tilia*, *Ulmus* and *Fraxinus* predominate. Owing to higher annual precipitation (up to 800 mm), the spectrum of agricultural land-use with cereals is complemented by extensive pasture areas and forage crops (lucerne, *Medicago sativa* L.). At altitudes of 50–1000 m, *H. mantegazzianum* predominantly occupies ruderalized, open habitats with good water supply in alluvial softwood forests along the Kuban River and its northern tributaries (Little Laba, Bjelaja, Selentschuk). Occasionally *H. mantegazzianum* forms dominant stands on abandoned crop fields and grasslands close to running waters.

Table 2.1. Altitudinal distribution of vegetation belts in the Western Greater Caucasus, prevailing agricultural land-use form and habitats with *H. mantegazzianum*. Numbers of vegetation relevés documenting particular vegetation types are shown. In total, this study is based on 94 relevés.

Altitude (m a.s.l.)	Vegetation (Dolukhanov, 1966; Henning, 1972)	Agricultural land use (Henning, 1972; Nakhutsrishvili, 1999; own surveys)	Alluvial softwood forest	Clearing	Arable land	Meadow	Pasture	Chalk grassland	Abandoned grassland	Livestock enclosure	Subalpine tall forb communities	Total
50–150	Potential *Stipa*-steppe, softwood forests at Kuban river and its tributaries	Arable land (grain crops)	11	3	1	1						16
> 150–600	Open woodland (with *Quercus petraea*, *Q. robur*)	Arable land (grain crops), pastures on hill tops	11	1	1		1					14
> 600–1000	Lowland hardwood forests with *Quercus* ssp., *Carpinus betulus*, *Acer* spp., *Tilia* spp., *Ulmus* spp., *Fraxinus* spp. (*Fagus orientalis*)	Arable land (grain crops), hay-meadows, pastures	7	1				4	4			16
> 1000–1400	Montane forests with *Fagus orientalis* (*Abies nordmanniana*, *Picea orientalis*)	Hay-meadows, pastures, less arable land		5				1	2	1	2	11
> 1400–2200	Montane forests with *Abies nordmanniana* and *Picea orientalis* (*Fagus orientalis*)	Meadows, pastures	3	9					4	4	17	37
> 2200–2400	Park-like forests with *Betula verrucosa*, *Acer trautvetteri*; Crook-stem forests of *Betula pubescens*, *B. medwedewii*, *B. litvinowii* mixed with subalpine meadows	Alpine pasturing systems (cattle, sheep)										
> 2400–3300	Alpine meadows	Alpine pasturing systems (cattle, sheep)										
		Total	32	19	2	1	1	5	10	5	19	94

Fig. 2.2. Vegetation type 1: tall-forb vegetation with *H. mantegazzianum* on a river bank consisting of sand and gravel near the confluence of the rivers Arkhyz and Psysch (1500 m a.s.l., July 2003). *Heracleum mantegazzianum* grows around the edges of an *Alnus incana*-alluvial forest. Photo: A. Otte.

In the lower montane belt (above 1000–1400 m a.s.l.) the forest fraction, which is dominated by *Fagus orientalis*, increases considerably. The occurrence and proportion of *Abies nordmanniana* (Stev.) Spach and *Picea orientalis* (L.) Link in the forests rise with increasing altitude. Under favourable climatic conditions, *Fagus* forests even reach to the upper forest limit (above 2000 m) or build up mixed forests with *A. nordmanniana* and *P. orientalis*. When pure coniferous forests are developed, they commonly reach up to 2200 m. Throughout the montane forest belt, stands of *H. mantegazzianum* occur in wet forest gaps, along open banks of mountain torrents (Figs 2.2 and 2.3), as well as on abandoned hay meadows, pastures and livestock enclosures. Extensive meadows and pastures with interspersed tree groups stretch from the valley upwards into the high montane region (Figs 2.4 and 2.5). These large islands in the forest can be clearly recognized on satellite images (Fig. 2.6). The lower, less steep grasslands are usually used as hay meadows and cut once or twice a year, whereas areas at higher altitudes are used as pastures (Fig. 2.7). The proportion of abandoned grassland in these steep areas, which are difficult to manage, is already high and still increasing. Starting with single individuals along forest and scrub edges, extensive tall-forb fringes with *H. mantegazzianum* disperse within abandoned grassland by water, wind and human activities from the montane belt down to the valley bottoms. Such single individuals exert a considerable seed pressure since they produce more seeds than individuals within dense stands (Hüls, 2005) and therefore potentially make a significant contribution to the spread of *H. mantegazzianum* in open habitats. This means that the spread of *H. mantegazzianum* at lower altitudes is facilitated by reduced grassland management (mowing, grazing) in

Fig. 2.3. Vegetation type 1: seedlings of *H. mantegazzianum* at the border of a river bank near the waterline (confluence of the rivers Arkhyz and Psysch, 1500 m a.s.l., July 2003). Photo: A. Otte.

the montane belt. A similar unhampered spread of *H. mantegazzianum* in a cultural landscape depopulated after World War II was documented by Müllerová *et al.* (2005) in the Czech Republic (see Pyšek *et al.*, Chapter 3, this volume). Comprehensive studies from Germany (see Thiele *et al.*,

Fig. 2.4. Vegetation type 2: montane ruderal tall-forb vegetation with *H. mantegazzianum* on Jurassic limestone below Lago Naki (1200 m a.s.l., July 2003), where *H. mantegazzianum* occurs in clearings of extensively used or abandoned grassland within the mixed montane forest belt composed of *Fagus orientalis* and *Abies nordmanniana*. Photo: A. Otte.

Fig. 2.5. Vegetation type 2: park-like mosaic landscape dominated by grassland in the mixed montane forest belt composed of *Fagus orientalis* and *Abies nordmanniana*. *Heracleum mantegazzianum* occupies ecotones around hedgerows and deciduous forests. From these sites *H. mantegazzianum* spreads quickly if agricultural land-use (here hay-making) is given up (below Lago Naki, 1200 m a.s.l., July 2003). Photo: A. Otte.

Chapter 8, this volume; Thiele and Otte, 2006) show that *H. mantegaz-zianum* can be considered as an indicator species for the abandonment of grassland.

The timberline lies on average between 2000 and 2200 m a.s.l., that of *Pinus sylvestris* L. reaches up to 2650 m (Plesnik, 1972). It should be noted that owing to human impact, the current timberline is about 300–400 m lower than it was in the 19th century (Dolukhanov, 1966; Nakhutsrishvili, 2003a). The subalpine belt (above 2200–2400 m) is characterized by park-like, low-growing maple forests (*Acer trautvetteri* Medwedev). Under moist conditions, a crook-stem (krummholz) zone with low-growing birches (*Betula pubescens* Ehrh., *B. medwedewii* Regel) or beech (*Fagus orientalis*) occurs. The transition area between the upper timberline and the subalpine zone harbours the natural habitats of the tall-forb communities of the 'mammoth flora' with *H. mantegazzianum* (Levier, 1894; Radde, 1899; Déchy, 1906; Engler, 1913; Rikli, 1914; Nakhutsrishvili, 1999) gradually giving way to high mountain meadows and pastures, which are pure grass and sedge stands in the alpine belt (2400–3300 m). In the subalpine and alpine belts, traditional mountain pasture systems with relatively high livestock densities of young cattle and sheep predominate (see photographs in Déchy, 1906), whereas today winter fodder comes from large-scale meadows in the valleys. The subnival zone extends up to 3800–4000 m a.s.l.

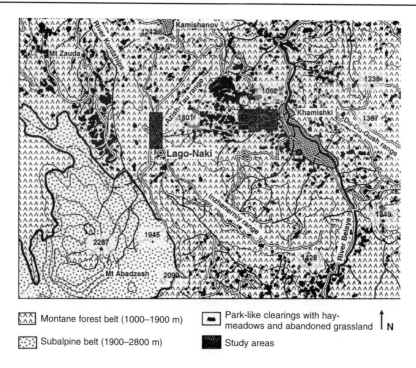

⟨ᴧᴧᴧ⟩ Montane forest belt (1000–1900 m)   ■ Park-like clearings with hay-
                                                  meadows and abandoned grassland   ↑ N

⟨∴∴⟩ Subalpine belt (1900–2800 m)   ■ Study areas

Fig. 2.6. Park-like clearings with grassland and old fields within the montane forest belt (Lago Naki, N 44°7′47.2″/ E 40°2′9.7″), District Tulski, Krasnodar Territory, Russia (redrawn after Kuznetsov and Truschin (2001)).

On the southern side of the Western Greater Caucasus, facing the Black Sea, the alpine zone extends about 100 m higher than on the northern slope, whereas the timberline lies between 2000 and 2200 m. Birch species like *Betula litwinowii* Doluch. and *B. megrelica* Sosn. join the vegetation building the subalpine krummholz, while *Quercus pontica* C. Koch occurs together with beech and birch. Between 2000 and 2200 m, *Abies nordmanniana*, *Fagus orientalis*, *Picea orientalis* and *Acer pseudoplatanus* L. build up mixed forests. The upper border of the deciduous forest belt is at about 1500 m and below this follows a Colchic-Central European deciduous forest belt with extensive mixed beech forests (Walter, 1974).

Environmental conditions of tall forb vegetation

Nakhutsrishvili (1999) describes the favourable environmental conditions for tall-forb vegetation in the subalpine belt of the Caucasus as follows: 'optimal air and soil temperature, negligible daily temperature fluctuations, high air humidity, high solar radiation and rich soils'. The tall-forb vegetation is composed of tall (2–3 m) perennials, mostly dicotyledonous plants, which are characterized by shoots without rosettes, short tap roots and rhizomes. Plant

Fig. 2.7. Vegetation type 3: ruderalized subalpine meadows with *H. mantegazzianum* in the montane forest belt of *Abies nordmanniana.* It should be noted, that these 'meadows' are pastures. It is likely that the observation of a herd of 30 aurochs in the Laba catchment area (Radde, 1899) took place in a similar setting (Lago Naki, 1660 m a.s.l., July 2003). Photo: A. Otte.

diversity of tall-forb vegetation is high and it is noticeable that *H. mantegaz-zianum* within these species-rich communities never dominates and its cover usually does not exceed 25% (Fig. 2.8). Tappeiner and Cernusca (1996, 1998) studied the spatial distribution of biomass in Caucasian tall-forb communities. They carried out detailed measurements of canopy structure, microclimate, energy fluxes and CO_2 exchange in a *Heracleum sosnowskyi* Mand. site (mountain research station Kasbegi, 2200 m a.s.l., on a north-facing slope bordered by two streams). Their results show that in mountain ecosystems structural properties of the canopy (dense leaf layers, accumulation of attached dead plant material) may enhance the development of a warm micro-environment. A high portion of leaf area in the upper canopy (inverted pyramid stands) allows the optimization of structural characteristics favouring a high carbon gain. Owing to strong similarity between giant hogweeds with respect to size and plant architecture, the effects of *H. sosnowskyi* on stand structure and micro-environment are probably also valid for *H. mantegazzianum.* In the latter, a high plasticity with respect to the length and orientation of petioles enables the species to position its leaves horizontally, away from the ground, and to expose a large leaf area in the upper canopy layer (Hüls, 2005). Further requisites for canopy photosynthesis are high soil nutrient availability and rapid nutrient cycling (Tappeiner and Cernusca, 1998). Therefore, *H. mantegaz-zianum* is not only an obligate light-demanding species, which easily eclipses its lower growing competitors, but it also requires habitats with good nutrient availability. On the other hand, *H. mantegazzianum* may create suitable

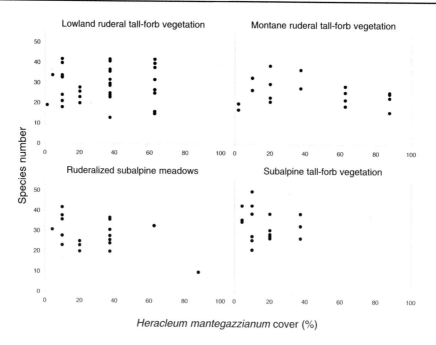

Fig. 2.8. Species densities (per 25 m^2) of stands of four vegetation types (cf. Table 2.2) in relation to the cover of *H. mantegazzianum*.

microclimatic conditions for other species, which may increase local species density. The observations of Nakhutsrishvili (2003a, 2003b) and Onipchenko (2002) that *H. mantegazzianum* stands on subalpine grasslands are species-poor (< 15 species per 25 m^2) cannot be confirmed by our study, where on average about 29 species per 25 m^2 were found (cf. Table 2.2). The reason for this discrepancy could be a bias towards stands dominated by *H. mantegazzianum* in the studies of Nakhutsrishvili (1999, 2003a, 2003b) and Onipchenko (2002). Nakhutsrishvili (1999) in particular based the characterization of associations on so-called leading species (*H. mantegazzianum* among them) and their dominance. According to our analyses, cover percentages of *H. mantegazzianum* are highest at medium altitudes (700–1400 m) and decrease towards the subalpine belt (Fig. 2.9).

In our 38 relevés from the high montane to subalpine belt (1300–1950 m a.s.l.), *H. mantegazzianum* with a cover exceeding 50% was only recorded twice. However, these two cases represented former livestock enclosures, which are among the most fertile habitats of alpine pasturing systems with respect to nitrogen and phosphorus because of fertilization by animal droppings. Usually the vegetation has been destroyed by animals so that in addition to *H. mantegazzianum*, other competitive species such as *Rumex alpinus* L. can successfully establish on the eutrophicated open ground following the abandonment of pastures. Such a stand dominated by *H. mantegazzianum* was used as a representative tall-forb community in the above-mentioned study

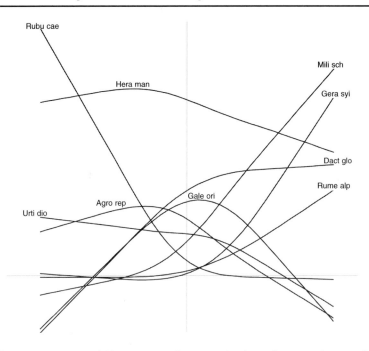

Rubu cae

Mili sch

Hera man

Gera syi

Dact glo

Rume alp

Gale ori

Agro rep

Urti dio

Fig. 2.9. Response curves of *H. mantegazzianum* and selected companion species along the first axis of DCA ordination of 94 relevés from the native distribution range in the Western Greater Caucasus. The first axis represented an altitudinal gradient from 50 to 2200 m a.s.l. The maximum predicted cover-abundance of *H. mantegazzianum* was in the class '2b' (15–25% cover). Abbreviations of species names: Agro rep = *Agropyron repens*; Dact glo = *Dactylis glomerata*; Gale ori = *Galega orientalis*; Gera syl = *Geranium sylvaticum*; Hera man = *Heracleum mantegazzianum*; Mili sch = *Milium schmidtianum*; Rubu cae = *Rubus caesius*; Rume alp = *Rumex alpinus*; Urti dio = *Urtica dioica*.

of structural characteristics (Tappeiner and Cernusca, 1996) and compared with hay meadows, lightly and heavily stocked pastures, and *Rhododendron* scrub communities. Normally, the cover of *H. mantegazzianum* in high montane to subalpine communities is below 25%. Due to the occurrence of many grasses (e.g. *Dactylis glomerata* L., *Milium schmidtianum* K. Koch) and tall forbs (e.g. *Cephalaria gigantea* (Ledeb.) Bobrov, *Geranium sylvaticum* L., *R. alpinus*, and others) these stands can be regarded as species rich (33.5 ± 7.5 species/25 m², mean ± SD, *n* = 18; cf. Table 2.2, Fig. 2.8). However, overgrazing of such communities favours further spread of *H. mantegazzianum* within the community and results in the development of stands dominated by this species. Most relevés with cover of *H. mantegazzianum* exceeding 50% are from montane altitudes of 700–1440 m a.s.l., where *H. mantegazzianum* spreads in abandoned ruderal grassland (meadows, pastures). In parallel with the highest cover of *H. mantegazzianum* run the highest cover values of *Agropyron repens* (L.) P. Beauv., an indicator of disturbance and ruderalization of farmland and marginal areas (Fig. 2.9). In the lowlands of the Kuban depression (70–600 m a.s.l.), stands dominated by

Table 2.2. Overview of vegetation types with *H. mantegazzianum* in the Western Greater Caucasus. Vegetation types were differentiated according to altitude and groups of differential species (denoted as d1–d8) that show a preference for certain vegetation types. Relevés from the lower montane belt (vegetation types 1, 2) represent nitrophilous tall-forb vegetation with (type 1) and without (type 2) alluvial forest species. Relevés from the sub-alpine belt (vegetation types 3, 4) represent ruderalized subalpine meadows (type 3) and subalpine tall-forb communities (type 4). Relevés of the latter type are from the southern slope of the main ridge of the Western Greater Caucasus, within the floristic region Colchis. At the head of the table are the number of relevés per vegetation type (set to 100%), altitude range, cover of trees, shrubs, and herbs and the number of species per 25 m^2 plot. Values in the body of the table represent species frequencies within vegetation types. Although placed in stands of *H. mantegazzianum*, individuals of this species were lacking in four plots. These relevés were kept in the analysis since they still represent the typical vegetation of *H. mantegazzianum* stands.

No. of vegetation type	1	2	3	4
Relevés (n)	37	19	20	18
Altitude range (m a.s.l.)	70–600	710–1440	1300–1800	1620–1950
Average cover of tree layer (%)	35	1	5	
Average cover of shrub layer (%)	6	1		
Average cover of herb layer (%)	75	89	88	100
Average number of species (5 x 5 m)	30	25	29	34
Heracleum mantegazzianum Somm. et Levier	**97**	**100**	**85**	**100**

1 Nitrophytic tall-forb vegetation (Galio-Urticetea Passarge ex Kopecký 1969)
1.1 Lowland ruderal tall forb vegetation

	1	2	3	4
d1 *Rubus caesius* L.	**81**	11	10	
Erigeron annuus (L.) Pers.	57			
Lythrum salicaria L.	30			
Artemisia vulgaris L.	27	11		
Torilis japonica (Houtt.) DC.	27	11		
Euphorbia stricta L.	22	16		
Petasites albus (L.) Gaertn.	22		5	
d2 *Agropyron repens* (L.) P. Beauv.	**54**	**68**		
Ranunculus repens L.	51	37	5	
Galium aparine L.	46	42	5	
Calystegia sepium (L.) R. Br.	46	26		
Stachys sylvatica L.	38	63	5	6
Geum urbanum L.	38	32		
Mentha longifolia (L.) Huds.	35	32	5	
Glechoma hederacea L.	35	21		
d3 *Urtica dioica* L.	49	32	50	
Chaerophyllum bulbosum L.	46	42	20	
Aegopodium podagraria L.	41	11	30	
Inula helenium L.	30	32	45	6
Telekia speciosa (Schreb.) Baumg.	16	21	25	

Table 2.2. *Continued*

No. of vegetation type	1	2	3	4
1.2 Montane ruderal tall-forb vegetation				
d4 *Galega orientalis* Lam.	14	**68**	60	
Poa trivialis L.	8	68	15	
Petasites hybridus (L.) P. Gaertn.,				
B. Mey. et Scherb.	16	47	40	
Carex sylvatica Huds.	8	37	30	
d5 *Dactylis glomerata* L.	11	**79**	80	78
Symphytum asperum Lepech.	16	53	90	50
Chaerophyllum aureum L.	8	53	70	89
Hesperis matronalis L.	11	5	50	61
2 Subalpine meadows and subalpine tall-forb vegetation				
(Mulgedio-Aconitetea Hadač & Klika in Klika & Hadač 1944)				
2.1 Ruderalized subalpine meadows				
d6 *Milium schmidtianum* K. Koch			70	94
Cephalaria gigantea (Ledeb.) Bobrov			50	94
Poa longifolia Trin.		5	55	56
Heracleum ponticum (Lipsky)				
Schischk. ex Grossh.	3	5	50	83
Aconitum orientale Mill.			40	50
Campanula latifolia L.		11	35	83
Veratrum lobelianum Bernh.		11	20	72
Rumex alpestris Jacq.			15	72
d7 *Rumex alpinus* L.		5	**55**	17
Lamium album L.		11	45	17
2.2 Subalpine tall forb vegetation				
d8 *Geranium sylvaticum* L.			10	**83**
Cirsium abkhasicum (Petr.) Grossh.				83
Vicia grossheimii Ekvtim.	3	5		67
Inula grandiflora Willd.				56
Asperula taurina L.			15	50
Betonica macrantha K. Koch			15	44
Silene vulgaris (Moench) Garcke			10	44
Euphorbia squamosa Willd.	3			44
Bistorta officinalis Delarbre	3		15	39
Senecio propinquus Schischk.	3		15	33
Woronowia speciosa (Alboff) Juzepczuk				33
Other species				
Lapsana communis agg. L.	43	74	70	83
Festuca gigantea (L.) Vill.	35	42	50	83
Brachypodium sylvaticum (Huds.)	27	42	45	11
P. Beauv.				
Vicia sepium L.	14	26	10	11
Geum latilobum Sommier et Levier	5	16	35	44
Rumex obtusifolius L.	8	37	15	
Angelica pachyptera Ave-Lall.	19	5	25	
Geranium gracile Ledeb. ex Nordm.			45	22

H. mantegazzianum are confined to abandoned open soils of former farm-lands and glades along rivers. In the shade of alluvial softwood forests, *H. mantegazzianum* occurs with considerably reduced abundance (< 25% cover) in ruderal stands of *Rubus caesius* L. and *Urtica dioica* L.; these species increase owing to their shade tolerance, whereas *H. mantegazzianum* decreases in shaded habitats.

Main Vegetation Types with *Heracleum mantegazzianum* and their Phytosociological Classification

Species composition of communities with *H. mantegazzianum* in the Greater Caucasus is primarily related to altitude (Fig. 2.10). The first axis of the

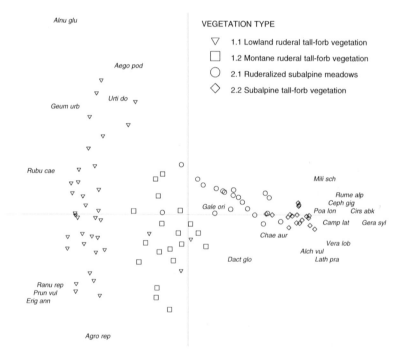

Fig. 2.10. DCA ordination diagram (axes 1 and 2) of 94 vegetation relevés with *H. mantegazzianum* from the native distribution range in the Western Greater Caucasus. Sites and selected species are plotted. Vegetation types (cf. Table 2.2) are presented as symbols. Abbreviations of species names: Aego pod = *Aegopodium podagraria*; Agro rep = *Agropyron repens*; Alch vul = *Alchemilla vulgaris* agg.; Alnu glu = *Alnus glutinosa*; Camp lat = *Campanula latifolia*; Ceph gig = *Cephalaria gigantea*; Chae aur = *Chaerophyllum aureum*; Cirs abk = *Cirsium abkhasicum*; Dact glo = *Dactylis glomerata*; Erig ann = *Erigeron annuus*; Gale ori = *Galega orientalis*; Gera syl = *Geranium sylvaticum*; Geum urb = *Geum urbanum*; Lath pra = *Lathyrus pratensis*; Mili sch = *Milium schmidtianum*; Poa lon = *Poa longifolia*; Prun vul = *Prunella vulgaris*; Ranu rep = *Ranunculus repens*; Rubu cae = *Rubus caesius*; Rume alp = *Rumex alpinus*; Urti dio = *Urtica dioica*; Vera lob = *Veratrum lobelianum*.

ordination represents the altitudinal amplitude covered by *H. mantegaz-zianum*. Vegetation relevés from lower altitudes (70–600 m), representing plant communities with characteristic species such as *Alnus glutinosa* (L.) P. Gaertn., *Geum urbanum* L., *R. caesius*, *Erigeron annuus* (L.) Pers. and *U. dioica*, are positioned at the left side of the ordination diagram, whereas those from the subalpine altitudinal belt (1620–1950 m), with *Milium schmid-tianum, Cephalaria gigantea, Poa longifolia* Trin. and *Geranium syl-vaticum*, are situated on the right. The second axis represents a gradient of land use (arable land, meadow, pasture, abandoned arable land and grassland, forest, no land use). The greatest heterogeneity (scatter) was found among relevés from the lowlands, where *H. mantegazzianum* occurs in a wide spectrum of habitats ranging from near-pristine alluvial forests (*Alnus glutinosa*) to old fields (*Agropyron repens, Ranunculus repens* L., *E. annuus*). With increasing altitude, agricultural land use is restricted to grassland systems (Table 2.1). As a consequence of this combination of abiotic and biotic conditions, relevés from medium to subalpine altitudes scatter only narrowly around the second axis and are thus floristically more homogeneous (Fig. 2.10).

Vegetation relevés with *H. mantegazzianum* from the planar altitude to lower montane altitude (70–1400 m a.s.l.) represent nitrophilous tall-forb vegetation (class Galio-Urticetea Passarge ex Kopecký 1969); Table 2.2, vegetation types 1, 2). Plant communities of the high montane coniferous forest belt (1500 m) belong to the phytosociological class of subalpine meadows and subalpine tall-forb vegetation (Mulgedio-Aconitetea Hadač & Klika in Klika & Hadač 1944; Table 2.2, vegetation type 3, 4). The basic group of species common to both classes is, besides *H. mantegazzianum*, those preferring habitats with high air humidity including *Lapsana communis* L., *Festuca gigantea* (L.) Vill., *Brachypodium sylvaticum* (Huds.) P. Beauv., *Vicia sepium* L. and *Geum latilobum* Sommier et Levier.

Within the class of nitrophilous tall-forb vegetation (Table 2.2, vegetation types 1, 2), relevés with *H. mantegazzianum* belong to the order *Glechometalia hederaceae* Tx. in Tx. et Brun-Hool 1975 and the alliance Aegopodion Tx. 1967. The co-occurring species are ruderal tall forbs, e.g. *U. dioica, Chaerophyllum bulbosum* L., *Aegopodium podagraria* L., *Inula helenium* L. (all belonging to a group of differential species designated as d3 in the Table), *Agropyron repens, Galium aparine* L., *Calystegia sepium* (L.) R. Br., *Glechoma hederacea* L. and *Geum urbanum* (all d2). Differential species of vegetation type 1 in the foreland of the Caucasus (70–600 m) are alluvial forest species (d1: *R. caesius, Lythrum salicaria* L., *Euphorbia stricta* L., *Petasites albus* (L.) Gaertn.) and ruderals (*Erigeron annuus, Artemisia vulgaris* L., *Torilis japonica* (Houtt.) DC.), which disappear with increasing altitude (Figs 2.2 and 2.3). Consequently, vegetation type 2 (710–1440 m) is mainly negatively characterized by the lack of species group d1, which is compensated by the emergence of tall forbs with montane distribution (*Galega orientalis* Lam., *Chaerophyllum aureum* L., *Symphytum asperum* Lepech.), grasses and sedges (*Poa trivialis* L., *Dactylis glomerata, Carex sylvatica* Huds.; Figs 2.4 and 2.5).

Subalpine meadows and subalpine tall-forb vegetation (vegetation types 3, 4) are floristically discrete; character species of the class Mulgedio-Aconitetea

Fig. 2.11. Vegetation type 3: ruderalized subalpine meadows with *H. mantegazzianum* and *Rumex alpinus* in the montane forest belt of *Abies nordmanniana* on a former livestock enclosure (Lago Naki, 1660 m a.s.l., July 2003). Photo: A. Otte.

(Table 2.2), are according to Onipchenko (2002), *Milium schmidtianum, Geranium sylvaticum, Campanula latifolia* L. and *Rumex alpestris* Jacq. (d6). Subalpine meadows with *H. mantegazzianum* can be separated into ruderalized subalpine meadows (vegetation type 3, d7) with *R. alpinus* and *Lamium album* L. (Figs 2.7 and 2.11) and subalpine tall-forb communities (vegetation type 4, d8) with *Cirsium abkhasicum* (Petr.) Grossh., *Vicia grossheimii* Ekvtim., *Inula grandiflora* Willd., *Asperula taurina* L., *Betonica macrantha* K. Koch, and others (Figs 2.12 and 2.13). According to Onipchenko (2002), who did phytosociological studies within the alpine vegetation of the Teberda Reserve, both vegetation types should be classified within the order Rumicetalia alpini Mucina in Karner & Mucina 1993 and the alliance Rumicion alpini Rübel ex Klika in Klika & Hadač 1944. The vegetation relevés representing vegetation type 4 were collected on the southern slope of the main Greater Caucasus ridge, thus within the floristic region Colchis. Their distinctness is reflected by the occurrence of a calciphilous species *Woronowia speciosa* (Alboff) Juzepczuk.

The differentiation of the vegetation in response to habitat conditions is shown in Fig. 2.14: natural subalpine meadows are characterized by high altitude, high radiation, high proportion of grasses and low management impact, whereas ruderal tall-forb vegetation of lower altitudes is characterized by the light shade of a tree layer, a high proportion of sedges and rushes and a high proportion of arable land use. It is remarkable that species densities are higher when *H. mantegazzianum* stands are associated with agricultural manage-

Fig. 2.12. Vegetation type 4: subalpine tall-forb vegetation with *H. mantegazzianum* on the south-eastern side of the Western Greater Caucasus within the montane belt (Adler district of Sochi, 1710 m a.s.l., August 2002). Note remnants of subalpine forests above the site at the top right edge. Photo: N. Portenier.

ment such as crop fields and pastures, which lead to an open stand structure. Lower species densities are found when *H. mantegazzianum* occurs in rarely used or less disturbed habitats (alluvial forests, subalpine tall-forb vegetation; Fig. 2.8).

Pristine vegetation with *H. mantegazzianum*, i.e. sites not influenced by human activities, cannot be documented in the present study. The original subalpine sites of the species at the forest line in the transition zone to alpine grassland vegetation have been lost due to the displacement of the treeline by up to 400 m in altitude into the montane belt. However, the documented stands at the south-eastern side of the mountain ridge of the Western Greater Caucasus at altitudes above 1700 m a.s.l., which contain more than 30% of endemic species, are very distinct with respect to their floristic composition. Owing to a high frequency of ruderal species such as *R. alpinus*, *L. album*, *Anthriscus nemorosa* (Bieb.) Spereng. etc., these communities should be placed within the alliance Rumicion alpini, a unit that with respect to its floristic compositions represents a group of natural and anthropogenic ruderal plant communities. To find natural vegetation with *H. mantegazzianum*, it would be worthwhile to systematically search the krummholz ecotone along the northwestern and south-eastern sides of the Greater Caucasus.

Fig. 2.13. Vegetation type 4: Subalpine tall-forb communities with *H. mantegazzianum* (*Senecio othonnae*, *Cephalaria gigantea* and *Cirsium abkhasicum*) occur often on wet steep slopes where pasturing is abandoned (Adler district of Sochi, 1710 m a.s.l., August 2002). Photo: N. Portenier.

Conclusions

In this chapter we have summarized the present knowledge on the occurrence of *H. mantegazzianum* within the primary distribution range of the Western Greater Caucasus. In its native range, the species shows a wide ecological amplitude and consequently occurs in a variety of habitats along an altitudinal range of between 50 and about 2000 m. At lower elevations, *H. mantegazzianum* is an element of nitrophilous tall-forb vegetation (class Galio-Urticetea) and occurs both in alluvial forests and old fields. At higher altitudes the species grows in subalpine meadows and subalpine tall-forb vegetation (class Mulgedio-Aconitetea). It is noticeable that in the latter vegetation type, *H. mantegazzianum* usually covers less than 25% of the ground and does not attain dominance in these species-rich plant communities. However, pristine sites

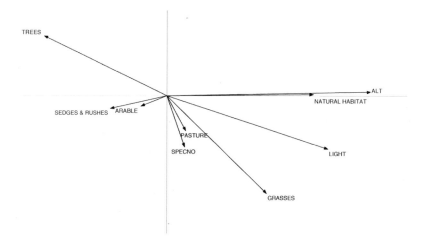

Fig. 2.14. DCA ordination diagram (axes 1 and 2) of 94 vegetation relevés with
H. mantegazzianum from the native distribution range in the Western Greater Caucasus.
Vectors of environmental variables significantly correlated with ordination axes are plotted.
Abbreviations: TREES/SEDGES&RUSHES/GRASSES = cover percentages of the
respective plant types; LIGHT = light supply; ALT = altitude; SPECNO = number of vascular
plant species per 25 m^2; ARABLE/PASTURE/NATURAL HABITAT = frequencies of the
respective (former) land-use regimes.

with *H. mantegazzianum* in the Greater Caucasus may have been lost due to
anthropogenic displacement of the treeline into the montane belt. No pristine
habitats could be documented in this study.

Acknowledgements

We thank D. Geltman and N. Portier for their help during data collection, P.
Pyšek, D. Simmering and two anonymous referees for insightful comments on
an earlier version of the manuscript, and M.J.W. Cock for language revision.
The study was supported by the project 'Giant Hogweed (*Heracleum mantegazzianum*) a pernicious invasive weed: developing a sustainable strategy for
alien invasive plant management in Europe', funded within the 'Energy,
Environment and Sustainable Development Programme' (grant no. EVK2-CT-
2001-00128) of the European Union 5th Framework Programme.

References

Barkman, J.J., Doing, H. and Segal, S. (1964) Kritische Bemerkungen und Vorschläge zur quan-
titativen Vegetationsanalyse. *Acta Botanica Neerlandica* 13, 394–419.
Braun-Blanquet, J. (1965) *Plant Sociology: the Study of Plant Communities*, 1st edn. Hafner,
London.

Déchy, M. v. (1906) *Kaukasus. Reisen und Forschungen im kaukasischen Hochgebirge*, Vol. II. Reimer, Berlin.

Dolukhanov, A.G. (1966) Vegetation. In: Gerasimov, I.P. (ed.) *Caucasus*. Nauka, Moscow, pp. 223–251. [In Russian.]

Engler, A. (1913) Über die Vegetationsverhältnisse des Kaukasus auf Grund der Beobachtungen bei einer Durchquerung des westlichen Kaukasus. *Abhandlungen des Botanischen Vereins für Brandenburg* 55, 1–26.

Feodorov, A.A. (1952) The history of high mountain flora of the Caucasus in the Quaternary period as an example of autochtonous development of the tertiary floristic basis. In: Lavrenko, E. (ed.) *Materiali po Chetvertichnomu Periodu SSSR* 3. Moscow, pp. 49–86. [In Russian.]

Gagnidze, R.I. (1974) *Botanical and Geographical Analysis of the Florocoenotic Complexes of Tall Herb Vegetation of the Caucasus*. Metsniereba, Tbilisi. [In Russian.]

Grossheim, A.A. (1948) *Plant Cover of the Caucasus*. Moskovskoe Obschestvo Ispytatelej Prirody Publication, Moscow. [In Russian.]

Henning, I. (1972) Die dreidimensionale Vegetationsanordnung in Kaukasien. *Erdwissenschaftliche Forschung* 4, 182–204.

Hüls, J. (2005) Populationsbiologische Untersuchung von *Heracleum mantegazzianum* Somm. & Lev. in Subpopulationen unterschiedlicher Individuendichte. PhD thesis, University of Giessen, Germany.

Kharadze, A.L. (1960) An endemic hemixerophilous element of the Greater Caucasus uplands. *Problemy Botaniki Pochvovedenija* 5, 115–126. [In Russian.]

Kuznetsov, Y. and Truschin, D. (2001) *The General Geographic Atlas 'Region Krasnodar and Republic of Adygeja'*, 2nd edn. TsEVKF, Moskow. [In Russian.]

Lepš, J. and Šmilauer, P. (2003) *Multivariate Analysis of Ecological Data Using CANOCO*. Cambridge University Press, Cambridge.

Levier, E. (1894) *A travers le Caucasus*. Attinger Frères, Neuchatel, Switzerland.

Mandenova, I.P. (1950) *Caucasian Species of the Genus Heracleum*. Georgian Academy of Sciences, Tbilisi. [In Russian.]

Müllerová, J., Pyšek, P., Jarošík, V. and Pergl, J. (2005) Aerial photographs as a tool for assessing the regional dynamics of the invasive plant species *Heracleum mantegazzianum*. *Journal of Applied Ecology* 42, 1042–1053.

Nakhutsrishvili, G. (1999) The vegetation of Georgia (Caucasus). *Braun-Blanquetia* 15, 5–74.

Nakhutsrishvili, G. (2003a) High mountain vegetation of the Caucasus region. *Ecological Studies* 167, 93–103.

Nakhutsrishvili, G. (2003b) Kaukasus und Alpen: Ein Vergleich der Vegetation. In: Psenner, R., Borsdorf, A. and Grabherr, G. (eds) *Forum Alpinum 2002. The Nature of the Alps*. Österreichische Akademie der Wissenschaften, Wien, Austria, pp. 82–87.

Onipchenko, V.G. (2002) Alpine vegetation of the Teberda Reserve, the Northwestern Caucasus. *Veröffentlichungen des Geobotanischen Institutes der ETH, Stiftung Rübel, Zürich* 130, 1–168.

Plesnik, P. (1972) Obere Waldgrenze in den Gebirgen Europas von den Pyrenäen bis zum Kaukasus. *Erdwissenschaftliche Forschung* 4, 73–92.

Radde, G. (1899) *Die Vegetation der Erde. III Grundzüge der Pflanzenverbreitung in den Kaukasusländern von der unteren Wolga über den Manytsch-Scheider bis zur Scheitelfläche Hocharmeniens*. Engelmann, Leipzig, Germany.

Rikli, M. (1914) *Natur- und Kulturbilder aus den Kaukasusländern von Teilnehmern der schweizerischen naturwissenschaftlichen Studienreise, Sommer 1912*. Orell Füssli, Zürich, Switzerland.

Tappeiner, U. and Cernusca, A. (1996) Microclimate and fluxes of water vapour, sensible heat and carbon dioxide in structurally differing subalpine plant communities in the Central

Caucasus. *Plant, Cell and Environment* 19, 403–417.

Tappeiner, U. and Cernusca, A. (1998) Model simulation of spatial distribution of photosynthesis in structurally differing plant communities in the Central Caucasus. *Ecological Modelling* 113, 201–223.

ter Braak, C.J.F. and Šmilauer, P. (1998) *CANOCO Reference Manual and User's Guide to CANOCO for Windows.* Centre for Biometry, Wageningen, The Netherlands.

Thiele, J. and Otte, A. (2006) Analysis of habitats and communities invaded by *Heracleum mantegazzianum* Somm. et Lev. (Giant Hogweed) in Germany. *Phytocoenologia* 36, 281–312.

Walter, H. (1974) *Vegetationsmonographien der einzelnen Großräume VII. Die Vegetation Osteuropas, Nord- und Zentralasiens.* Fischer, Stuttgart, Germany.

3

Historical Dynamics of *Heracleum mantegazzianum* Invasion at Regional and Local Scales

PETR PYŠEK,[1,2] JANA MÜLLEROVÁ[1] AND VOJTĚCH JAROŠÍK[1,2]

[1]*Institute of Botany of the Academy of Sciences of the Czech Republic, Průhonice, Czech Republic;* [2]*Charles University, Praha, Czech Republic*

Fashionable country gentlemen had some cultivated wild gardens in which they innocently planted the Giant Hogweed throughout the land
(Genesis, 1971)

Introduction

For several reasons, historical data on the occurrence of *Heracleum mantegazzianum* Sommier & Levier are fairly detailed, especially in countries with a strong floristic tradition. Such data allow a retrospective analysis of this species' spread. This species is attractive enough to be recorded by botanists, because of its alien origin and tendency to spread, and above all its conspicuousness (Fig. 3.1). In parts of the continent, such as Central Europe, it is also taxonomically unproblematic, hence easily recognizable by amateur botanists who are the main collectors of floristic data (Pyšek, 1991). The taxonomic problems referred to earlier (see Jahodová *et al.*, Chapter 1, this volume) are of little significance in the Czech Republic as the vast majority of the observations concern this species (Holub, 1997).

Probably the most systematically gathered data on the occurrence of *H. mantegazzianum* in Europe are those for the Czech Republic (Pyšek, 1991, 1994; Pyšek *et al.*, 1998a) and the UK (Collingham *et al.*, 2000), countries with a strong floristic tradition (Preston *et al.*, 2002; Pyšek *et al.*, 2002). Such data allow the invasion dynamics of *H. mantegazzianum* over the last 150 years to be reconstructed. These two data sets are complementary because they: (i) come from climatically different geographical areas of Europe (continental and oceanic climate); and (ii) were collected using different methods, i.e. collation of localities from the literature in the Czech Republic, and repeated

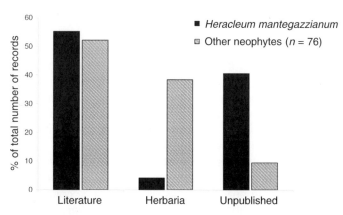

Fig. 3.1. Sources of floristic data for *H. mantegazzianum* differ from those for other neophytes, based on a sample of 76 neophytes recorded in the Czech Republic (based on 28,330 floristic records). The percentage of literature sources is very similar (55.6% and 52.2% of all records of *H. mantegazzianum* and all neophytes, respectively). However, *H. mantegazzianum* is avoided by herbarium collectors because of its large stature, which accounts for why only 4.1% of the records (25 of the 603 in total) are from herbaria, compared to 38.4% for an average neophyte. Interestingly, 245 unpublished records (40.6% of the total) were supplied by botanists, which greatly exceeds the average number for other neophytes (35.2 records).

mapping in different periods in the UK. That these different approaches yield similar results, as shown below, is an indication that conclusions drawn about the history of invasion at regional/geographical scale are robust.

Furthermore, floristic data not only make it possible to reconstruct the invasion, and describe the pattern of species abundance in the landscape and its development over time, but also provide information on species ecology and the temporal changes in the spectrum of habitats occupied during the course of the invasion (Pyšek, 1991). The present chapter reviews information on the historical dynamics of *H. mantegazzianum* and aims to: (i) describe the dynamics of the invasions of the Czech Republic and UK; (ii) compare the pattern of spread at local and regional/geographical scales; and (iii) place the invasion potential of *H. mantegazzianum* into a wider context of plant invasions by comparing its rate of spread recorded in these two countries with that recorded for important invasive species elsewhere.

Reconstruction of the Invasion of the Czech Republic

Previous work (e.g. Pyšek, 1991; Pyšek and Prach, 1995; Delisle *et al.*, 2003) shows how floristic data, systematically gathered over an area for a long time, may be used to reconstruct the pattern of a species invasion on a regional geographical scale. The history of the invasion by *H. mantegazzianum* is well described for the Czech Republic, in terms of the overall dynamics of spread (Pyšek, 1991) and changes in species ecology during the course of the invasion (Pyšek, 1994; Pyšek and Prach, 1993; Pyšek *et al.*, 1998a). This species was first introduced in the Slavkovský les region, in the western part of the country, as a garden ornamental, reportedly in 1862. In 1877 it was first collected in the wild, as an escape from cultivation, and documented by a herbarium specimen (Holub, 1997). It soon became popular among gardeners, which is assumed to have contributed to its spread. Up to 1950, only a few localities were known for the entire country, and their spatial distribution (Pyšek, 1991: Fig. 1a) indicates that at least some localities in the east must have resulted from humans translocating the species.

The same pattern is apparent in the UK. In general, the distribution of an invasive species spreading across a landscape is expected to show autocorrelation because the probability that an area will be colonized is a function of its distance from neighbouring populations. However, Collingham *et al.* (2000) found no evidence of a significant spatial autocorrelation using distribution data for *H. mantegazzianum* in the UK, although this species is believed to spread its fruits by both wind and water (see Moravcová *et al.*, Chapter 5, this volume). Although local aggregations of this species are obvious regionally and nationally, the majority of records in the UK are isolated. Collingham *et al.* (2000) conclude that this pattern reflects the importance of long-distance anthropogenic dispersal in the intial phase of the invasion in the UK. This suggests that in this country, as in the Czech Republic, *H. mantegazzianum* might have been spread as a garden ornamental over long distances in the early stages of invasion. This results in the existence of new sites, which subsequently act as foci for further spread (Pyšek, 1991).

Nevertheless, the current distribution in the Czech Republic (232 grid squares occupied in 1996; Williamson *et al.*, 2005) still indicates that the species has spread from the original place of introduction in the western part of the country, because its abundance decreases from west to east. The effect of distance from the source is highly significant and the form of the relationship is the same as it was at the end of the 1980s, i.e. some 20 years after the beginning of the period of rapid spread, but the effect of time, significant then, is no longer so when recent data are analysed (Fig. 3.2).

Based on the plot of the cumulative number of localities over time, the start of the exponential phase of invasion was 1943 (for details, see Pyšek and Prach, 1993). Williamson *et al.* (2005) used a more recent data set of the cumulative number of grid squares occupied, instead of the number of localities. This study also revealed a distinct lag phase, defined as the period before the log plot of the cumulative number of squares occupied against time becomes straight. Such a lag phase was only found in 19 species of neophytes of the 63 examined in the Czech Republic. This paper gives the beginning of the exponential phase as 1936, which is similar to the previous estimate of Pyšek and Prach (1993). Both estimates are close enough to conclude that the lag phase, from the first record outside cultivation to the beginning of exponential spread, lasted 60–70 years in the Czech Republic. Nevertheless, a comparison of these papers indicates that the exact date of the transition from the lag to the exponential phase depends on the method used to analyse the data (Pyšek and Prach, 1993; Williamson *et al.*, 2005).

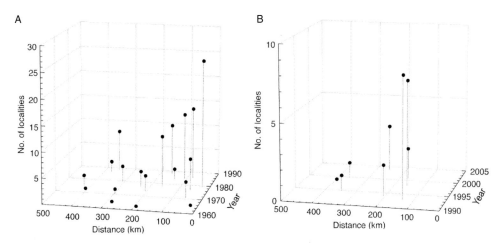

Fig. 3.2. (A) Effects of the distance from the locality of the original introduction in 1862 and of time on the frequency of occurrence of *H. mantegazzianum* in the Czech Republic, at the end of the 1980s, approximately 20 years after the onset of rapid spread in this country (Pyšek, 1991; Pyšek and Prach, 1993). Data are the number of localities recorded by annual floristic summer schools organized by the Czech Botanical Society, during which a region is systematically sampled and standardized data are obtained. Both year ($F = 27.13$; $df = 1$, 17; $P < 0.001$) and distance ($F = 5.59$; $df = 1$, 17; $P < 0.05$) were significant predictors in the multiple regression derived in 1989: LN (NUMBER + 1) = $-194.2 + 0.099$ YEAR + 33.06 (1/DISTANCE); $F = 14.72$; $df = 2$, 16; $R^2 = 0.65$. See Pyšek (1991) for details; additional localities are: 15 – Svitavy 1965; 16 – Lanškroun 1970 (Kovář *et al.*, 1996); 17 – Humpolec 1974 (Skalický and Štech, 2000); 18 – Blovice 1986 (Nesvadbová and Sofron, 1996); 19 – Tábor 1988 (Štech, 2005). Note that the regression coefficients differ from the original data in Pyšek (1991), as five more data points are included. (B) Fifteen years later, the relationship is only significant for distance; deletion test on the effect of year, using ANCOVA with the two time periods (1963–1989; 1990–2004) as a factor, and year and distance as covariates: $F = 0.04$; $df = 1$, 22; NS. The average abundance in the 1990–2004 period is significantly higher ($F = 23.85$; $df = 1$, 24; $P < 0.001$) than at the end of the 1980s. The effect of distance remains highly significant ($F = 18.91$; $df = 1$, 24; $P < 0.001$) and the same as in 1989 (deletion test in ANCOVA on a different slope of distance in each time period: $F = 0.28$; $df = 1$, 23; NS). Data for the latter period: 1 = Mělník 1993 (Hrouda *et al.*, 1996); 2 = Světlá nad Sázavou 1994 (Čech, 2003); 3 = Břeclav 1995 (Danihelka and Grulich, 1996); 4 = Česká Lípa 1998 (Kubát *et al.*, 1998); 5 = Nový Jičín 1999 (Grulich, 2003a); 6 = Kroměříž 2000 (Grulich, 2003b); 7 = České Budějovice 2001 (Lepší *et al.*, 2005); 8 = Kostelec nad Orlicí 2004 (Kaplan, 2005).

Heracleum mantegazzianum started to spread exponentially in the Czech Republic after colonizing very few localities (Pyšek and Prach, 1993). This implies that the lag phase, the time necessary for adaptation to a new region, is not too important in this species; it started to spread in the second half of the 20th century and as for most invasive aliens, changes in the landscape were the main trigger (Williamson *et al.*, 2005). This seems to accord with the small genetic difference between populations from Europe and the Caucasus Mountains (see Jahodová *et al.*, Chapter 1, this volume).

Pyšek (1991) estimated the rate of spread of *H. mantegazzianum* on a geographical scale in the Czech Republic, based on the slope of a semi-log plot of the cumulative number of grid squares occupied by the species up to a certain date. The value of 0.0835 is very similar to that obtained by Williamson *et al.* (2005) for *H. mantegazzianum* in the Czech Republic, using more recent cumulative grid square data. These authors used $\log_{10}$ base and arrived at the value of 0.0327, which is very close to 0.0363, obtained if the $\log_e$ based value given by Pyšek (1991) is transformed to $\log_{10}$ base.

Environmental Factors and Habitat Preferences During the Invasion

In the course of the invasion of the Czech Republic, this species shifted to lower altitudes; at the beginning of the 1970s, 28.5% of the localities were above 600 m a.s.l., but only 14.7% in 1990 (Pyšek, 1994). This suggests that the inherent preferences for a cooler climate that *H. mantegazzianum* acquired at higher altitudes in the Caucasus Mountains affected its ability to invade warmer areas at the beginning, but this constraint was overcome and is no longer present. Moreover, the above frequency distribution of altitudes in the Czech Republic is similar to a recent distribution of altitudes at which the localities of *H. mantegazzianum* are recorded, indicating that its occurrence is no longer affected by altitude (Pyšek, 1994).

Changes in the representation of habitats occupied in the course of the invasion of the Czech Republic are rather profound, indicating that the habitat preferences of *H. mantegazzianum* changed as the invasion progressed (Fig. 3.3). Because of its popularity as a garden ornamental, this species was more or less confined initially to parks and gardens and their immediate surroundings. Later on the importance of these habitats decreased. That 'semi-natural' habitats (Fig. 3.3) were occupied to the same degree from the very start of the invasion indicates that *H. mantegazzianum* was able to invade such less disturbed habitats and some grasslands, wetlands and scrub. The beginning of the exponential spread in the 1930s–1940s (Pyšek and Prach, 1993; Williamson *et al.*, 2005) is associated with an increase in the proportion of riparian corridors and other linear habitats (roads, railways), suitable for efficient seed dispersal (see Moravcová *et al.*, Chapter 5, this volume). Major changes in habitat spectra are detectable in the 1970s, when profound landscape changes are thought to have encouraged the spread of neophytes in the country in general (Pyšek *et al.*, 1998b; Williamson *et al.*, 2005). In this period, *H. mantegazzianum* started to spread into urban habitats and linear dispersal along corridors, such as roads and railways, became relatively more important than along rivers and water courses (Fig. 3.3). That is, large rivers in particular acted as efficient dispersal vectors in the early stages of the invasion, but later on the species started to spread to more distant areas not associated with rivers (Pyšek, 1994). Although rivers are in general a good dispersal vector for invasive plants (Pyšek and Prach, 1994), the major invasive species in the Czech flora differ significantly in their affinity to riparian habitats and

H. mantegazzianum is less confined to riparian habitats than, for example, *Impatiens glandulifera* and two *Fallopia* species (Pyšek and Prach, 1993). In general, *H. mantegazzianum* is able to invade a rather large spectrum of habitats, and the rate of invasion and dates of first records do not differ substantially among these habitats (Pyšek and Prach, 1993). This suggests that the nature of recipient habitats is less important than may be the case for most neophytes, and indicates that once *H. mantegazzianum* enters a habitat, it spreads exponentially regardless of the characteristics of the recipient vegetation (Pyšek, 1994). This is strongly supported by ecological studies on its seed bank (see Moravcová *et al.*, Chapter 5, this volume; Krinke *et al.*, 2005), reproductive biology (Moravcová *et al.*, 2005 and Chapter 5, this volume) and habitat occupation on both local (Müllerová *et al.*, 2005) and national (Pyšek, 1994) scales; these studies indicate that this species is only slightly limited by the character of the invaded vegetation or specific site conditions – once there are suitable habitats in the landscape it spreads at a constant and rather high rate.

In the mid 1990s, urban sites (including dumps and deposits in open landscapes) and linear habitats (such as roads, paths and railways) were frequently reported as habitats for *H. mantegazzianum*, accounting for 29.2% and 29.3% of the total records (*n* = 679 year/habitat records from 603 localities, with some localities assigned to more than one habitat type). The species was also fairly frequently recorded in various less disturbed 'semi-natural' habitats (15.6%) and riparian habitats (14.4%). Over the entire invasion history, 7.5% of the records came from parks and gardens. These numbers are cumulative,

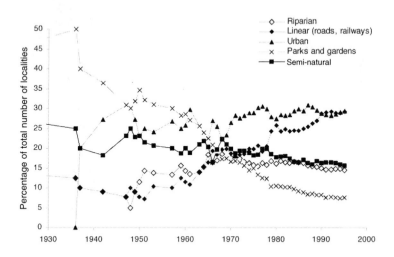

Fig. 3.3. Changes in habitat preferences of *H. mantegazzianum* during the course of its invasion in the Czech Republic. Habitat preference is expressed as the percentage of the total number of localities reported up to the given year, which are for a particular habitat. 'Semi-natural' habitats include less disturbed sites such as scrub, grassland, wetlands, forest and their margins (for details of the classification of habitats based on floristic records, see Pyšek *et al.*, 1998b). Based on data from Pyšek (1994) and updated.

therefore slightly biased if inferring current habitat spectra, but correspond reasonably well to where the species is currently found. Taking only records from the last decade (1986–1995, n = 344) yields very similar percentages of 28.8% (urban sites), 33.7% (linear habitats), 14.8% ('semi-natural' habitats), 12.5% (riparian habitats) and 5.8% (parks and gardens).

These figures correspond reasonably well with those of a detailed investigation of habitat preferences at a local scale in the Slavkovský les region (Pyšek and Pyšek, 1995). Taking the percentage of a habitat invaded by *H. mantegazzianum* as a measure, 40.4% of path margins, 38.2% of road ditches and 38.0% of the area adjacent to water courses are occupied by more or less dense populations of this species. In addition, 30.7% of willow scrub and 7.9% of forest margins are also invaded, but only 2.3% of dry grassland and 2.2% of wetlands (Pyšek and Pyšek, 1995).

Invasion Dynamics at a Local Scale, Analysed Using Aerial Photographs

At the local scale, regression models that establish a relationship between the area invaded from an invasion focus and time have been useful in quantifying invasion patterns (Higgins and Richardson, 1996). Data from an analysis of historical aerial photographs showing the invasion of the Slavkovský les region, Czech Republic, by *H. mantegazzianum* (Müllerová *et al.*, 2005) provide an insight into the invasion history of a noxious alien species on a local scale. This species is easily detectable on aerial photographs taken at flowering and early fruiting, from June to August (Fig. 3.4). These data document the invasion from the beginning, which is rarely possible for other alien plants, and therefore allow an analysis of the rate of spread and a study of the species' population dynamics.

Mean rate of areal spread over 50 years, calculated for nine sites (Müllerová *et al.*, 2005), was 1261 m^2/year, and of linear spread 10.8 m/year. Absence of a correlation between linear and areal rates of spread indicates that *H. mantegazzianum* did not spread as an advancing front but that long-distance dispersal (Higgins and Richardson, 1999; Hulme, 2003) played an important role in the invasion. The direct effect of the rate of invasion on invaded area was larger than that of residence time (defined as the time for which a species has been present at a locality, see Rejmánek, 2000; Pyšek and Jarošík, 2005), but the total, direct and indirect, effect of residence time was only slightly less than that of the rate of invasion (Fig. 3.5). As the invasion proceeded, the populations spread from linear habitats into the surroundings, i.e. a pattern similar to that observed at the geographical/national scale. Flowering intensity did not exhibit any significant trend over time (Müllerová *et al.*, 2005).

A

B

C

Fig. 3.4. Series of aerial photographs showing *H. mantegazzianum* invading one of the localities in the Slavkovský les region, where the species was first introduced as a garden ornamental in 1862 and escaped from cultivation in 1877. Locality Žitný, showing increase in the area occupied by *H. mantegazzianum* from (A) 1962 to (B) 1973 and (C) 1991. Photographs were taken at flowering and plants of *H. mantegazzianum* appear as white dots (for details, see Müllerová *et al.*, 2005).

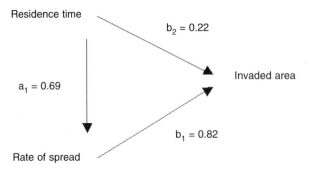

Fig. 3.5. Path model and path coefficients of the invaded area as a function of residence time and rate of spread. Based on aerial photographs of nine sites in the Slavkovský les region, Czech Republic, invaded by *H. mantegazzianum*, taken on 11 sampling dates between 1947 (before the invasion started) and 2002. The area covered by the plant in a 60 ha section of landscape was measured digitally, and used to obtain information on invaded habitats, year of invasion, flowering intensity and structure of patches. Invaded area was regressed on residence time (time since the beginning of invasion), and regression slopes were used to measure the rate of spread. Residence time directly affects the rate of spread and both the residence time and the rate of spread directly affect the invaded area. The direct effect of the rate of invasion on invaded area (0.82) was larger than that of residence time (0.22), but the total effect (direct and indirect) of residence time was only slightly less (0.79) than that of the rate of invasion (0.82). Based on data from Müllerová *et al.* (2005).

Rate of *Heracleum mantegazzianum* Spread Compared to that of Other Important Invasive Species

Since spread is the stage of invasion that is most easily modelled and most accessible to quantitative analysis (Williamson, 1996), there are many studies of the rate of spread of one or a few species of alien plants (reviewed in Pyšek and Hulme, 2005). Although some obvious limitations need to be kept in mind, such as multiple origins (the spread may start from several places, not just one), effect of boundaries (Williamson *et al.*, 2005) and effect of scale (Pyšek and Hulme, 2005), it is possible to make a rough comparison of the rate of spread of *H. mantegazzianum* with that of other species.

Measured by the regression slope of the increase in the cumulative number of localities over time, the rate of spread of *H. mantegazzianum* is similar to that of other important invasive species in the Czech Republic, if the whole period of invasion is considered. However, considering only the exponential phase of the invasion, the rate during this phase (slope: 1.107, $\log_e$ base) was 152% of that of *Impatiens glandulifera* (0.070), 248% of *Fallopia sachalinensis* (0.043) and 263% of *F. japonica* (0.040) (Pyšek and Prach, 1993; calculation based on their Table 3). Nevertheless, Williamson *et al.* (2005) provide estimates of spread for 31 species that can be compared. During 1971–1995, the period over which the section of the $\log_{10}$ base plot of the cumulative number of occupied mapping quadrats against time is straight, the number of squares occupied by *H. mantegazzianum* doubled every 9.21 years. This value indicates an average (geometric mean: 8.75 years; range 2.04–39.02) rate of spread (Williamson *et al.*, 2005).

Rates of areal spread are difficult to compare as they crucially depend on the scale of the study (Pyšek and Hulme, 2005). Some indication can be obtained by using a relative value, i.e. the multiplication of invaded area over time as recorded in the local-scale study of Müllerová *et al.* (2005). *Heracleum* increased its area from the first time it was recorded at a site to the date of the largest invaded area recorded 26.7 ± 37.9 times (mean $\pm$ SD, $n = 9$); the average mean residence time in a site was 33.7 ± 5.5 years. During the invasion of New Mexico by species of *Tamarix*, part of the invasion of the south-western USA by these species, which is one of the most spectacular invasions reported in the literature (Zavaleta, 2000), the invaded area increased 20.5 times from 1915 to 1925, and 4.6 times from 1925 to 1960 (Robinson, 1965). Thus the *Heracleum* invasion in the study area proceeded at a rate similar to that of most aggressive invasive species in other parts of the world.

What can be Inferred from the Comparison of Invasion at Local and National Scales?

Since detailed information on the course of the invasion of the Czech Republic by *H. mantegazzianum* is available (Pyšek, 1991), it is possible to compare the rate of spread at local and national scales. Plotting the cumulative number of this species' records against time over the whole period of the invasion yields a doubling time of 13.2 years for localities (analysed in Pyšek, 1991) and 14.3 for mapping squares (analysed in Williamson *et al.*, 2005), and these values do not differ significantly from the doubling time at the local scale of 13.9 (Müllerová *et al.*, 2005). This indicates that *Heracleum* spread nationally at the same rate as locally in the region of its introduction to the country, and that the constraints to its spread imposed by landscape features and availability of suitable habitats were similar at both scales (see Pyšek and Hulme, 2005, also for statistical details of the comparison of spreading rates).

Invasive species of plants do not always have the same rate of spread at local and geographical scales. Data that can be compared are extremely rare. *Heracleum mantegazzianum*, however, can be compared with the *Mimosa pigra* invasion in Australia. Lonsdale (1993) compared the rate of increase of this species at local and geographical scales. At the former scale, measured by invaded area, the average doubling time over a 6-year period was 1.2 years (which indicates that this invasion was an order of magnitude faster than that of *H. mantegazzianum* in the Slavkovský les). Across the region as a whole, the doubling time for numbers of infestation was much slower, 6.7 years, probably because of the separation of suitable habitats by eucalypt savannas that are less readily colonized.

Conclusions

Studies of the dynamics of the *H. mantegazzianum* invasion allow some conclusions to be drawn that may apply to a wider geographical area of Europe.

1. This species exhibited a distinct lag phase of between 60 and 70 years, depending on the method of estimation, in the Czech Republic. Exponential phase of spread was associated with distinct changes in habitat preferences. Water courses and other riparian habitats were initially the major dispersal routes, but once the species started to spread beyond the river corridors, other linear habitats such as roads and railways became important. In the course of the invasion, the species became less confined to high altitudes and invaded warmer areas.

2. During its invasions, *H. mantegazzianum* spread at rates comparable to that of some of the most important and spectacular invaders in other parts of the world. A rigorous comparison of rates of spread is, however, limited by the variety of measures used to express and ways used to calculate this characteristic, as well as by the effect of scale at which the data were recorded.

3. The rate of spread was similar in at least two European countries (Czech Republic and the UK), which differ markedly in climate, indicating that environmental constraints imposed by the landscapes of the countries invaded may be of little importance. In addition, the fact that this species spread at a similar rate in different habitats in the Czech Republic also supports this conclusion.

Acknowledgements

We thank Mark Williamson and Claude Lavoie for helpful comments on the manuscript, Tony Dixon for improving our English, and Jan Pergl for technical assistance. The study was supported by the project 'Giant Hogweed (*Heracleum mantegazzianum*) a pernicious invasive weed: developing a sustainable strategy for alien invasive plant management in Europe', funded within the 'Energy, Environment and Sustainable Development Programme' (grant no. EVK2-CT-2001-00128) of the European Union 5th Framework Programme. PP and VJ were supported by institutional long-term research plans no. AV0Z60050516, from the Academy of Sciences of the Czech Republic, no. 0021620828, from Ministry of Education of the Czech Republic and by the Biodiversity Research Centre no. LC 06073.

References

Čech, L. (ed.) (2003) Results of the floristic summer school of the Czech Botanical Society in Světlá nad Sázavou (30.6–4.7.1997). *Zprávy České botanické společnosti* 38/Příloha 2, 42–88. [In Czech.]

Collingham, C.Y., Wadsworth, R.A., Huntley, B. and Hulme, P.E. (2000) Predicting the spatial distribution of non-indigenous riparian weeds: issues of spatial scale and extent. *Journal of*

Applied Ecology 37 (suppl. 1), 13–27.

Danihelka, J. and Grulich, V. (eds) (1996) 34. Floristic summer school of the Czech Botanical Society in Břeclavi II. *Zprávy České botanické společnosti* 31/Příloha 1, 1–125. [In Czech.]

Delisle, F., Lavoie, C., Jean, M. and Lachance, D. (2003) Reconstructing the spread of invasive plants: taking into account biases associated with herbarium specimens. *Journal of Biogeography* 30, 1033–1042.

Grulich, V. (ed.) (2003a) Results of the floristic summer school of the Czech Botanical Society in Nový Jičín (4–10 July 1999). *Zprávy České botanické společnosti* 38/Příloha 2, 89–174. [In Czech.]

Grulich, V. (ed.) (2003b) Results of the floristic summer school of the Czech Botanical Society in Kroměříž (10–16 July 2000). *Zprávy České botanické společnosti* 38/Příloha 2, 89–174. [In Czech.]

Higgins, S.I. and Richardson, D.M. (1996) A review of models of alien plant spread. *Ecological Modelling* 87, 249–265.

Higgins, S.I. and Richardson, D.M. (1999) Predicting plant migration rates in a changing world: the role of long-distance dispersal. *American Naturalist* 153, 464–475.

Holub, J. (1997) *Heracleum* – hogweed. In: Slavík, B., Chrtek, J. and Tomšovic, P. (eds) *Flora of the Czech Republic 5*. Academia, Praha, pp. 386–395. [In Czech.]

Hrouda, L., Mandák, B. and Hadinec, J. (eds) (1996) Materials on the flora of the Kokořínsko a Mělnicko regions. *Příroda* 7, 7–109. [In Czech.]

Hulme, P.E. (2003) Biological invasions: winning the science battles but losing the conservation war? *Oryx* 37, 178–193.

Kaplan, Z. (ed.) (2005) Results of the floristic summer school of the Czech Botanical Society in Kostelec nad Orlicí (4–10 July 2004). *Zprávy České botanické společnosti* 40, Suppl. 2005/1, 1–76. [In Czech.]

Kovář, P., Jirásek, J. and Grundová, H. (eds) (1996) Floristic summer schools of CBS in Svitavy (11–17.7.1965) and Lanškroun (2–10.7.1970). *Zprávy České botanické společnosti* 31/Suppl. 1996/2, 1–74. [In Czech.]

Krinke, L., Moravcová, L., Pyšek, P., Jarošík, V., Pergl, J. and Perglová, I. (2005) Seed bank of an invasive alien, *Heracleum mantegazzianum*, and its seasonal dynamics. *Seed Science Research* 15, 239–248.

Kubát, K., Ondráček, Č. and Machová, I. (eds) (1998) Results of the floristic summer school in Česká Lípa. *Severočeskou Přírodou*, Suppl. 11, 19–81. [In Czech.]

Lepší, M., Lepší, P. and Štech, M. (eds) (2005) Results of the floristic summer school of CBS in České Budějovice 2001 (1–7.7.2001). *Zprávy České botanické společnosti* 40, Suppl. 2005/2, 71–135. [In Czech.]

Lonsdale, W.M. (1993) Rates of spread of an invading species: *Mimosa pigra* in northern Australia. *Journal of Ecology* 81, 513–521.

Moravcová, L., Perglová, I., Pyšek, P., Jarošík, V. and Pergl, J. (2005) Effects of fruit position on fruit mass and seed germination in the alien species *Heracleum mantegazzianum* (*Apiaceae*) and the implications for its invasion. *Acta Oecologica* 28, 1–10.

Müllerová, J., Pyšek, P., Jarošík, V. and Pergl, J. (2005) Aerial photographs as a tool for assessing the regional dynamics of the invasive plant species *Heracleum mantegazzianum*. *Journal of Applied Ecology* 42, 1042–1053.

Nesvadbová, J. and Sofron, J. (1996) Floristic course of CBS in Blovice (5.7–12.7.1986). *Sborník Západočeského Muzea Plzeň, Příroda* 94, 23–48. [In Czech.]

Preston, C.D., Pearman, D.A. and Dines, T.D. (2002) *New Atlas of the British and Irish Flora*. Oxford University Press, Oxford, UK.

Pyšek, P. (1991) *Heracleum mantegazzianum* in the Czech Republic: the dynamics of spreading from the historical perspective. *Folia Geobotanica et Phytotaxonomica* 26, 439–454.

Pyšek, P. (1994) Ecological aspects of invasion by *Heracleum mantegazzianum* in the Czech Republic. In: de Waal, L.C., Child, L.E., Wade, P.M. and Brock, J.H. (eds) *Ecology and Management of Invasive Riverside Plants*. Wiley, Chichester, UK, pp. 45–54.

Pyšek, P. and Hulme, P.E. (2005) Spatio-temporal dynamics of plant invasions: linking pattern to process. *Ecoscience* 12, 302–315.

Pyšek, P. and Jarošík, V. (2005) Residence time determines the distribution of alien plants. In: Inderjit (ed.) *Invasive Plants: Ecological and Agricultural Aspects*. Birkhäuser Verlag-AG, Basel, Switzerland, pp. 77–96.

Pyšek, P. and Prach, K. (1993) Plant invasions and the role of riparian habitats – a comparison of four species alien to central Europe. *Journal of Biogeography* 20, 413–420.

Pyšek, P. and Prach, K. (1994) How important are rivers for supporting plant invasions? In: de Waal, L.C., Child, E.L., Wade, P.M. and Brock, J.H. (eds) *Ecology and Management of Invasive Riverside Plants*. Wiley, Chichester, UK, pp. 19–26.

Pyšek, P. and Prach, K. (1995) Invasion dynamics of *Impatiens glandulifera* – a century of spreading reconstructed. *Biological Conservation* 74, 41–48.

Pyšek, P. and Pyšek, A. (1995) Invasion by *Heracleum mantegazzianum* in different habitats in the Czech Republic. *Journal of Vegetation Science* 6, 711–718.

Pyšek, P., Kopecký, M., Jarošík, V. and Kotková, P. (1998a) The role of human density and climate in the spread of *Heracleum mantegazzianum* in the Central European landscape. *Diversity and Distributions* 4, 9–16.

Pyšek, P., Prach, K. and Mandák, B. (1998b) Invasions of alien plants into habitats of Central European landscape: an historical pattern. In: Starfinger, U., Edwards, K., Kowarik, I. and Williamson, M. (eds) *Plant Invasions: Ecological Mechanisms and Human Responses*. Backhuys, Leiden, The Netherlands, pp. 23–32.

Pyšek, P., Sádlo, J. and Mandák, B. (2002) Catalogue of alien plants of the Czech Republic. *Preslia* 74, 97–186.

Rejmánek, M. (2000) Invasive plants: approaches and predictions. *Austral Ecology* 25, 497–506.

Robinson, T.W. (1965) Introduction, spread and areal extent of saltcedar (*Tamarix*) in the western states. *Studies of Evapotranspiration. Geological Survey Professional Paper* 491-A, 1–12.

Skalický, V. and Štech, M. (2000) *Výsledky floristického kursu ČSBS v Humpolci 1974*. Česká botanická společnost, Praha.

Štech, M. (ed.) (2005) Results of the floristic summer school of the CSBS in Tábor 1988 (2–9.7.1988). *Zprávy České botanické společnosti* 40, Suppl. 2005/2, 3–70. [In Czech.]

Williamson, M. (1996) *Biological Invasions*. Chapman and Hall, London.

Williamson, M., Pyšek, P., Jarošík, V. and Prach, K. (2005) On the rates and patterns of spread of alien plants in the Czech Republic, Britain and Ireland. *Ecoscience* 12, 424–433.

Zavaleta, E. (2000) Valuing ecosystem services lost to *Tamarix* invasion in the United States. In: Mooney, H.A. and Hobbs, R.J. (eds) *The Impact of Global Change on Invasive Species*. Island Press, Washington, DC, pp. 261–300.

4 Reproductive Ecology of *Heracleum mantegazzianum*

IRENA PERGLOVÁ,[1] JAN PERGL[1] AND PETR PYŠEK[1,2]

[1]*Institute of Botany of the Academy of Sciences of the Czech Republic, Průhonice, Czech Republic;* [2]*Charles University, Praha, Czech Republic*

> *Botanical creature stirs, seeking revenge*
>
> (Genesis, 1971)

Introduction

Reproduction is the most important event in a plant's life cycle (Crawley, 1997). This is especially true for monocarpic plants, which reproduce only once in their lifetime, as is the case of *Heracleum mantegazzianum* Sommier & Levier. This species reproduces only by seed; reproduction by vegetative means has never been observed.

As in other *Apiaceae*, *H. mantegazzianum* has unspecialized flowers, which are promiscuously pollinated by unspecialized pollinators. Many small, closely spaced flowers with exposed nectar make each insect visitor to the inflorescence a potential and probable pollinator (Bell, 1971). A list of insect taxa sampled on *H. mantegazzianum* (Grace and Nelson, 1981) shows that Coleoptera, Diptera, Hemiptera and Hymenoptera are the most frequent visitors.

Heracleum mantegazzianum has an andromonoecious sex habit, as has almost half of British *Apiaceae* (Lovett-Doust and Lovett-Doust, 1982); together with perfect (hermaphrodite) flowers, umbels bear a variable proportion of male (staminate) flowers. The species is considered to be self-compatible, which is a typical feature of *Apiaceae* (Bell, 1971), and protandrous (Grace and Nelson, 1981; Perglová *et al.*, 2006). Protandry is a temporal separation of male and female flowering phases, when stigmas become receptive after the dehiscence of anthers. It is common in umbellifers. Where dichogamy is known, 40% of umbellifers are usually protandrous, compared to only about 11% of all dicotyledons (Lovett-Doust and Lovett-Doust, 1982). Although protandry has traditionally been considered to be a mechanism of avoiding or reducing selfing, it is itself unlikely to guarantee outcrossing. However, when it is strongly developed, the male and female phases of a plant may be completely separated in time so that outcrossing is assured (Webb, 1981; Snow and Grove, 1995).

Interspecific hybrids between *H. mantegazzianum* and *H. sphondylium* L. are reported from Great Britain (McClintock, 1975) and Germany (Ochsmann, 1996). Hybrids are found in sites where both species grow together, although they are not numerous (Grace and Nelson, 1981; Stewart and Grace, 1984). This was studied and it was found that only experimental crosses in which *H. sphondylium* was the female parent were successful (Stewart and Grace, 1984).

The aim of this chapter is to summarize the knowledge about the reproductive ecology of *H. mantegazzianum*, including the results obtained during the triennial European project GIANT ALIEN. During this project, an extensive study of flowering phenology and seed production of wild populations of *H. mantegazzianum* was conducted at ten localities in the Slavkovský les Protected Landscape Area in the Czech Republic (Fig. 4.1, Table 4.1). This region is where the species was first introduced to this country in the second half of the 19th century and from where it started to spread (Pyšek, 1991; see Pyšek *et al.*, Chapter 3, this volume). The rapid spread was probably facilitated by the fact that after World War II inhabitants were displaced and part of the region became a military area until the 1960s. This led to a lack of appropriate management and a specific disturbance regime; military activities are rather specific in that they occur in 'natural' parts of the landscape, which are less affected under standard

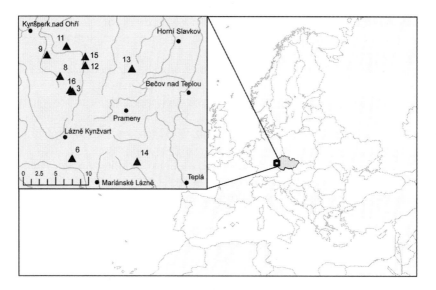

Fig. 4.1. The study area in the Slavkovský les Protected Landscape Area, which is the region where *H. mantegazzianum* was first introduced into the Czech Republic. Currently the region is still heavily infested. Most of the region is in the Ore Mountains and is formed from granite. Total size of the protected area is 617 km^2, altitudinal range is 373–983 m a.s.l. (Kos and Maršáková, 1997), January temperature ranges from –5.1°C (average mimimum) to –0.2°C (average maximum), July temperature from 10.5 to 21.5°C, respectively. Annual sum of precipitation is 1094 mm (Mariánské Lázně meteorological station, 50-year average).

Table 4.1. Geographical location, altitude (m a.s.l.) and population size of *H. mantegazzianum* estimated from aerial photographs taken at ten study sites in the Slavkovský les Protected Landscape Area, Czech Republic. Estimates were made for 60 ha sections of landscape (taken from Müllerová *et al.*, 2005).

Site no.	Name	Latitude	Longitude	Altitude (m)	Population (m^2)
3	Žitný I	50°03.754'	12°37.569'	787	99,121
6	Lískovec	49°59.156'	12°38.721'	541	8174
8	Potok	50°04.660'	12°35.953'	643	39,774
9	Dvorečky	50°05.982'	12°34.137'	506	24,817
11	Arnoltov	50°06.801'	12°36.147'	575	47,170
12	Krásná Lípa I	50°05.685'	12°38.546'	597	–
13	Litrbachy	50°06.009'	12°43.777'	800	4711
14	Rájov	49°59.704'	12°54.933'	753	5198
15	Krásná Lípa II	50°06.306'	12°38.393'	596	7945
16	Žitný II	50°03.837'	12°37.304'	734	–

land use. There are still very few people in this protected landscape, which consists mainly of extensive wetlands, pastures and spruce plantations. Nowadays, the area of the Slavkovský les is still invaded to a large extent (see Pyšek *et al.*, Chapter 3, this volume; Müllerová *et al.*, 2005). The field study was complemented by detailed studies of flowering phenology and selfing in the experimental garden of the Institute of Botany, Průhonice, Czech Republic (50°0.071' N, 14°33.5281' E; 310 m a.s.l.). Furthermore, information on the age at which *H. mantegazzianum* reproduces, gathered in both its native (Western Greater Caucasus) and invaded (Czech Republic) areas, is presented.

Description of the Pattern and Timing of Flowering

Flowering plants of *H. mantegazzianum* have a distinct architecture. The inflorescences are compound umbels of four orders. The main flowering shoot develops as a leafy stem that terminates in a primary (first-order) umbel, also called 'terminal'. Lateral shoots, which are produced on the stem, terminate in secondary (second-order) umbels and can be found in a satellite position, surrounding the primary umbel (hereafter also called 'satellites'), or in a branch position below them on the stem (hereafter also called 'branches'). Third-order umbels may arise on shoots branching from secondary shoots (in both satellite and branch position) and fourth-order umbels on shoots branching from tertiary shoots (Fig. 4.2). Under favourable conditions, strong plants can produce several other shoots, which arise from the base of the flowering stem at ground level (further referred to as 'basal branches'). The character of the terminal umbels of these basal branches varies and in terms of umbel size, fruit size, fecundity and proportion of male flowers is intermediate between typical first- and second-order umbels.

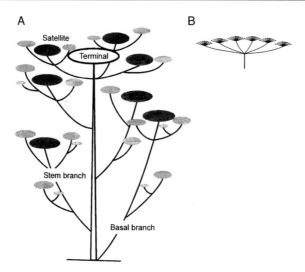

Fig. 4.2. (A) Schematic representation of the ordering of umbels and their position within the hierarchical inflorescence system of *H. mantegazzianum*. Umbel orders: primary ☐, secondary ◼, tertiary ▦ and quaternary ◉. (B) Each compound umbel consists of umbellets that bear a large number of small, closely packed flowers. Taken from Perglová *et al.* (2006), published with permission from the Czech Botanical Society.

Each compound umbel consists of umbellets (umbellules), simple umbels that bear a large number of small, closely packed flowers. Flowers are either hermaphrodite or male (staminate); the latter usually contain reduced stylopodia (Bell, 1971) and no, or a shrunken, style. The proportion of male flowers increases in higher-order umbels, while the terminal umbel usually contains only hermaphrodite flowers. If present, male flowers are located in the central part of umbellets. Within the same umbel, the proportion of male flowers seems to be the same in all umbellets (Perglová *et al.*, 2006), despite the increasing percentage of male flowers towards the centre of the umbels reported for *Zizia aurea* (Michaux) Fernald and *Thaspium barbinode* (Michaux) Nutt. (Bell, 1971). Fourth-order umbels usually consist only of male flowers (Perglová *et al.*, 2006).

Male sterility was observed in the experimental garden in Průhonice. A plant transplanted from a natural stand at the seedling stage and grown in a garden bed bore only physiologically female flowers with stamens, filaments of which remained unrolled, and the anthers remained closed and did not dehisce. Pistils were fully functional and the fruits were set after fertilization. The same phenomenon is described for wild *Daucus carota* L. plants by Braak and Kho (1958).

Within-flower and within-umbel phenology

A study of flowering phenology conducted on plants growing in the experimental garden in Průhonice, Czech Republic (Perglová *et al.*, 2006) revealed that in an individual flower, flowering starts by the sequential expansion and dehiscence of the five stamens, which takes usually 1 and sometimes 2 days. Anthers are ready to shed pollen almost immediately after expansion is completed. Within an umbellet, the outer flowers are first to flower and flowering continues to the centre, where pollen is shed 3 (in the case of the terminal umbels) to 6 (secondary umbels) days later (Fig. 4.3A). In umbellets located in the centre of an umbel, the onset of flowering can be 1 day later than in peripheral umbellets.

In contrast, stigma receptivity is well synchronized throughout the whole umbel and lasts 1–2, maximum 3 days. Receptivity can be recognized visually – the stigmas are fully elongated with a fresh glistening appearance at the tip of the initially dome-shape style, which spreads and becomes bulbous

Fig. 4.3. (A) The flowers of an umbellet open centripetally over a period of several days. (B) When the stigmas are receptive the styles are fully elongated, separate and with a fresh glistening appearance at the tip. Photo: I. Perglová.

Fig. 4.4. The overlap in anther dehiscence and stigma receptivity in flowers of the same umbellet. Photo: P. Pyšek.

(Fig. 4.3B). Flowers in the centre of umbellets can be male and do not have a female phase.

Between anther dehiscence and stigma receptivity, there is a neutral phase of variable length, depending on the position of a flower in an umbellet and position of an umbellet in the umbel. In the outer flowers of peripheral umbellets, the neutral phase may last up to 6 days because they are the first flowers within an umbel to shed pollen. The neutral phase of the outer flowers of central umbellets is usually 1 day shorter. Neutral phase of central flowers, both in peripheral and central umbellets, lasts 2 days at most or there is no neutral phase or even an overlap in anther dehiscence of these flowers, which are often male, with the receptivity of other flowers in the same umbel (Fig. 4.4).

Such overlaps only occur in some umbels and only a small proportion of the late dehiscing anthers are usually involved (Perglová *et al.*, 2006). Consequently, stigmas are not covered by a mass of pollen from the same umbel but geitonogamous selfing can occur.

Within-plant phenology

Umbels of different orders flower in sequence. The terminal (primary) umbel is the first to flower, followed by secondary, and later tertiary and quaternary umbels on satellites and branches. By way of an example, Fig. 4.5 shows the course of flowering of different umbels and the flowering phases of one plant growing in the experimental garden. In some umbels of higher orders (mainly quaternary), the female phase and fruit development do not occur after the male phase, because those umbels contain only male flowers and thus wither after anther dehiscence. At the umbel level, male phase is defined as a phase in which at least some flowers dehisce anthers while other flowers can be

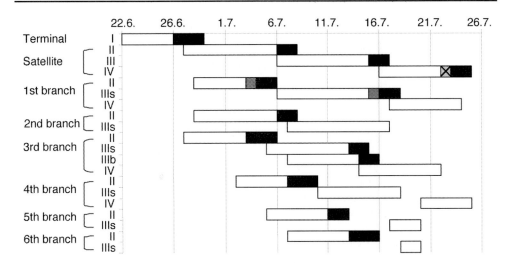

Fig. 4.5. Phenological pattern of a single plant. The sequence of ☐ male, ✕ neutral and ■ female phases throughout the vegetation period is shown for each umbel. Overlap in anther dehiscence and female receptivity within an umbel is indicated by ▨. Numbers I, II, III, IV refer to umbel order, letters s, b to the location of tertiary umbels on branches (satellite or branch position).

already in the neutral phase. The neutral phase of an umbel is thus defined as the stage before stigma receptivity when none of the flowers contain dehiscing anthers. However, neutral phases are uncommon. Male and female phases of umbels on the same plant can overlap and a study of the flowering phenology of 100 plants at ten localities in the Slavkovský les area revealed that such overlaps are common; at least a short overlap between some umbels was observed in 99% of plants (Perglová *et al.*, 2006).

Evidence for overlaps is also provided by Stewart and Grace's (1984) study of hybridization between *H. mantegazzianum* and *H. sphondylium*. They report complete protandry within an umbel and an overlap between female and male phases of primary and secondary umbels in only two plants out of the nine studied. However, the plants were transplanted to a greenhouse in the spring of the same year in which they flowered. The results therefore might have been affected by the plants being smaller and possibly not producing tertiary umbels, which reduce the possibilities of overlaps between male and female phases. Stewart and Grace (1984) did not include plants growing under natural conditions in their study.

An opportunity for geitonogamous selfing usually increases if a plant produces basal branches, because these often exhibit asynchronous flowering (umbels do not flower at the same time as other umbels of the same order). Basal branches are usually vigorous and branched, bearing umbels of higher orders. However, several plants observed at the Slavkovský les study sites produced late umbels on short basal shoots (shoot length up to about 20 cm, umbel diameter 10–25 cm), which consisted of physiologically female flowers with shrunken anthers (I. Perglová, Průhonice, 2003, personal observation). At that time, pollen from quaternary umbels (usually containing only male flowers) was still available; such plants exploited the last opportunity and resources to produce fruit and did not invest in the production of pollen, which would have been wasted, as there were no or very few receptive stigmas.

Timing of flowering at the plant and population level

In plants destined to flower, the first signs of the development of a flowering stem become visible in early June in the Slavkovský les (Perglová *et al.*, 2006), late May in Giessen in Germany (Otte and Franke, 1998) and approximately 1 month earlier in the UK (Tiley *et al.*, 1996). A rapid stem elongation culminates in the opening of a terminal bud, which contains terminal and satellite umbels. Then the terminal umbel expands, opens its flowers and starts to flower.

In the Slavkovský les protected area in 2002, flowering began within a period of 1 week (from 20 to 27 June) at all ten localities, despite differences in exposure and altitude (Table 4.1). The peak of flowering, expressed as the average date on which the primary and secondary umbels of 30 randomly selected plants at each locality flowered, was between 27 June and 6 July. The duration of flowering of an individual plant (time from beginning of male phase in the terminal umbel to end of female phase in the last umbel on the plant) in the Slavkovský les was on average 36 days (range of averages for individual localities: 31–41 days). The maximum observed duration of flowering was 60 days and duration increased with the number of umbels on a plant. The terminal umbel flowered on average for 10 days and its fruits were ripe on average 44 days after the beginning of flowering (Perglová *et al.*, 2006). In the second half of August, the majority of all fruits were ripe and started to be released.

Potential for Selfing

The complete separation of male and female phases in many *Apiaceae* is effective in promoting outcrossing. In contrast, a weak protandry does not appear to be an effective outcrossing mechanism and is more easily understood in terms of sexual selection and optimal allocation of resources to maternal and paternal functions (Lovett-Doust, 1980; Webb, 1981). Although species of *Apiaceae* are considered to be fully self-compatible (Bell, 1971), the potential for selfing need not be determined only by the degree of protandry. In some *Apiaceae*, selfing seems to be limited by a genetic mechanism: maternal control before fertilization (i.e. partial self-incompatibility), late-acting self-incompatibility or inbreeding depression very shortly after fertilization. These mechanisms are probably responsible for the low selfed seed set in the endangered species *Eryngium alpinum* L. (Gaudeul and Till-Bottraud, 2003) and in *Trachymene incisa* Rudge subsp. *incisa* (Davila and Wardle, 2002).

Table 4.2. Fruit production of umbellets subjected to controlled crossing. For each plant (*n* = 8), there were four different pollination treatments, each on three umbellets of the terminal umbel: 1. manual outcrossing, 2. manual selfing, 3. autonomous selfing and 4. natural open pollination (as a control). Fruit number per umbellet (mean ± SD) for each treatment is shown, letters indicate significant differences.

Treatment	Fruit set/umbellet	
Open pollination	92.0 ± 19.5	a
Manual outcrossing	87.7 ± 21.3	a
Manual selfing	87.8 ± 24.2	a
Autonomous selfing	3.8 ± 6.3	b

To determine whether selfing is possible in *H. mantegazzianum* and whether there is a self-incompatibility mechanism and inbreeding depression, controlled crosses were made in the experimental garden in Průhonice (I. Perglová *et al.*, unpublished). There were no significant differences in fruit production among open pollinated, manually outcrossed (i.e. bagged and hand-pollinated by a mixture of pollen from other plants) and manually selfed (i.e. bagged and hand-pollinated with pollen from the same plant) umbellets (Table 4.2), indicating that artificial pollination was effective and pollen from the same and other plants was equally successful. *Heracleum mantegazzianum* is self-compatible and selfing does not seem to be limited by any genetic mechanism, such as maternal control before fertilization or inbreeding depression shortly after fertilization. There was an almost full fruit set in both hand- and open-pollinated umbellets. In contrast, fruit production of autonomously selfed umbellets (i.e. umbellets bagged to exclude pollinators and not subjected to manual pollination) was very low (Table 4.2), suggesting that pollen transfer by pollinators is needed for a standard fruit set. However, as the fruits produced by these umbellets must have arisen from near-flower fertilization, plants can reproduce even in the absence of pollinators (and successfully colonize new sites following the long-distance dispersal of a single propagule).

Experimental crosses on *H. mantegazzianum* and *H. sphondylium* were also made by Stewart and Grace (1984) in their study of interspecific hybridization of these two species. Only a negligible fruit set (1%) was obtained when pollen was transferred between flowers within the terminal umbel, which is consistent with rare overlap in anther dehiscence and female receptivity in the same umbel. However, they were able to realize high levels of selfing (68% fruit set) in two plants with incomplete separation of staminate and pistillate phases between umbels.

Implications of Self-compatibility for the Invasion

Self-pollination was identified as advantageous in some colonizing species (Brown and Burdon, 1987; Rejmánek *et al.*, 2005) and selfing may lead to

acceleration of the rates of spread (Lewis, 1973; Daehler, 1998). Of the introduced species included in a study of Western Australian members of *Apiaceae*
(Keighery, 1982), all the naturalized species were capable of autogamy and
self-fertile. They all possessed attractive inflorescences, and were pollinated by
a variety of native and introduced insects. Controlled pollination experiments
on 17 invasive alien plant species in South Africa revealed that 100% of them
were either self-compatible or apomictic (Rambuda and Johnson, 2004).

The ability to self is advantageous for successful colonization following
long-distance dispersal of a single propagule, because there is no need to wait
for a sexual partner (Baker's law; Baker, 1955). Once a plant has successfully
established, selfing transmits proved genes of a plant, which was able to
survive at that site. Nevertheless, theoretical models suggest that an optimal
mating system for a sexually reproducing invader in a heterogeneous landscape is to be able to modify selfing rates according to local conditions. In early
stages of invasions, when populations are small, plants should self to maximize
fertility. Later, when populations are large and pollinators and/or mates are
not limiting, outcrossing is more beneficial because it generates increased
genetic polymorphism (Pannell and Barrett, 1998; Rejmánek *et al.*, 2005).

Heracleum mantegazzianum is fully self-compatible, as indicated by the
fact that selfed fruit set was not lower than that of naturally pollinated flowers,
and it does not suffer from inbreeding depression at the germination stage (I.
Perglová, unpublished results). The study of flowering phenology showed that
overlaps between male and female flowering phases allow for geitonogamous
(i.e. between-flowers) pollination. This indicates that plants of *H. mantegazzianum* are probably highly self-fertile if isolated or growing in very sparse populations where pollinators transport pollen within a single plant. This has very
important implications for the invasion because even a single isolated plant of
H. mantegazzianum, resulting from a long-distance dispersal event, is capable
of founding a new population. However, when the species grows in abundant
and dense populations, it is likely to produce predominantly outcrossed progeny
because of the high incidence of pollinators moving between plants. However,
natural frequencies of self-fertilization can be detected only by a genetic study
of seed progeny and determination of the selfing rate in natural populations.

Fecundity

The first mention indicating the high fecundity of *H. mantegazzianum* was in
the paper by Sommier and Levier (1895) in which the species was described.
They mention a plant grown in Geneva, which bore no less than 10,000
flowers. Nevertheless, before evaluating what this and other reports mean in
terms of the fecundity, it is useful to describe the morphology of flowers of *H.
mantegazzianum*. Every flower with a fertilized ovule can produce two winged
mericarps (for simplicity, the morphologically correct term 'mericarp' is
replaced by 'fruit' in this chapter and refers to the unit of generative reproduction and dispersal). Thus, the number of flowers recorded by Sommier and
Levier transforms into a potential fecundity of more than 20,000 fruits.

However, as some umbels (mainly those of higher orders) contain male flowers (see the section 'Description of the pattern and timing of flowering'), fruit set is most likely lower.

Surprisingly, a more precise estimate of fecundity was made no earlier than 100 years later by Pyšek *et al.* (1995). Before then, there were several reports (Williamson and Forbes, 1982; Brondegaard, 1990; Table 4.3), which were insufficiently documented in terms of where the numbers come from and how they have been obtained, yet are frequently cited. The only exception might have been the thesis of Warde (1985, cited by Caffrey, 1999), which is unfortunately not easily available. Another report of 1500–18,000 fruits, cited by Tiley *et al.* (1996) is claimed to have come from Neiland (1986), but there is no estimate of fecundity in that study. This suggests that tracing the origin of reports on *H. mantegazzianum* fecundity is as difficult as that reported for the presumed longevity of seeds in the soil (see Moravcová *et al.*, Chapter 5, this volume).

Pyšek *et al.* (1995) record the first estimate of fecundity based on several plants, which provides details of the method of assessment. It reports an average fruit set of 16,140 in the Czech Republic (Table 4.3). Ochsmann (1996) records an average of 9696 fruits in Germany. Similar average values are reported from Scotland (15,729 fruits; Tiley and Philp, 2000) and again from the Czech Republic (20,671 fruits; Perglová *et al.*, 2006). Caffrey (1999) reports numbers from Ireland that are markedly higher than those of previous reports: on average 41,202 fruits, and maximum 107,984 fruits! However, the method of assessment is described as 'a total count of seed numbers was recorded'. From this it is unclear how the values were derived. The number of umbels (the primary umbel, nine secondary and 14 tertiary) this author gives for that most fecund plant is by no means exceptional and is similar to that of plants studied in the Czech Republic. It is therefore likely that in Caffrey's study (Caffrey, 1999) the number of fruits was derived from the number of flowers including male flowers. Be it flowers or fruits on which the assessment was based, it is unclear how the figures were obtained; given the number, counting individual fruits is unlikely.

Tiley and Philp (1997), who 'selected the apparently largest plants' for their estimates, published a maximum number of 81,500 flowers per plant, and the fruit set of a plant with even slightly more umbels as 'only' 52,800. This is clear evidence that the number of flowers is not a good estimate of fecundity.

Probably the highest estimate of fecundity in the literature is 120,000 fruits cited by Ochsmann (1996), reportedly coming from Dodd *et al.* (1994). However, Dodd *et al.* (1994) is not the primary source as these authors refer to Tiley and Philp (1994), which is a chapter in the same book, but that gives no estimate! In any case, Ochsmann's citation is imprecise because the note in Dodd *et al.* (1994) refers to 60,000 flowers, not 120,000 fruits. As cited above (and further illustrated by Perglová *et al.*, 2006), the assumption of fruit set being twice the number of flowers is unrealistic.

The most detailed study of fecundity, based on a large sample of plants, is provided by Perglová *et al.* (2006). Their study estimated the fecundity of 98

Table 4.3. Summary of published data on fecundity of *H. mantegazzianum*. The sources are arranged according to the year of publication.

Source	No. of fruits	No. of plants	No. of localities	Country	Method of assessment
Sommier and Levier (1895)	> 10,000[1]	1	1	Switzerland	n.g.
Williamson and Forbes (1982)	'up to 5,000 or more'	n.g.	n.g.	n.g.	n.g.
Warde (1985) cited by Tiley *et al.* (1996)	14,000–29,000	n.g.	n.g.	Ireland	n.g.
Brondegaard (1990)	5000–6000; max. 27,000	n.g.	n.g.	n.g.	n.g.
Pyšek *et al.* (1995)	avg. 16,140; max. 25,894	8	1	Czech Republic	estimated from fruit mass
Ochsmann (1996)	avg. 9695; max. 28,908	33	n.g.	Germany	n.g.
Tiley and Philp (1997)[2]	81,500; 65,000[1]	2	1	Scotland	estimated from sample counts[3]
	52,800	1	1	Scotland	estimated from fruit mass
Caffrey (1999)	avg. 41,202; min. 1516; max. 107,984	80	8	Ireland	'a total count of seed numbers was recorded'[4]
Tiley and Philp (2000)	avg. 15,729	4	1	Scotland	estimated from fruit mass
Perglová *et al.* (2006)	avg. 20,671; min. 7545; max. 46,470	98	10	Czech Republic	estimated from fecundity class and umbel diameter[5]

[1] Number of flowers.
[2] Three largest plants were sampled, two for flower and one for fruit number.
[3] Flowers in the terminal umbel were counted individually, flower counts for the remaining umbels were derived from the number of rays per umbel and the number of flowers per ray, which were counted on four rays.
[4] Formulation from the original source, see text for interpretation.
[5] See text for details.
n.g. = not given.

plants growing at ten sites in the area of Slavkovský les, Czech Republic (see Table 4.1 for characteristics of the sites). All umbels that developed on these plants were classified according to umbel order, umbel position (satellite, branch) and proportional fecundity (0, 1–25%, 26–75% and >75% of flowers set fruit; Table 4.4). For each combination of umbel order and fecundity class, fruit set was estimated based on data from an additional 100 umbels and regressions based on umbel diameter (Perglová *et al.*, 2006). Nearly 81% of the terminal umbels were placed in the highest fecundity class (>75% of flowers set fruits), while tertiary umbels often set no fruits (Table 4.4). Of the quaternary umbels, 99% did not produce fruit; they did not contribute to the fruit set and their flowers only served as pollen donors. This is consistent with the increase in proportion of male flowers in higher order umbels (see 'Description of the pattern and timing of flowering'). An average plant produced 20,671 ± 5130 fruits (mean ± SD) (Table 4.5) and the maximum estimated number of fruits was 46,470. Almost half of the fruits were produced by the terminal umbel. The fecundity of individual plants significantly increased with the diameter of the flowering stem (Perglová *et al.*, 2006).

The study of Perglová *et al.* (2006) also provides detailed information on the architecture of *H. mantegazzianum* plants, based on the number of umbels of a particular order per plant and on their size (Table 4.5). In the area of Slavkovský les, a typical flowering plant bears a terminal, 4 satellites, 3–4 branches, 17 tertiary and 3 quaternary umbels. This is the first information about the architecture of *H. mantegazzianum* based on a large sample of plants; previously published estimates of fecundity did not provide information on the number of umbels (Caffrey, 1999), considered only primary and secondary umbels (Pyšek *et al.*, 1995) or are based on only a few plants (Tiley *et al.*, 1996; Tiley and Philp, 1997).

To summarize the issue of fecundity in *H. mantegazzianum*, it can be concluded that: (i) the hierarchical structure of flowering organs, typical of umbellifers, and varying proportions of male and female flowers made reports of the number of fruits, which did not take these factors into account, rather unreliable; (ii) in addition, values were cited without checking the original sources, which created a similar myth to the one regarding seed longevity in this species (see Moravcová *et al.*, Chapter 5, this volume). *Heracleum mantegazzianum* is particularly prone to becoming the subject of exaggeration; and (iii) a detailed inspection of the literature, together with our own field estimates, indicates that the maximum reported fruit sets are overestimated. It is very doubtful whether an individual plant of *H. mantegazzianum* is able to produce over 100,000 fruits. Values of 10,000–20,000 seem to be appropriate for Europe, with the maximum occasionally reaching around 50,000 fruits.

Life Span and Age of Flowering Plants in Native and Invaded Ranges, and Under Different Management Regimes

In the above sections, the flowering pattern of *H. mantegazzianum* is described in terms of days; the timing of flowering in terms of years determines

Table 4.4. Distribution of umbel types in fecundity classes, defined as the percentage of flowers that produce fruits. Percentage of the total number of umbels in each class is presented. For example, no terminal was completely infertile, while in 80.6%, more than 75% of the flowers produced fruits. Based on 98 plants growing in the Slavkovský les Protected Landscape Area in 2002. I–IV: umbel order.

Umbel type	Fecundity class				Total no. of umbels
	0	1–25%	26–75%	76–100%	
Termina I	0	17.3	2.0	80.6	98
Satellite II	10.4	77.4	12.2	0	442
Satellite III	87.7	4.1	8.1	0	751
Satellite IV	95.0	1.7	3.3	0	180
Branch II	13.6	73.2	13.3	0	369
Branch III	82.8	7.3	9.9	0	899
Branch IV	99.0	0	1.0	0	305

its life span. Literature reports on the flowering strategy of *H. mantegazzianum* are neither consistent nor detailed. This section summarizes knowledge on the age of flowering plants and whether the species is a monocarpic (flowering only once in its lifetime) or polycarpic perennial plant (flowering repeatedly). Tiley *et al.* (1996) suggest that *H. mantegazzianum* is usually monocarpic and 2–5 years old when flowering. This conclusion is based, among others, on a study of interbreeding between *H. mantegazzianum* and *H. sphondylium*, in which Stewart and Grace (1984) found that plants of the former species flower at the age of 2–5 years. Unfortunately, their plants were grown under artificial conditions in cultivation. A possible effect of growing conditions was observed in the experimental garden at Průhonice; some of the plants grown in a garden bed, with a suitable substrate and regularly watered, flowered in the second year, which was never observed in the field, nor in a native or invaded distribution range (Pergl *et al.*, 2006). Information on life span is fairly limited in literature from the native distribution range, but *H. mantegazzianum* is reported to be polycarpic in Russia (Shumova, 1972). Furthermore, the possibility of repeated flowering in the subsequent years is reported for plants damaged before they finished flowering (Tiley *et al.*, 1996).

Before reporting the age of flowering plants, it is useful to clarify the position of *H. mantegazzianum* in the continuum between monocarpy to polycarpy. In field studies carried out since 2002, *H. mantegazzianum* was never observed to flower repeatedly (Pergl *et al.*, 2006). Nevertheless, some species in this genus, such as the European species *H. sphondylium* (Stewart and Grace, 1984) are polycarpic, as is *H. persicum* Desf. ex Fischer, which is invasive in Scandinavia (see Jahodová *et al.*, Chapter 1, this volume; Nielsen *et al.*, 2005). Another invasive congener, *H. sosnowskyi* Manden, is also monocarpic (Nielsen *et al.*, 2005). That both monocarpy and polycarpy exist in closely related species of the genus indicates that an occasional shift to polycarpic behaviour is not excluded and exploration of this possibility deserves attention. In addition, the majority of interspecific hybrids between

Table 4.5. The number of umbels, their size and estimate of fecundity of plants growing in the Slavkovský les Protected Landscape Area in 2002. The umbels are classified according to their order and position in the case of secondary umbels. Ten plants from each of ten sites were assessed; two plants were damaged before the fruit matured and were excluded from the analysis. Taken from Perglová *et al.* (2006), published with permission from the Czech Botanical Society.

	Terminal	II. order umbel – satellite	II. order umbel – branch	III. order umbel	IV. order umbel	Total
No. of umbels/plant [median (min–max)]*	1	4.3 (2–8)	3.5 (1–10)	17.3 (2–43)	2.8 (1–39)	27.5 (5–98)
Fecundity/umbel [avg ± SD]**	9216.0 ± 481.7	1288.0 ± 614.8	1157.0 ± 429.1	32.0 ± 30.1	n.a.	20,671.0 ± 5129.8
Umbel size [cm] avg ± SD (min–max)**	61.7 ± 3.7 (44–85)	36.8 ± 3.3 (20–56)	36.3 ± 4.8 (5–62)	17.8 ± 2.3 (2–36)	7.9 ± 1.8 (1–17)	–

* Calculated as a median of site medians.
** Calculated as averages of site averages and among-sites standard deviations.

H. mantegazzianum and *H. sphondylium* are polycarpic (Stewart and Grace, 1984), which indicates a potential for polycarpy in the former species.

To clarify the reported possibility that *H. mantegazzianum* is capable of flowering the following year if damaged before it finishes flowering (Tiley *et al.*, 1996), all the umbels produced during the course of the growing season were removed from 20 flowering plants grown in the experimental garden at Průhonice. None of the plants survived and flowered the following year (see Pyšek *et al.*, Chapter 7, this volume). From field observations, a study of age structure (Pergl *et al.*, 2006) and this experiment, it can be concluded that *H. mantegazzianum* is strictly monocarpic and dies after flowering. It is possible that the reported survival of flowering plants can be attributed to plants forming rather dense clumps, which makes the identification of individual plants difficult. A new individual emerging next to a dead stem can be easily considered as resprouting from the rootstock and lead to the wrong conclusion, as was probably the case of Morton (1978).

Since the reproduction of *H. mantegazzianum* depends exclusively on fruit production, which occurs only once in its lifetime, the right timing of flowering is a crucial point in the life history of this species. Its monocarpic behaviour thus opens questions about the timing of flowering and its relation to fruit set, and whether the life span varies across distribution ranges (native/invaded) and habitat types. Monocarpic species have a single opportunity to reproduce, and need to trade-off postponing flowering until the next season, which allows the accumulation of more resources and setting more fruits in spite of the increased risk of death, or flowering as soon as possible with the resources currently available (Metcalf *et al.*, 2003).

The age at flowering of native plants in the Western Greater Caucasus and invasive plants in the Czech Republic, in managed (pastures) and unmanaged habitats, is recorded in Pergl *et al.* (2006). Unmanaged sites in the Caucasus can be considered as natural habitats of *H. mantegazzianum* (see Otte *et al.*, Chapter 2, this volume), while unmanaged sites in the Slavkovský les study area in the Czech Republic are abandoned pastures, forest clearings, meadows and abandoned former villages (for details of the history of the invasion of this region, see Müllerová *et al.*, 2005). Pergl *et al.* (2006) found that age at flowering in unmanaged habitats is significantly different between distribution ranges. Plants from the native range flowered later (median age 4 years) than those from the invaded range (3 years). Within the invaded range, plants from managed sites needed significantly more time (5 years) to flower than those from unmanaged sites. But the oldest, a 12-year-old flowering plant, was found in an unmanaged site in the Czech Republic; this can be attributed to the harsh conditions at that site. However, the general pattern is that plants from pastures flowered later than those from unmanaged sites in both distribution ranges (the mean age at flowering in pastures was 5 years), although the difference in the native range was not significant. Additional analysis by Pergl *et al.* (2006) showed that the delay in the time of flowering can be a result of higher altitude in the native distribution range, which affects the length of the growing season and consequently the time needed for accumulation of resources.

The timing of flowering in relation to the reproductive effort of plants in their native and invaded distribution ranges was also studied by Pergl *et al.* (2006), who found no relationship between the age of flowering plants and potential fecundity, based on an architecture score, calculated from the number of umbels of different orders and their importance for fruit set. This implies that once a plant accumulates a certain minimum level of resources it flowers. This strategy is the same in both distribution ranges (Pergl *et al.*, 2006). Complementary information on the age at flowering is provided by C. Nielsen *et al.* (unpublished data), who investigated the effect of the length of the management regime on the population structure of *H. mantegazzianum*. Their data are similar to those from the Czech Republic. At four study sites, grazed for 2–8 years, plants flowered from the third to fifth year. The trend in the data indicates that the proportion of older plants in the population increased with the duration of grazing, but further research is needed to confirm this.

From the overall pattern shown by *H. mantegazzianum* it is concluded that although it is able to flower in the second year, this only occurs in the favourable conditions in experimental gardens or cultivated fields (Stewart and Grace, 1984; Pergl *et al.*, 2006). The youngest flowering plant found under natural conditions was 3 years old; this is also the most common age of flowering plants in the invaded distribution range where, nevertheless, they can be as much as 12 years old (Pergl *et al.*, 2006). This suggests that *H. mantegazzianum* is remarkably plastic and when growing in unsuitable conditions can wait until the needed resources are accumulated and then reproduce.

Acknowledgements

Thanks to Lenka Moravcová for helpful consultations, Zuzana Sixtová, Ivan Ostrý and Šárka Jahodová for technical assistence, and Tony Dixon for improving our English. The study was supported by the project 'Giant Hogweed (*Heracleum mantegazzianum*) a pernicious invasive weed: developing a sustainable strategy for alien invasive plant management in Europe', funded within the 'Energy, Environment and Sustainable Development Programme' (grant no. EVK2-CT-2001-00128) of the European Union 5th Framework Programme, by institutional long-term research plans no. AV0Z60050516 from the Academy of Sciences of the Czech Republic and no. 0021620828 from the Ministry of Education of the Czech Republic, and grant no. 206/05/0323 from the Grant Agency of the Czech Republic.

References

Baker, H.G. (1955) Self-compatibility and establishment after 'long-distance' dispersal. *Evolution* 9, 347–348.

Bell, C.R. (1971) Breeding systems and floral biology of the *Umbelliferae* or evidence for specialization in unspecialized flowers. In: Heywood, V.H. (ed.) *The Biology and Chemistry of the Umbelliferae*. Academic Press, London, pp. 93–107.

Braak, J.P. and Kho, Y.O. (1958) Some observations on the floral biology of the carrot (*Daucus carota* L.). *Euphytica* 7, 131–139.

Brondegaard, V.J. (1990) Massenausbreitung des Bärenklaus. *Naturwissenschaftliche Rundschau* 43, 438–439.

Brown, A.H.D. and Burdon, J.J. (1987) Mating systems and colonizing success in plants. In: Gray, A.J., Crawley, M.J. and Edwards, P.J. (eds) *Colonization, Succession and Stability*. Blackwell, Oxford, UK, pp. 115–131.

Caffrey, J.M. (1999) Phenology and long-term control of *Heracleum mantegazzianum*. *Hydrobiologia* 415, 223–228.

Crawley, M.J. (1997) *Plant Ecology*, 2nd edn. Blackwell Publishing, Oxford, UK.

Daehler, C.C. (1998) Variation in self-fertility and the reproductive advantage of self-fertility for an invading plant (*Spartina alterniflora*). *Evolutionary Ecology* 12, 553–568.

Davila, Y.C. and Wardle, G.M. (2002) Reproductive ecology of the Australian herb *Trachymene incisa* subsp. *incisa* (*Apiaceae*). *Australian Journal of Botany* 50, 619–626.

Dodd, F.S., de Waal, L.C., Wade, P.M. and Tiley, G.E.D. (1994) Control and management of *Heracleum mantegazzianum* (giant hogweed). In: de Waal, L.C., Child, L.E., Wade, P.M. and Brock, J.H. (eds) *Ecology and Management of Invasive Riverside Plants*. Wiley, Chichester, UK, pp. 111–126.

Gaudeul, M. and Till-Bottraud, I. (2003) Low selfing in a mass-flowering, endangered perennial, *Eryngium alpinum* L. (*Apiaceae*). *American Journal of Botany* 90, 716–723.

Grace, J. and Nelson, M. (1981) Insects and their pollen loads at a hybrid *Heracleum* site. *New Phytologist* 87, 413–423.

Keighery, G.J. (1982) Reproductive strategies of Western Australian *Apiaceae*. *Plant Systematics and Evolution* 140, 243–250.

Kos, J. and Maršáková, M. (1997) *Chráněná území České Republiky*. Agentura ochrany přírody a krajiny, Praha.

Lewis, H. (1973) The origin of diploid neospecies in Clarkia. *American Naturalist* 107, 161–170.

Lovett-Doust, J. (1980) Floral sex ratios in andromonoecious *Umbelliferae*. *New Phytologist* 85, 265–273.

Lovett-Doust, J. and Lovett-Doust, L. (1982) Life-history patterns in British *Umbelliferae*: a review. *Botanical Journal of the Linnean Society* 85, 179–194.

McClintock, D. (1975) *Heracleum* L. In: Stace, C.A. (ed.) *Hybridization and the Flora of the British Isles*. Academic Press, London, p. 270.

Metcalf, J.C., Rose, K.E. and Rees, M. (2003) Evolutionary demography of monocarpic perennials. *Trends in Ecology and Evolution* 18, 471–480.

Morton, J.K. (1978) Distribution of giant cow parsnip (*Heracleum mantegazzianum*) in Canada. *Canadian Field Naturalist* 92, 182–185.

Müllerová, J., Pyšek, P., Jarošík, V. and Pergl, J. (2005) Aerial photographs as a tool for assessing the regional dynamics of the invasive plant species *Heracleum mantegazzianum*. *Journal of Applied Ecology* 42, 1042–1053.

Neiland, M.R.M. (1986) The distribution and ecology of Giant Hogweed (*Heracleum mantegazzianum*) on the river Allan, and its control in Scotland. Thesis, University of Stirling, UK.

Nielsen, C., Ravn, H.P., Cock, M. and Nentwig, W. (eds) (2005) *The Giant Hogweed Best Practice Manual. Guidelines for the Management and Control of an Invasive Alien Weed in Europe*. Forest and Landscape Denmark, Hørsholm, Denmark.

Ochsmann, J. (1996) *Heracleum mantegazzianum* Sommier & Levier (*Apiaceae*) in Deutschland. Untersuchungen zur Biologie, Verbreitung, Morphologie und Taxonomie. *Feddes Repertorium* 107, 557–595.

Otte, A. and Franke, R. (1998) The ecology of the Caucasian herbaceous perennial *Heracleum mantegazzianum* Somm. et Lev. (Giant Hogweed) in cultural ecosystems of Central Europe. *Phytocoenologia* 28, 205–232.

Pannell, J.R. and Barrett, S.C.H. (1998) Baker's Law revisited: reproductive assurance in a metapopulation. *Evolution* 52, 657–668.

Pergl, J., Perglová, I., Pyšek, P. and Dietz, H. (2006) Population age structure and reproductive behavior of the monocarpic perennial *Heracleum mantegazzianum* (*Apiaceae*) in its native and invaded distribution range. *American Journal of Botany* 93, 1018–1027.

Perglová, I., Pergl, J. and Pyšek, P. (2006) Flowering phenology and reproductive effort of the invasive alien plant *Heracleum mantegazzianum*. *Preslia* 78, 265–285.

Pyšek, P. (1991) *Heracleum mantegazzianum* in the Czech Republic – the dynamics of spreading from the historical perspective. *Folia Geobotanica & Phytotaxonomica* 26, 439–454.

Pyšek, P., Kučera, T., Puntieri, J. and Mandák, B. (1995) Regeneration in *Heracleum mantegazzianum* – response to removal of vegetative and generative parts. *Preslia* 67, 161–171.

Rambuda, T.D. and Johnson, S.D. (2004) Breeding systems of invasive alien plants in South Africa: does Baker's rule apply? *Diversity and Distributions* 10, 409–416.

Rejmánek, M., Richardson, D.M., Higgins, S.I., Pitcairn, M.J. and Grotkopp, E. (2005) Ecology of invasive plants: state of the art. In: Mooney, H.A., Mack, R.M., McNeely, J.A., Neville, L., Schei, P. and Waage, J. (eds) *Invasive Alien Species: Searching for Solutions*. Island Press, Washington, DC, pp. 104–161.

Shumova, E.M. (1972) The morphology of main shoot of giant hogweed during the juvenile stage. *Doklady Tskha* 180, 235–242. [In Russian.]

Snow, A.A. and Grove, K.F. (1995) Protandry, a neuter phase, and unisexual umbels in a hermaphroditic, neotropical vine (*Bomarea acutifolia*, Alstroemeriaceae). *American Journal of Botany* 82, 741–744.

Sommier, S. and Levier, E. (1895) Decas Umbelliferarum Novarum Caucasi. *Nuovo Giornale Botanico Italiano* 2, 79–81.

Stewart, F. and Grace, J. (1984) An experimental study of hybridization between *Heracleum mantegazzianum* Somm. & Levier and *H. sphondylium* L. subsp. *sphondilium* (*Umbelliferae*). *Watsonia* 15, 73–83.

Tiley, G.E.D. and Philp, B. (1994) *Heracleum mantegazzianum* (Giant Hogweed) and its control in Scotland. In: de Waal, L.C., Child, L.E., Wade, P.M. and Brock, J.H. (eds) *Ecology and Management of Invasive Riverside Plants*. John Wiley and Sons, Chichester, UK, pp. 101–109.

Tiley, G.E.D. and Philp, B. (1997) Observations on flowering and seed production in *Heracleum mantegazzianum* in relation to control. In: Brock, J.H., Wade, M., Pyšek, P. and Green, D. (eds) *Plant Invasions: Studies from North America and Europe*. Backhuys Publishers, Leiden, The Netherlands, pp. 123–137.

Tiley, G.E.D. and Philp, B. (2000) Effects of cutting flowering stems of Giant Hogweed *Heracleum mantegazzianum* on reproductive performance. *Aspects of Applied Biology* 58, 77–80.

Tiley, G.E.D., Dodd, F.S. and Wade, P.M. (1996) *Heracleum mantegazzianum* Sommier & Levier. *Journal of Ecology* 84, 297–319.

Webb, C.J. (1981) Andromonoecism, protandry, and sexual selection in *Umbelliferae*. *New Zealand Journal of Botany* 19, 335–338.

Williamson, J.A. and Forbes, J.C. (1982) Giant Hogweed (*Heracleum mantegazzianum*): its spread and control with glyphosate in amenity areas. *Proceedings 1982 British Crop Protection Conference – Weeds*. British Crop Protection Council, Farnham, UK, pp. 967–972.

5

Seed Germination, Dispersal and Seed Bank in *Heracleum mantegazzianum*

LENKA MORAVCOVÁ,[1] PETR PYŠEK,[1,2] LUKÁŠ KRINKE,[3]
JAN PERGL,[1] IRENA PERGLOVÁ[1] AND KEN THOMPSON[4]

[1]Institute of Botany of the Academy of Sciences of the Czech Republic,
Průhonice, Czech Republic; [2]Charles University, Praha, Czech Republic;
[3]Museum of Local History, Kladno, Czech Republic; [4]University of
Sheffield, Sheffield, UK

*Royal beast did not forget. Soon they escaped, spreading their seed,
preparing for an onslaught, threatening the human race*

(Genesis, 1971)

Introduction

Since reproduction of *Heracleum mantegazzianum* Sommier & Levier is
exclusively by seed (Tiley *et al.*, 1996; Moravcová *et al.*, 2005; Krinke *et al.*,
2005), a detailed knowledge of its seed ecology is crucial for understanding this
species' invasive behaviour. This chapter summarizes available information on
seed dormancy, pattern of germination, seed bank formation and dynamics, as
well as the first solid data on the longevity of seed in the soil.

Heracleum mantegazzianum has oval-elliptical broadly winged mericarps
which are connected into pairs by carpophore (Fig. 5.1A) and split when
mature (Holub, 1997). The mericarps are 6–18 mm long and 4–10 mm wide,
and each contains one seed. The embryo is rudimentary (Martin, 1946) and
surrounded by oily endosperm. Mature fruits have a strong resinous smell
(Tiley *et al.*, 1996). For simplicity the unit of generative reproduction and dispersal is termed 'fruit' throughout this chapter rather than the morphologically
correct 'mericarp', and the term 'seed' is used when referring to germination.

Dormancy Breaking Mechanisms

It has long been known that seeds of *H. mantegazzianum* do not germinate
after dry storage (Grime *et al.*, 1981) and cold stratification is necessary for germination (Nikolaeva *et al.*, 1985; Tiley *et al.*, 1996; Otte and Franke, 1998).

A

B

Fig. 5.1. (A) *Heracleum mantegazzianum* has oval-elliptical broadly winged mericarps which are connected into pairs by carpophore. (B) In the autumn, seeds are released from plants forming dense stands, as in the largest locality of the Slavkovký les study region, Czech Republic. Photo: P. Pyšek.

Under natural conditions in the Slavkovský les study area, Czech Republic, seeds germinate early in spring after snow melting (March to April) (see Pergl *et al.*, Chapter 6, this volume; Krinke *et al.*, 2005). Seeds of *H. mantegazzianum* exhibit morphophysiological dormancy in the sense of Nikolaeva *et al.* (1985) and Baskin and Baskin (2004); ripe seeds have an underdeveloped embryo which is physiologically dormant. For a seed to germinate, embryo growth needs to be completed and its physiological dormancy broken. Both these processes occur in cold and wet conditions of autumn and winter stratification in the field; corresponding laboratory conditions are temperature within the range of 1–6°C (Moravcová et al., 2005). Gibberellic acid does not stimulate the germination of freshly ripe seeds (L. Moravcová, unpublished results). From all this information and according to the results of Nikolaeva et al. (1985), this type of morphophysiological dormancy (further MPD) in the sense of Baskin and Baskin (2004) resembles a deep complex MPD. The same type of MPD was recorded in other species of Apiaceae, e.g. Heracleum sphondylium L. (Stokes, 1952a, b; Nikolaeva, 1985) or Anthriscus sylvestris (L.) Hoffm. (Lhotská, 1978; Baskin et al., 2000). Nevertheless, information on embryo growth is needed to prove this for H. mantegazzianum.

Germination Characteristics Related to the Position on the Plant

Heracleum mantegazzianum bears umbels at various positions and orders (for a scheme of the plant architecture see Moravcová et al., 2005: Fig. 1; and Perglová et al., Chapter 4, this volume) and characteristics of seeds are affected by where on the plant they are produced. The position of a seed or fruit on a mother plant has been shown to affect seed mass, morphology, germination and dormancy characteristics (for a survey see Baskin and Baskin, 1998; Gutterman, 2000). For Apiaceae, seed mass and/or germination has been shown to depend on umbel position in many species (e.g. Ojala, 1985; Thomas et al., 1978, 1979; Hendrix, 1984a; Hendrix and Trapp, 1992). Compared to the other members of its family, H. mantegazzianum is not unusual with respect to seed position on the mother plant. Moravcová et al. (2005) studied the influence of fruit position on the mother plant on fruit mass, germination percentage and rate of germination in H. mantegazzianum. The data were collected in the Slavkovský les Protected Area in the west of the Czech Republic (for details on this region, see Perglová et al., Chapter 4, this volume). Seeds were collected from the terminal (i.e. primary) umbel, secondary umbels in satellite positions and secondary umbels in branch positions, separately from the centre and margin of each sampled umbel (see Fig. 4.1). The overall mean mass of a single fruit was 13.1 mg (Table 5.1) and corresponded to the range 4.6–23.2 mg given by Tiley et al. (1996) for H. mantegazzianum. Fruits from terminal inflorescences were heavier than those from satellites and branches, and those produced in the centre of an umbel were heavier than those from the margin (Moravcová et al., 2005). Fruits from terminals weighed on average 15.9 mg, whereas those from satellites and branches weighed 11.7 mg, and the fruits from the centre were significantly heavier than those from the margins, being 13.3 and 12.9 mg, respectively (Table 5.1). Overall, the determination of fruit mass in H. mantegazzianum follows the same rules as in other Apiaceae (Hendrix, 1984a, b; Thompson, 1984; Hendrix and Sun, 1989), but the variation is several orders of magnitude lower than in some other species.

The mean percentage germination at 8–10°C (after 2 months of cold stratification at 2–4°C) found by Moravcová et al. (2005) was 91%, and varied among the seven sites studied, but was not affected by fruit position on a plant.

Table 5.1. Fruit mass in *H. mantegazzianum* depends on the position of fruit on a mother plant. Data are means ± SD (mg) and *n* (sample size) for fruit sampled from combinations of umbel types on mother plant (terminal, satellite, branch) and positions within an umbel (centre or margin) from eight plants in each of seven sites in the Slavkovský les region, West Bohemia, Czech Republic. There were five replicates of each treatment. Data from Moravcová *et al.* (2005).

Fruit position within umbel	Umbel type (fruit position on the plant)							
	Terminal	*n*	Satellite	*n*	Branch	*n*	Mean	*n*
Centre	16.15 ± 2.75	270	11.93 ± 3.47	270	11.65 ± 3.12	265	13.25 ± 3.74	805
Margin	15.60 ± 2.59	265	11.52 ± 3.44	275	11.66 ± 3.3	270	12.90 ± 3.66	810
Mean	15.88 ± 2.67	535	11.73 ± 3.46	545	11.66 ± 3.12	535	13.08 ± 3.70	1615

Such a high percentage germination seems to be usual within the context of the *Apiaceae* family, where seeds usually germinate readily once dormancy is broken (Baskin and Baskin, 1990, 1991; Baskin *et al.*, 1995); a value of 94% was reported for *Anthriscus sylvestris* (Baskin *et al.*, 2000). In the study of Moravcová *et al.* (2005), cold and dark conditions mimicked closely the natural situation under the soil surface during spring. Moreover, the high percentage germination recorded in the laboratory corresponds with that obtained in a garden burial experiment, where about 90% of seed stored in the soil germinated over the first winter (Fig. 5.5 and Moravcová *et al.*, 2006). Given its fecundity (see Perglová *et al.*, Chapter 4, this volume), *H. mantegazzianum* exerts enormous pressure of highly germinable propagules in invaded sites (see Pyšek *et al.*, Chapter 19, this volume).

These results (Moravcová *et al.*, 2005) have practical implications as mechanical control often focuses on cutting terminal umbels or whole stems at flowering time (see Pyšek *et al.*, Chapter 7, this volume). Regeneration then occurs via higher-order umbels that produce fruit with the same capability to germinate as those produced on the terminal and low-order umbels. Similarly, the ability to produce a standard fruit in terms of weight, even on umbels of higher orders, contributes to successful regeneration after the loss of flowering tissues due to control efforts (see Pyšek *et al.*, Chapter 7, this volume).

The percentage of seeds of *H. mantegazzianum* which germinate does not depend on where on the plant the fruit is produced, but the rate at which they germinate does (Moravcová *et al.*, 2005). Large seeds germinated faster than small seeds; germination rate increased with increasing fruit mass and this pattern was consistent for all plants at each site (Fig. 5.2). Since seeds from terminals were heavier than those from branches, the former germinated sooner than the latter. Heavy seeds germinated faster than light seeds, but the difference was only obvious at the beginning of the experiment (Moravcová *et al.*, 2005). Faster germination of heavier seed adds to the ecological advantage resulting from their size; heavier seeds produce bigger seedlings (Harper, 1977) and this was also reported for other species within *Apiaceae* (Thomas *et al.*, 1979; Thomas, 1996).

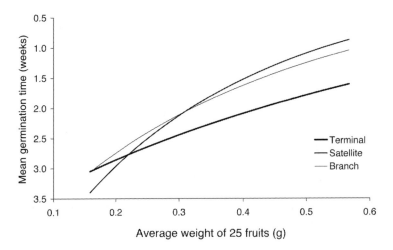

Fig. 5.2. The rate of germination of *H. mantegazzianum* seed increases with increasing fruit mass and the relationship is valid for all three umbel types (terminal, satellite, branch). Data are for a randomly chosen plant ($n = 715$ seeds). Terminal: rate = $1/\exp(-3.345 + 3.852\text{mass})^{(1/2.45)}$; satellite: rate = $1/\exp(-4.296+8.171\text{mass})^{(1/2.45)}$; branch = $1/\exp(-3.763 + 6.441\text{mass})^{(1/2.45)}$. $\chi^2 = 1301.0$; $df = 5$; $P < 0.001$. The y-axis is reversed so the seeds that germinate first appear above those that germinate last for a given fruit mass. For most of the ranges in fruit mass, the germination rate of a particular fruit mass is higher for satellites and branches than terminals, but the average germination rate for terminals is higher than for satellites and branches because fruit from terminal umbels is, on average, much heavier than that from the other two umbel types. Hence seeds produced on terminals germinate faster. Seeds were stratified for 2 months at 2–4°C and then germinated at 8–10°C. Germination was recorded weekly for 6 months. From Moravcová *et al.* (2005).

Neither the characteristics of individual umbels (duration of flowering, size) nor those of whole plants (fecundity, age, height, basal diameter) had an effect on germination characteristics. The only significant relationship found was a negative one between fruit mass and plant height (Moravcová *et al.*, 2005).

Germination at Different Temperature Regimes

In the Czech Republic, seeds of *H. mantegazzianum* germinate exclusively in spring; no seedlings were found in the field in the autumn (Krinke *et al.*, 2005; Moravcová *et al.*, 2005; see Pergl *et al.*, Chapter 6, this volume). This conclusion is based on results obtained in the Slavkovský les study region, but may be considered as valid for Central European areas with similar climatic conditions. In the above region, seeds in the field germinate in March and April, when temperatures are still below 10°C (Fig. 5.6).

Preliminary results of laboratory experiments carried out at the Institute of Botany, Průhonice (Moravcová *et al.*, 2006) allow field observations to be linked with germination patterns revealed under controlled conditions.

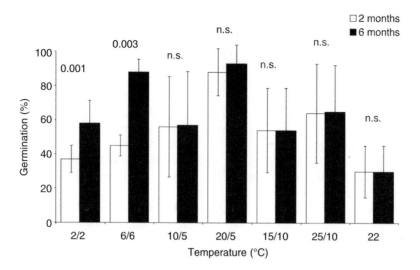

Fig. 5.3. Germination percentage of *H. mantegazzianum* seeds (mean ± SE) depends on the temperature regime and on the time provided for germination. Seeds cold-stratified for 2 months before the experiment were used for germination. Seven temperature regimes were used: 2, 6, 10/5, 20/5, 15/10, 25/10 and 22°C (in alternating day/night temperatures the day and night lasted 12 h). Percentage of seed that germinated was ascertained after 2 and 6 months. Differences between times at a particular regime are indicated on top of bars (t-test for paired comparisons). From Moravcová *et al.* (2006).

Although the stratified seeds of *H. mantegazzianum* were able to germinate under various temperature regimes (Fig. 5.3), the best germination was achieved either at a low temperature of 6°C (88% of seed germinated) or at an alternating temperature of 20/5°C (93%). At higher temperatures all germination occurred in the first 2 months, but then it virtually ceased and the percentage of germinated seeds did not increase up to the 6th month. On the other hand, seeds at low constant temperatures of 2°C and 6°C germinated gradually and for a long time; between 2 and 6 months, the percentage of germinated seeds increased by 20% and 43%, respectively (Fig. 5.3). These results suggest that higher spring and summer temperatures either induce seed dormancy or rather prevent the process of breaking dormancy which can only be completed under cold temperatures.

The Persistent Seed Bank of *Heracleum mantegazzianum*: in Pursuit of an Urban Myth

The development of opinions on the persistence of the soil seed bank in *H. mantegazzianum* is an interesting story, illustrating how dangerous it can be to accept information without checking original sources. Manchester and Bullock (2000) mention *H. mantegazzianum* among plants that have so far caused a serious problem in the UK and conclude that 'due to extensive seed banks and possible long-term viability of seeds, any control programme would need to have follow-up monitoring and control for at least 7 years after the initial control measures using herbicide or cutting'. To support this statement, they refer to Dodd *et al*. (1994) and Collingham *et al*. (2000). The paper of Dodd *et al*. (1994) is, however, a general review of the control of *H. mantegazzianum*, and bases the statement that the species produces 'an extensive seed bank in the vicinity of the parent plant' on Morton (1978), who refers explicitly to the longevity of dry seeds, and Lundström (1989). The latter paper is more ambiguous, claiming that 'seed viability of 15 years is possible' (Lundström, 1989). Collingham *et al*. (2000) make no direct reference to seed persistence, they only refer to the *Biological Flora of British Isles* for *H. mantegazzianum* (Tiley *et al*., 1996).

Tiley *et al*. (1996) are reasonably clear about seed persistence in the soil. They note that the seed biology of *H. mantegazzianum* is very similar to that of the native *H. sphondylium* – once seeds are adequately chilled, they germinate quite well at 5°C in the dark – and therefore a persistent seed bank is unlikely. They conclude that 'field observations indicate that most if not all shed seeds germinate in the following year', although only unpublished observations are cited in support of this. Once more Morton (1978) and Lundström (1989) are cited as sources of data on seed longevity, but again it is not clear whether Lundström is referring to dry seeds or to seeds in the soil.

Astute readers will have begun to detect a pattern here. Wherever the possibility of a persistent seed bank in *H. mantegazzianum* is raised, the ultimate source appears to be Lundström (1989). Lundström and Darby (1994) is a good example:

> The plant is characterised by rapid growth and the production of large numbers of viable seeds, which remain capable of germinating over 7–8 years. Because of these factors, vigorous control measures are required over a number of years in order to control and eradicate the plant (Dodd *et al*., 1994).

The circle seems to close here, because the information on seed persistence in Dodd *et al*. (1994) comes from Lundström (1989)...

So, what does Lundström (1989) have to say? Very little, in fact: 'It cannot be avoided [even after control with herbicides] that new seedlings emerge for several years when the stand has been in the same place for several decades. There are data suggesting that germinability could extend up to fifteen years.' There are at least two ambiguities here. It is hardly surprising that new seedlings emerge after an attempt to control a large stand by herbicides. Such control is rarely complete, and new seeds may always be dispersed from outside. It is also now clear why those who cite Lundström are vague about whether he is referring to dry seeds or to seeds buried in the soil – Lundström is himself vague. Even if he is referring to buried seeds, it is most likely that the 'suggestive data' come from the sporadic appearance of seedlings in the field, which could be recently dispersed.

In fact, the similarity of *H. mantegazzianum* and *H. sphondylium*, a species without a long-persistent seed bank (Thompson *et al*., 1997), might have made researchers more suspicious about the claimed longevity of

H. mantegazzianum seed in the soil. Further, indirect evidence that the period is actually much less has been around for some time – Andersen and Calov (1996) sampled the soil beneath a stand of *H. mantegazzianum* that had been eliminated by 7 years of sheep grazing and, as they found no viable seeds, concluded that longevity in the soil is certainly less than 7 years.

Seed Bank Type, its Size and Seasonal Dynamics

What is clear from the above is that no-one has previously conducted a serious study of seed bank dynamics or controlled burial experiments with *H. mantegazzianum* seeds. To clarify the behaviour of this species in terms of seed bank type and its dynamics, several studies were carried out within the framework of the GIANT ALIEN project. The amount of seeds in the seed bank, its vertical distribution and seasonal changes were studied by Krinke *et al.* (2005) at seven sites in the Slavkovský les region, Czech Republic. The study spanned two growing periods with censuses made in autumn after most seeds were released, in spring before seed germination, and in summer after spring germination and before seed release in the following autumn. The seeds were classified into three categories (dormant, non-dormant and dead) after collection (for explanation see Fig. 5.4).

The total number of seeds significantly increased with mean density of flowering plants at a site. The numbers of living, dead and total seeds were high in autumn, remained at the same level until spring, but from spring to summer, they all decreased. The number and proportion of dormant seeds was significantly higher in autumn than in spring or summer, and the number of non-dormant seeds was highest in spring (Fig. 5.4A). Proportions of dormant, non-dormant and dead seeds exhibited considerable seasonal dynamics. The percentage of dead seeds consistently increased from autumn to the following summer (Fig. 5.4A). The percentage of living seeds in the total seed bank decreased during winter from 55.9% in the autumn sample to 41.7% in spring to 14.8% in summer (Fig. 5.4B). The percentage of non-dormant seeds among living seeds was 0.3% in autumn, over winter it increased to 87.5% in the spring sample, and decreased to 3.0% in summer (Fig. 5.4B). After massive fruit release in autumn, nearly all living seeds (99.7%) were dormant. As almost no non-dormant seeds were found in autumn, this supports the observations that germination and population recruitment from seedlings in this species occur exclusively in spring (Krinke *et al.*, 2005; see Pergl *et al.*, Chapter 6, this volume).

Of the total variation in seed bank size, about four-fifths was attributed to variation among sites, and one-fifth to that within sites. Expressed per m^2, the average value pooled across localities was 6719 ± 4119 (mean ± SD) in autumn, 4907 ± 2278 in spring and 1301 ± 1036 in summer for the total number of seeds, and 3759 ± 2906, 2044 ± 1198 and 192 ± 165, respectively, for living seeds (Krinke *et al.*, 2005 and their Table 5). These data (Krinke *et al.*, 2005) represent the first quantitative estimate of a seed bank in *H. mantegazzianum*, because numbers reported previously were based on

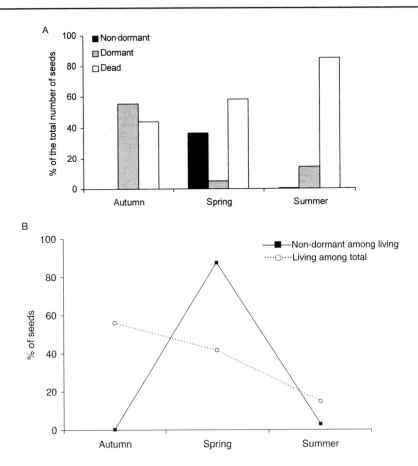

Fig. 5.4. (A) Changes in representation of dormant, non-dormant and dead seeds in the seed bank of *H. mantegazzianum* from autumn (after seed release) through spring (before germination) to the following summer (before new seeds are shed). Mean values shown are pooled across nine localities in the Slavkovský les region, Czech Republic. (B) The proportion of non-dormant among living seed is close to zero in autumn, reaches a peak in spring after dormancy has been broken by cold and wet stratification over winter, and decreases to very low values in summer, after the vast majority of non-dormant seed germinated in spring. The proportion of living seed among the total number of seed steadily decreases in the course of the 'seed-cycle year', as part of the seed population gradually decays. Based on data from Krinke *et al.* (2005). Seeds germinated up to 1 month were considered as non-dormant, non-germinated living seeds were considered as dormant and decayed seeds found in the soil sample or seeds found dead after germination were considered as dead.

estimates from seedlings germinating in the field (Andersen and Calov, 1996) or on multi-species seed bank studies (Thompson *et al.*, 1997). Such numbers of seeds found per m² of the soil exceed the average value in the family *Apiaceae* by an order. Only two species of *Apiaceae* (*Ammi majus* L. and *Torilis japonica* (Houtt.) DC.) exhibit seed density values comparable with

H. mantegazzianum (Thompson *et al.*, 1997). The reproductive potential of *H. mantegazzianum* is enormous and seems to be a crucial feature making invasion possible to the extent observed in the region (Krinke *et al.*, 2005) and elsewhere in Europe (Ochsmann, 1996; Tiley *et al.*, 1996).

Quantitative data of Krinke *et al.* (2005) allow some extrapolations to the landscape level. From knowledge of *H. mantegazzianum* population size in the largest study site (99,000 m^2) and the average number of non-dormant seeds present in the spring, it can be calculated that each year in spring there are 386 million seeds, ready to germinate, in a single site.

So which seed bank type best fits what we now know about *H. mantegazzianum*? The species was considered to have a transient soil seed bank, i.e. missing from the seed bank or present only in the surface layer (Thompson *et al.*, 1997). However, Krinke *et al.* (2005) classified the soil seed bank of *H. mantegazzianum* as a short-term persistent soil seed bank *sensu* Thompson *et al.* (1997) as in their samples 95% of seeds were concentrated in the upper soil layer and some living seeds were also present in lower soil layers. Moreover, the data reported in the following sections clearly indicate that seeds of *H. mantegazzianum* do persist in the soil for some years and that a short-term persistent seed bank is the case here.

Seasonal and Long-term Survival of Buried Seed and Timing of Germination

In another experiment, Moravcová *et al.* (2006) monitored the survival of buried seeds and seasonal timing of germination in *H. mantegazzianum* over 2 years. They found that the depletion of the seed bank during the first winter following burial was very fast; of seeds buried in September 2002 at the experimental garden of the Institute of Botany in Průhonice, Czech Republic, 91.4% germinated (or decayed) by May and 91.8% by October of the following year. During the second winter, the proportion of germinated (or decayed) seeds increased to 97.4% in May 2004 (Fig. 5.5). Only a small proportion of seeds survived in a viable state in the soil over the first winter following burial. In May and October 2003, dormant seeds made up 8.4% and 8.2%, respectively, of the total. A year later, the proportions recorded were 2.6% in May 2004 and 3.3% in October 2004 (Fig. 5.5).

These results perfectly correspond with those of Krinke *et al.* (2005) and with results of long-term survival experiments of buried seeds in the Czech Republic (Fig. 5.7). Of seeds (of the same seed batch) buried at ten different localities in the Czech Republic (five replications of 100 seeds), selected in order to represent a range of climates and soil types, on average 8.8% survived 1 year, 2.7% 2 years and 1.2% remained viable and dormant after 3 years of burial. These preliminary results provide evidence that *H. mantegazzianum* forms a seed bank, with some proportion of seeds surviving longer than 1 year. From the results of seed longevity mentioned above (Moravcová *et al.*, 2006), it can be supposed that the seeds of *H. mantegazzianum* will not persist in soil for a long time. Such species with seeds

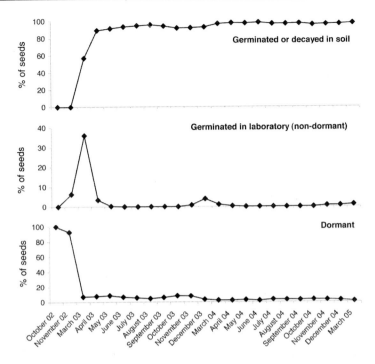

Fig. 5.5. Seasonal pattern of seed bank depletion in the course of two vegetation seasons. Percentage of dormant and non-dormant seeds among the total buried at the experimental garden of the Institute of Botany, Průhonice, Czech Republic. Each number is a mean of ten replicates. Seeds collected earlier in the same year were buried at a depth of 5–8 cm at the beginning of October 2002 and followed for two subsequent vegetation seasons, until March 2005. They were taken from the soil every month, except when the soil was frozen, and those that germinated or decayed were recorded; those that did not were tested for dormancy by germination in the laboratory at 10/5°C and for viability by tetrazolium. Based on data from Moravcová *et al.* (2006).

persisting in soil for at least 1 year, but less than 5 years, form in the sense of Thompson *et al.* (1997) a short-term persistent seed bank. The experiment continues in order to determine the maximum persistence of seeds of *H. mantegazzianum* in the soil, which can only be reliably determined by this kind of experiment.

Breaking of Dormancy: Towards a Threshold or Dormancy/Non-dormancy Cycle?

Combined together, the results of the experiments reported above make it possible to outline the pattern of seed germination and survival in the soil in *H. mantegazzianum*. Seeds in the field start to germinate early in spring (March in the study area in the western part of the Czech Republic) and most

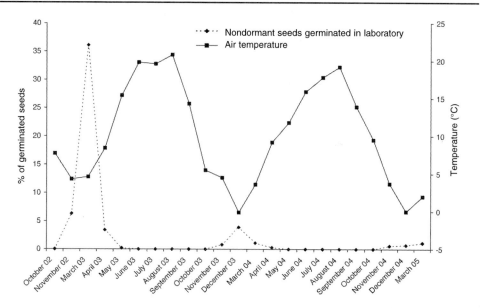

Fig. 5.6. Percentage of non-dormant seeds that germinated in the laboratory at 10/5°C, immediately after removal from the soil. Pattern recorded over two vegetation seasons is related to mean air temperature for months when seeds were tested. See Fig. 5.5 for details of data. Based on data from Moravcová *et al.* (2006).

of them do so up to the end of April or beginning of May (Figs 5.5 and 5.6). Despite considerable effort, no new seedlings were found in invaded sites later in the season and also, no buried seeds germinated in the soil after then (Moravcová *et al.*, 2006). However, the buried seeds removed from the soil started to germinate in the laboratory in November (Fig. 5.6). This corresponds to the fact that seeds need about 2 months of cold stratification below 8°C to break dormancy (Krinke *et al.*, 2005; Moravcová *et al.*, 2005); in November, this requirement is met for the seeds buried in the soil.

However, in the field seeds do not germinate in the autumn because of unsuitable conditions; in the laboratory they do. Obviously, buried seeds start to germinate in the laboratory approximately 2 months after the outside air temperature has dropped below 10°C (Fig. 5.6). But no buried seeds germinated before, i.e. in the course of the preceding late spring, summer and autumn when day temperature was above 10°C. This confirms the experimentally detected fact that although stratified seeds of *H. mantegazzianum* are able to germinate at a broad range of temperatures (Fig. 5.3), after being exposed to above 6°C for a longer time (up to 2 months) they cease germination and another cold period is needed to restart the germination process (Fig. 5.6).

It appears that seeds that have not germinated in spring (ca. 9% of seeds buried in the previous autumn) re-enter or retain dormancy during the high summer temperatures and break dormancy again during the following cold autumn and winter period (Fig. 5.5 and 5.6). It is hard to say whether the seeds in soil go through the annual dormancy/non-dormancy cycles which have been known for many annuals (Baskin and Baskin, 1998) and also for some perennials. The conditional dormancy/non-dormancy cycles are reported, for example, for a perennial *Rumex obtusifolius* L. (van Assche and Vanlerberghe, 1989), or biennials *Verbascum blattaria* L., *V. thapsus* L. (Vanlerberghe and van Assche, 1986; Baskin and Baskin, 1981) and *Oenothera biennis* L. (Baskin and Baskin, 1991) and dormancy/non-dormancy cycles are known also for perennials such as *Lychnis flos-cuculi* L. (Milberg, 1994a), *Primula veris* L. and *Trollius europaeus* L. (Milberg, 1994b), *Rhexia mariana* L. var. *interior* (Pennell) Kral & Bostick (Baskin *et al.*, 1999) or in sedges (Schütz, 1998). In *H. mantegazzianum* this question seems to be more complex. In burial experiments, the majority of seeds found dormant after 1 year in the soil had a morphologically fully developed embryo (Fig. 5.8C), which means that such seeds are only physiologically dormant; morphological dormancy must have been broken at that time. However, some seeds with an underdeveloped (Fig. 5.8A) or partially developed embryo (Fig. 5.8B) that had kept morphological (or morphophysiological) dormancy were also found (Moravcová *et al.*, 2006). Therefore, it seems more probable that dormancy in seeds of *H. mantegazzianum* is not broken completely in the first spring, but the breaking happens gradually and this process can take place only in months with sufficiently cold temperatures. This explains why the seeds staying dormant in soil have to wait until the next spring to germinate. That a small amount of seeds (about 1%) are able to survive in dormant state in the soil for at least 3 years (Fig. 5.7) suggests that the dormancy-breaking processes can take quite a long time in a small fraction of seeds and that the threshold is gradually achieved through accumulation of active temperatures during cold months.

Mechanisms of Dispersal

Fruits of *H. mantegazzianum* are elliptical, winged and dispersed mostly by wind, water and human activities. The majority of ripe fruits fall close to mother plants. For plants 2-m high, 60–90% of fruits fall within a radius of 4 m from the mother plant (Nielsen *et al.*, 2005). Clegg and Grace (1974) and Ochsmann (1996) argue that dispersal by wind could be important only over short distances. There is no direct evidence of dispersal by animals, but it can be supposed that adherence to animal skin could only play a role in short distance dispersal. Since at the landscape scale, long-distance dispersal and random events can play a crucial role in the dynamics of plant species, buoyancy can potentially affect the distance the species can reach. Clegg and Grace (1974) and Dawe and White (1979) report an ability to float up to 3 days for *H. mantegazzianum*, but L. Moravcová (unpublished results) found that 6-month-old fruits sink within 8 h. Such time, nevertheless, is likely to be sufficient for spreading a long distance, especially by fast-flowing streams. Other important dispersal vectors are humans, who spread fruit of *H. mantegazzianum* stuck to car tyres along roads, move them to new

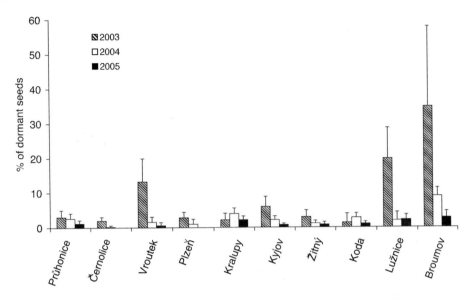

Fig. 5.7. The proportion of viable dormant seeds in the soil is rather low after the first year and rapidly decreases further. Survival over 3 years of seeds buried at ten localities in the Czech Republic in November 2002 is shown. Seeds were taken from the soil in October of the following 3 years and tested for viability by tetrazolium. Numbers are means of five replicates. Based on data from Moravcová *et al.* (2006).

locations with soil transport, or deliberately transport decorative umbels with dry fruit (Tiley *et al.*, 1996). Given that long-distance dispersal is important to the success of possible invasion (Pyšek and Hulme, 2005), dispersal by water and humans seem to be the most significant factors in this respect.

If suitable sites are available, high rate of spreading is realized at both local and regional scales. At the scale of the Czech Republic the number of localities doubled each 14 years during the exponential phase of invasion (see Pyšek *et al.*, Chapter 3, this volume). Müllerová *et al.* (2005) report an average rate of spread of about 10 m/year, and increase in the area invaded by more than 1200 m^2 each year in the Slavkovský les region, Czech Republic.

To illustrate the spreading from local populations to wider surroundings, aerial photographs can be explored (Müllerová *et al.*, 2005). Diaspore output of *H. mantegazzianum* populations can be calculated and evaluated by using additional data from experiments running in the sites analysed (Krinke *et al.*, 2005; Perglová *et al.*, Chapter 4, this volume; Pergl *et al.*, Chapter 6, this volume). Density of flowering plants as recorded from aerial photographs varied around 1.76 plants per m^2 at an average site (Müllerová *et al.*, 2005) and this value corresponds reasonably well to that recorded in permanent plots in the field (J. Pergl *et al.*, unpublished data). For the site harbouring the largest population of *H. mantegazzianum* (see Perglová *et al.*, Chapter 4, this volume, for the size of populations in individual sites), 14,164 flowering plants

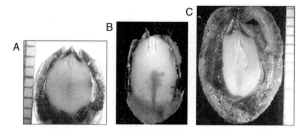

Fig. 5.8. Embryo in (A) a freshly harvested, (B) 2-months stratified and (C) 5-months stratified seed of *H. mantegazzianum*. Seeds were stratified in the soil. Photo: L. Moravcová.

were estimated to be present from aerial photographs (Fig. 3.4). Given the mean fecundity of 20,500 fruits per plant in the study area (see Perglová *et al.*, Chapter 4, this volume), the total fruit number per 60-ha area (size of the research plot used by Müllerová *et al.*, 2005) is over 290,000,000, representing an annual input of 484 fruits/m^2/year. Relating the total fruit set to the mean area actually infested at a site (31,946 m^2) gives 9089 fruits/m^2/year. These values can be compared with the number of seeds in seed banks, estimated in permanent plots: the mean value from autumn 2002, after the fruits were shed, was 3650 seeds/m^2 (Krinke *et al.*, 2005). Bearing in mind that the values derived from aerial photographs are rough estimates, they provide some idea of how large a proportion of fruits are spread outside the actual stands. The value of 484 seeds/m^2/year is a theoretical one since seeds are not dispersed evenly across the whole site. On the other hand, the amount produced by monitored populations (9089) greatly exceeds the value recorded in the field (3650); this difference indicates that a large proportion of fruits are spread into surroundings, making further population growth possible.

Conclusions

A combination of reproductive traits such as high fecundity, high germination capacity, opportunistic behaviour associated with limited effect of fruit position on a plant on germination characteristics, dormancy mechanisms together with short-term persistent soil seed bank and possibility of long-distance spread are likely to determine the ability of *H. mantegazzianum* to invade successfully new habitats in the secondary distribution range.

Acknowledgements

Thanks to Per Milberg and Norbert Hölzl for helpful comments on the manuscript, and to Zuzana Sixtová, Ivan Ostrý, Šárka Jahodová and Vendula Čejková for technical assistance. The study was supported by the project 'Giant Hogweed (*Heracleum mantegazzianum*) a pernicious invasive weed: developing a sustainable strategy for alien invasive plant management in

Europe', funded within the 'Energy, Environment and Sustainable Development Programme' (grant no. EVK2-CT-2001-00128) of the European Union 5th Framework Programme, by institutional long-term research plans no. AV0Z60050516 from the Academy of Sciences of the Czech Republic and no. 0021620828 from Ministry of Education of the Czech Republic, and grant no. 206/05/0323 from the Grant Agency of the Czech Republic.

References

Andersen, U.V. and Calov, B. (1996) Long-term effects of sheep grazing on giant hogweed (*Heracleum mantegazzianum*). *Hydrobiologia* 340, 277–284.

Baskin, C.C. and Baskin, J.M. (1998) *Seeds. Ecology, Biogeography and Evolution of Dormancy and Germination*. Academic Press, San Diego, California.

Baskin, C.C., Meyer, S.E. and Baskin, J.M. (1995) Two types of morphophysiological dormancy in seeds of two genera (*Osmorhiza* and *Erythronium*) with an arcto-tertiary distribution pattern. *American Journal of Botany* 82, 293–298.

Baskin, C.C., Baskin, J.M. and Chester, E.W. (1999) Seed dormancy and germination in *Rhexia mariana* var. *interior* (Melastomataceae) and eco-evolutionary implications. *Canadian Journal of Botany* 77, 488–493.

Baskin, C.C., Milberg, P., Andersson, L. and Baskin, J.M. (2000) Deep complex morpho-physiological dormancy in seeds of *Anthriscus sylvestris* (Apiaceae). *Flora* 195, 245–251.

Baskin, J.M. and Baskin, C.C. (1981) Seasonal changes in germination responses of buried seeds of *Verbascum thapsus* and *V. blattaria* and ecological implications. *Canadian Journal of Botany* 59, 1769–1775.

Baskin, J.M. and Baskin, C.C. (1990) Germination ecophysiology of the winter annual: *Chaerophyllum tainturieri*: a new type of morphophysiological dormancy. *Journal of Ecology* 78, 993–1004.

Baskin, J.M. and Baskin, C.C. (1991) Germination requirements of *Oenothera biennis* seeds during burial under natural seasonal temperature cycles. *Canadian Journal of Botany* 72, 779–782.

Baskin, J.M. and Baskin, C.C. (2004) A classification system for seed dormancy. *Seed Science Research* 14, 1–16.

Clegg, L.M. and Grace, J. (1974) The distribution of *Heracleum mantegazzianum* (Somm. & Levier) near Edinburgh. *Transactions of the Botanical Society of Edinburgh* 42, 223–229.

Collingham, Y.C., Wadsworth, R.A., Huntley, B. and Hulme, P.E. (2000) Predicting the spatial distribution of non-indigenous riparian weeds: issues of spatial scale and extent. *Journal of Applied Ecology* 37, 13–27.

Dawe, N.K. and White, E.R. (1979) Giant Cow Parsnip (*Heracleum mantegazzianum*) on Vancouver Island, British Columbia. *Canadian Field Naturalist* 93, 82–83.

Dodd, F.S., de Waal, L.C., Wade, P.M. and Tiley, G.E.D. (1994) Control and management of *Heracleum mantegazzianum* (Giant Hogweed). In: de Waal, L.C., Child, L.E., Wade, P.M. and Brock, J.H. (eds) *Ecology and Management of Invasive Riverside Plants*. Wiley, Chichester, UK, pp. 111–126.

Grime, J.P., Mason, G., Curtis, A.V., Rodman, J., Band, S.R., Mowforth, M.A., Neal, A.M. and Shaw, S. (1981) A comparative study of germination characteristics in a local flora. *Journal of Ecology* 69, 1017–1059.

Gutterman, Y. (2000) Maternal effects on seeds during development. In: Fenner, M. (ed.) *Seeds*.

The Ecology of Regeneration in Plant Communities, 2nd edn. CABI, Wallingford, UK, pp. 59–84.

Harper, J.L. (1977) *Population Biology of Plants.* Academic Press, London.

Hendrix, S.D. (1984a) Variation in seed weight and its effect on germination in *Pastinaca sativa* L. (*Umbelliferae*). *American Journal of Botany* 71, 795–802.

Hendrix, S.D. (1984b) Reactions of *Heracleum lanatum* to floral herbivory by *Depressaria pastinacella. Ecology* 65, 191–197.

Hendrix, S.D. and Sun, I.-F. (1989) Inter- and intraspecific variation in seed mass in seven species of umbellifer. *New Phytologist* 112, 445–451.

Hendrix, S.D. and Trapp, E.J. (1992) Population demography of *Pastinaca sativa* (*Apiaceae*): effects of seed mass on emergence, survival, and recruitment. *American Journal of Botany* 79, 365–375.

Holub, J. (1997) *Heracleum* – bolševník. In: Slavík, B., Chrtek jun., J. and Tomšovic, P. (eds) *Květena České republiky 5.* Academia, Praha, pp. 386–395.

Krinke, L., Moravcová, L., Pyšek, P., Jarošík, V., Pergl, J. and Perglová, I. (2005) Seed bank of an invasive alien, *Heracleum mantegazzianum,* and its seasonal dynamics. *Seed Science Research* 15, 239–248.

Lhotská, M. (1978) Contribution to the ecology of germination of the synanthropic species of the family *Daucaceae.* II. Genus *Anthriscus. Acta Botanica Slovaca Academiae Scientiarum Slovacae, Series A* 3, 157–165.

Lundström, H. (1989) New experience of the fight against the giant hogweed, *Heracleum mantegazzianum.* In: *Weeds and Weed Control, 30th Swedish Crop Protection Conference 2.* Swedish University of Agriculture Sciences, Uppsala, Sweden, pp. 51–58.

Lundström, H. and Darby, E. (1994) The *Heracleum mantegazzianum* (giant hogweed) problem in Sweden: suggestions for its management and control. In: de Waal, L.C., Child, L.E., Wade, P.M. and Brock, J.H. (eds) *Ecology and Management of Invasive Riverside Plants.* Wiley, Chichester, UK, pp. 93–100.

Manchester, S.J. and Bullock, J.M. (2000) The impacts of non-native species on UK biodiversity and the effectiveness of control. *Journal of Applied Ecology* 37, 845–864.

Martin, A.C. (1946) The comparative internal morphology of seeds. *American Midland Naturalist* 36, 513–660.

Milberg, P. (1994a) Annual dark dormancy cycle in buried seeds of *Lychnis flos-cuculi. Annales Botanici Fennici* 31, 163–167.

Milberg, P. (1994b) Germination ecology of the polycarpic grassland perennials *Primula veris* and *Trollius europaeus. Ecography* 17, 3–8.

Moravcová, L., Perglová, I., Pyšek, P., Jarošík, V. and Pergl, J. (2005) Effects of fruit position on fruit mass and seed germination in the alien species *Heracleum mantegazzianum* (*Apiaceae*) and the implications for its invasion. *Acta Oecologica* 28, 1–10.

Moravcová, L., Pyšek, P., Pergl, J., Perglová, I. and Jarošík, V. (2006) Seasonal pattern of germination and seed longevity in the invasive species *Heracleum mantegazzianum. Preslia* 78, 287–301.

Morton, J.K. (1978) Distribution of giant cow parsnip (*Heracleum mantegazzianum*) in Canada. *Canadian Field Naturalist* 92, 182–185.

Müllerová, J., Pyšek, P., Jarošík, V. and Pergl, J. (2005) Aerial photographs as a tool for assessing the regional dynamics of the invasive plant species *Heracleum mantegazzianum. Journal of Applied Ecology* 42, 1042–1053.

Nielsen, Ch., Ravn, H.P., Netwig, W. and Wade, M. (eds) (2005) *The Giant Hogweed Best Practice Manual. Guidelines for the Management and Control of Invasive Weeds in Europe.* Forest and Landscape, Hørsholm, Denmark.

Nikolaeva, M.G., Rasumova, M.V. and Gladkova, V.N. (1985) *Reference Book on Dormant Seed Germination.* Nauka, Leningrad. [In Russian.]

Ochsmann, J. (1996) *Heracleum mantegazzianum* Sommier et Levier (*Apiaceae*) in Deutschland: Untersuchungen zur Biologie, Verbreitung, Morphologie und Taxonomie. *Feddes Repertorium* 107, 557–595.

Ojala, A. (1985) Seed dormancy and germination in *Angelica archangelica* subsp. *archangelica* (*Apiaceae*). *Annales Botanici Fennici* 22, 53–62.

Otte, A. and Franke, R. (1998) The ecology of the Caucasian herbaceous perennial *Heracleum mantegazzianum* Somm. et Lev. (Giant Hogweed) in cultural ecosystems of Central Europe. *Phytocoenologia* 28, 205–232.

Pyšek, P. and Hulme, P.E. (2005) Spatio-temporal dynamics of plant invasions: linking pattern to process. *Ecoscience* 12, 302–315.

Schütz, W. (1998) Dormancy cycles and germination phenology in sedges of various habitats. *Wetlands* 18, 288–297.

Stokes, P. (1952a) A physiological study of embryo development in *Heracleum sphondylium* L. I. The effect of temperature on embryo development. *Annals of Botany* 16, 441–447.

Stokes, P. (1952b) A physiological study of embryo development in *Heracleum sphondylium* L. II. The effect of temperature on after-ripening. *Annals of Botany* 16, 571–576.

Thomas, T.H. (1996) Relationships between position on the parent plant and germination characteristics of seeds of parsley (*Petroselium crispum* Nym). *Plant Growth Regulation* 18, 175–181.

Thomas, T.H., Gray, D. and Biddington, N.L. (1978) The influence of the position of the seed on the mother plant on seed and seedling performance. *Acta Horticulturae* 83, 57–66.

Thomas, T.H., Biddington, N.L. and O'Toole, D.F. (1979) Relationship between position on the parent plant and dormancy characteristics of seeds of three cultivars of celery (*Apium graveolens*). *Physiologia Plantarum* 45, 492–496.

Thompson, J.T. (1984) Variation among individual seed masses in *Lomatium grayi* (*Umbelliferae*) under controlled conditions: magnitude and partitioning of the variance. *Ecology* 65, 626–631.

Thompson, K., Bakker, J.P. and Bekker, R.M. (1997) *The Soil Seed Bank of North West Europe: Methodology, Density and Longevity.* Cambridge University Press, Cambridge.

Tiley, G.E.D., Dodd, F.S. and Wade, P.M. (1996) Biological flora of the British Isles. 190. *Heracleum mantegazzianum* Sommier et Levier. *Journal of Ecology* 84, 297–319.

Van Assche, J.A. and Vanlerberghe, K.A. (1989) The role of temperature on the dormancy cycle of seeds of *Rumex obtusifolius* L. *Functional Ecology* 3, 107–115.

Vanlerberghe, K.A. and Van Assche, J.A. (1986) Dormancy phases in seeds of *Verbascum thapsus* L. *Oecologia* 68, 479–480.

6 Population Dynamics of *Heracleum mantegazzianum*

Jan Pergl,[1] Jörg Hüls,[2] Irena Perglová,[1] R. Lutz Eckstein,[2] Petr Pyšek[1,3] and Annette Otte[2]

[1]Institute of Botany of the Academy of Sciences of the Czech Republic, Průhonice, Czech Republic; [2]Justus-Liebig-University Giessen, Giessen, Germany; [3]Charles University, Praha, Czech Republic

Around every river and canal their power is growing

(Genesis, 1971)

Introduction

The aim of this chapter is to summarize existing knowledge on the dynamics of populations of *Heracleum mantegazzianum* Sommier & Levier in regions where it is an invasive species, with particular emphasis on survival, growth and reproduction. The data come from populations studied in the Czech Republic and Germany. Unfortunately, corresponding information on the population biology of *H. mantegazzianum* in its native distribution range in the Western Greater Caucasus is still incomplete, because population studies require long-term observations in a wide range of habitats and environmental conditions. Nevertheless, there is some information on the population biology of *H. mantegazzianum* in its native range, which can be compared with corresponding data from the invaded range, such as population age structure, which is a crucial characteristic of monocarpic plants (Pergl *et al.*, 2006).

Because population dynamics involve a variety of processes, this chapter is presented in sections dealing with seedling dynamics, relative growth rate of seedlings, population stage structure and mortality, analysis of life tables and effects of stand structure on population dynamics. Implications of these results for the long-term population dynamics are outlined. Together with other biological characteristics, which affect the population dynamics of *H. mantegazzianum* and are dealt with elsewhere (seed bank – Moravcová *et al.*, Chapter 5, this volume; Krinke *et al.*, 2005; reproductive biology – Moravcová *et al.*, 2005 and Chapter 5, this volume; timing of flowering – Perglová *et al.*, 2006 and Chapter 4, this volume), this provides the most complete picture of the population dynamics of this species. Knowledge of the population biology of *H. mantegazzianum* will: (i) provide insights into spread dynamics at various scales based on field data; (ii) help to determine what makes this species such a successful invader; and (iii) enable us to analyse the complete life cycle of this invasive herb with the aim of finding possible weak events in its life cycle that can be targeted for weed control. In addition, data on population dynamics can be used in models with the aim of making both generalized and site-specific predictions (see Nehrbass and Winkler, Chapter 18, this volume; Nehrbass *et al.*, 2006).

Population studies in the Czech Republic were conducted in the western part of the country, in the Slavkovský les Protected Landscape Area (see Moravcová *et al.*, 2005; Müllerová *et al.*, 2005 for characteristics of the region, and Perglová *et al.*, Chapter 4, this volume, for overview of the localities). *Heracleum mantegazzianum* was introduced to this area in 1862 as an ornamental species. The invasive behaviour of the species accelerated when the majority of the inhabitants of this area were moved out after World War II. This led to radical changes in land use, with considerable increase in the proportion of unmanaged habitats. Historical dynamics of the invasion of this region by *H. mantegazzianum* were reconstructed based on a series of aerial photographs covering a period of 50 years from World War II (start of the invasion at the studied sites) up to the present (Müllerová *et al.*, 2005). The history of the *H. mantegazzianum* invasion of the Czech Republic is described in detail elsewhere (Pyšek, 1991; Pyšek and Prach, 1993; Pyšek and Pyšek, 1995; Pyšek *et al.*, 1998; see Pyšek *et al.*, Chapter 3, this volume).

In Germany, population studies were conducted on paired open and dense stands of *H. mantegazzianum* in five populations (Hüls, 2005). These were situated at an altitude of 155–335 m in the low mountainous region of Hesse, central western Germany, which is characterized by a temperate sub-oceanic climate with annual average temperatures of 7.6–9.0°C and annual precipitation of 609–767 mm (for details see Hüls, 2005).

Seedling Dynamics

The seedling stage is the most vulnerable in the life cycle of a plant. Young seedlings suffer a high mortality due to attack by herbivores and pathogens, as well as from intra- and interspecific competition or unsuitable environmental conditions such as frost or drought (Harper, 1977; Crawley, 1997). Because reproduction in *H. mantegazzianum* is exclusively by seed (see Perglová *et al.*, Chapter 4 and Pyšek *et al.*, Chapter 7, this volume), seedlings represent the only means of colonization of new sites and subsequent population recruitment. Data on seedling dynamics are available for a wide range of sites, including dominant stands of *H. mantegazzianum*, with only a few ruderal species and a high proportion of bare ground in spring at the time of seedling emergence, to those covered with grass and with a sparse occurrence of *H. mantegazzianum*. These sites are the same as those at which the data on reproduction ecology and seed bank were collected (see Perglová *et al.*, Chapter 4 and Moravcová *et al.*, Chapter 5).

High percentages of the seeds of *H. mantegazzianum* germinate; Moravcová *et al.* (2005) give an average of 91% for seed collected at a range of study sites in the Slavkovský les region and germinated in laboratory conditions. Prior to germination, the seeds need cold and wet stratification to break the morphophysiological dormancy. Under experimental conditions (Moravcová *et al.*, 2005 and Chapter 5, this volume) this process takes about 2 months, but once dormancy is broken seeds germinate in conditions suitable for cold stratification. Under natural conditions at the study region the cold period lasts longer than the minimum of 2 months needed for stratification. Thus all non-dormant seeds are ready to germinate in early spring (see Moravcová *et al.*, Chapter 5, this volume; Krinke *et al.*, 2005). A study of the dynamics of seedling emergence conducted in 2002–2004 (J. Pergl *et al.*, unpublished) shows that massive germination occurred a few days after the snow melted. In 2002 the peak density of seedlings occurred during the first census on 5 April at six out of 11 localities. In 2003, some seedlings already had true leaves on 20 April, despite the presence of scarce snow cover; at the same time, the rosettes of previous year(s)' plants reached about 10 cm in diameter. The peak number of seedlings occurred about 10 days after the last snowfall (30 April 2003). A week later, the ground was overgrown by leaf rosettes of the older *H. mantegazzianum* individuals, which shaded out the seedlings of this and other plant species. No new seedlings with cotyledon leaves were observed after May. This accords well with the results of seed bank studies and germination experiments (see Moravcová *et al.*, Chapter 5, this volume; Krinke *et al.*, 2005), which show that the seed bank is depleted in early spring and germination under higher temperatures later in the season does not occur. The minimum morning ground temperatures recorded at the nearest meteorological station (Mariánské Lázně; Czech Hydrometeorological Institute) in 2002 reveal that in this year, the massive germination occurred 5 days before the temperatures increased and remained above freezing. A similar pattern is reported for the milder oceanic climate of the British Isles, where seedlings start to emerge from January to March (Tiley *et al.*, 1996).

Autumn emergence of seedlings was not observed in the study sites in the western part of the Czech Republic, although it is reported from Poland (Cwiklinski, 1973, cited in Tiley *et al.*, 1996), Ireland (Caffrey, 1999) and Scotland (Tiley and Philp, 1994). Nevertheless, the small proportion of non-dormant seed in the seed bank in autumn at some localities (Krinke *et al.*, 2005) indicates that germination in autumn is possible under suitable climatic conditions.

To estimate the dynamics of seedling emergence and their survival at each site, the number of seedlings was counted at one permanent plot in three consecutive seasons in early spring (Fig. 6.1). Five censuses were made in 2002, and three in 2003 and 2004. The maximum density of 3700 seedlings/m^2 was found in one plot in 2004. The mean values for the sites were (mean ± SD per m^2) 671.8 ± 439.2 (2002), 734.3 ± 441.7 (2003) and 1613.9 ± 1322.1 (2004). The seedling densities were not related to characteristics of the maternal population, such as total or adult density or to those characteristics of the seed bank, such as number of total, live or non-dormant seeds

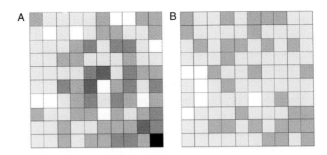

Fig. 6.1. Spatial distribution of seedlings in one of the permanent plots (1 × 1 m) in (A) early (5 April 2002 – first census) and (B) late (1 May 2002 – last census) spring. Each permanent plot was divided into 100 subplots in which the number of seedlings was counted. The censuses were carried out at weekly intervals until the beginning of May, when the leaf canopy of the rosettes closed and prevented further observations. Number of seedlings: □ 0, ▨ 1–3, ▨ 4–6, ▨ 7–8, ▨ 9–11, ■ 12–15.

(J. Pergl *et al.*, unpublished). The density of seedlings, defined as new plants with a lamina width of < 3.5 cm, in ten plots in a population close to Giessen, Germany, in 2003 was 504.0 ± 363.3 per m² (Hüls, 2005). Other available data for seedling densities come from Scotland, where Tiley *et al.* (1996) found 400 seedlings/m².

However, seedling survival is low. In the Czech Republic and Germany, less than 1% of the seedlings survive until the next spring (J. Pergl *et al.*, unpublished; Hüls, 2005). In Scotland there is a rapid decrease in seedling density from 400 seedlings/m² in spring to 33 surviving plants in autumn (Tiley *et al.*, 1996). Caffrey (1999) records survival of seedlings within the range of 1.2–13.7%. In another study, the survival rate of seedlings and immature plants from March to June was approximately 2.5% (Caffrey, 2001: estimated from his Fig. 2).

Seedling emergence and survival indicate that established populations are not seed limited. Seedling survival is related to microclimatic conditions, intraspecific competition and stochastic events. Seedling establishment is promoted by favourable conditions of open ground, where competition from co-occurring species, particularly grasses, is low. As stands of *H. mantegazzianum* are often visited by game animals (deer, hogs), seeking shelter and food, the soil surface is disturbed and the herb layer removed, which increases the probability of establishment and survival of seedlings and decreases competition from other species. In undisturbed grasslands and other vegetation types with a dense cover, the probability of seedling establishment and survival to the end of the first growing season is very low; in the dense cover of grasses, the average number of seedlings found in midsummer in a 1 m² plot next to a plant that flowered the previous year was 56 (J. Pergl and I. Perglová, unpublished). As *H. mantegazzianum* is a monocarpic perennial plant and dies after seed release, seedlings do not have to compete with the mother plant. Thus, small seedlings can take advantage of a safe place in close proximity to a dead, decaying flower stem, which creates a patch of disturbed ground even in otherwise compact vegetation. Suitable conditions for germination and survival are also found at wet sites, such as riverbanks, where there are sufficient nutrients, moisture and suitable patches of disturbed ground (Ochsmann, 1996; Tiley *et al.*, 1996).

The germination of *H. mantegazzianum* seeds very early in spring provides this species with an advantage. Seedlings are adapted to climatic conditions in the native distribution range, and hence are not sensitive to frost. Early germination allows the seedlings to cover patches of open ground and reach a sufficiently advanced stage before they are overgrown by adult plants of *H. mantegazzianum* or other species. The mortality of seedlings is high, but comparable with those generally observed in other plants (Harper, 1977).

Seedling RGR

To evaluate the growth potential of *H. mantegazzianum*, seedling relative growth rate (RGR) was measured using standard procedures (Grime and Hunt, 1975). Thus, it is possible to compare the RGR of *H. mantegazzianum* with that of native species and explore whether early germination is associated with fast growth and rapid accumulation of biomass.

In *H. mantegazzianum*, fruits are produced in umbels of various orders and positions within the plant (see Fig. 4.2). The final contribution of an individual umbel to the overall fitness of a plant is a function of its fruit set, the germination capacity of the seeds and their ability to survive to flowering. While germination is not, fruit mass is significantly affected by umbel position (Moravcová *et al.*, 2005). To determine the role of umbel position on population growth and offspring fitness, the RGR of seedlings from different umbels was compared (Perglová *et al.*, unpublished).

Ripe fruits were collected in 2004 at the site Žitný I (see Perglová *et al.*, Chapter 4, this volume; Müllerová *et al.*, 2005) from randomly selected individuals. They were collected separately from the primary umbels (hereafter referred to as 'terminals') and secondary umbels on satellites ('satellites'), stem branches ('stem branches') and branches growing from the base of the flowering stem at ground level ('basal branches'). One umbel per plant was sampled. After 1 month of storage, seeds were stratified in Petri dishes on wet sand (for details of stratification see Moravcová *et al.*, 2005). After emergence of the radicule, seedlings were placed in pots filled with sand, moved to a climate chamber (Fitotron, Sanyo) with a standard regime (day/night: 12 h/12 h, 22°C/15°C), watered with Rorison nutrient solution (Hendry and Grime, 1993) and harvested after 7 or 21 days. RGR was calculated according to Hoffmann and Poorter (2002) and Hunt *et al.* (2002).

Seedlings from basal branches showed considerably high variation in RGR (Table 6.1), which accords with the observation that terminal umbels on these branches are generally highly variable in terms of size, fruit

Table 6.1. Relative growth rate (RGR) of seedlings of *H. mantegazzianum* grown from seeds produced by different types of umbels. Different letters indicate significant differences ($P = 0.05$) tested using multiple t-tests with Bonferroni's correction.

Umbel position	Seedling RGR (g/g/day)		
	Mean	SD	
Terminal	0.186	0.0497	a
Satellite	0.184	0.0598	ab
Stem branch	0.155	0.0667	b
Basal branch	0.156	0.0803	ab

mass, fecundity and proportion of male flowers. Some terminal umbels on basal branches resemble primary umbels (terminals) in appearance, while others resemble secondary umbels (see Perglová *et al.*, Chapter 4, this volume). RGR differed significantly only between the seedlings grown from seed from terminals and stem branches, with the growth of the former being quicker (Table 6.1). RGR of seedlings from satellites and basal branches did not differ significantly from others. However, seedlings from satellites differed marginally significantly ($P = 0.1$) from those from stem branches. Unexpectedly, there was no difference in the RGR of seedlings from terminals and satellites, i.e. from the umbels of a different order that differed significantly in seed mass.

In order to consider the RGR values in the context of plant communities invaded by *H. mantegazzianum*, its growth rate was compared with that of co-occurring species or of those found in the close vicinity of invaded sites.

The following RGR values (using data from Grime and Hunt, 1975) are reported for common species in communities invaded by *H. mantegazzianum* (M. Hejda, unpublished data from the Slavkovský les region): *Urtica dioica* L. 0.314 g/g/day, *Dactylis glomerata* L. 0.187, *Galium aparine* L. 0.167, *Geum urbanum* L. 0.104, *Poa trivialis* L. 0.200, *Anthriscus sylvestris* (L.) Hoffm. 0.074, *Heracleum sphondylium* L. 0.083 and *Alopecurus pratensis* L. 0.184. Seedlings from terminal umbels of *H. mantegazzianum* have an RGR of 0.186 g/g/day, which is less than that of *Urtica dioica*, but higher than or similar to that of other common species. In contrast, the RGR of the native Apiaceae, *H. sphondylium* and *Anthriscus sylvestris*, is only half of that of *H. mantegazzianum*.

The RGR of *H. mantegazzianum* seedlings, grown from seeds produced by terminal umbels, is the 34th highest in the set of 117 herbaceous species studied in the UK (Fig. 6.2). This relatively high RGR allows seedlings to establish themselves in dense *Heracleum* stands within the short period between germination and when they are overgrown by leaf rosettes. Also this RGR is similar to that of grasses, which seem to be the most important competitors in neighbouring areas that are not invaded.

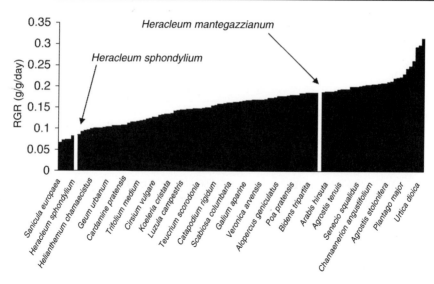

Fig. 6.2. Position of invasive *H. mantegazzianum* and its native congener *H. sphondylium* in a continuum of 117 herbaceous species ranked according to increasing seedling RGR. Data for the other species are from Grime and Hunt (1975). Only one species in five is labelled by name. For *H. mantegazzianum*, the value reported is that of seedlings developing from seeds from terminal umbels (0.186 g/g/day).

Population Structure, Mortality and Flowering

What can the structure of populations tell us about the species' life history? Can we determine the reasons why *H. mantegazzianum* is such a successful invasive species? Can an understanding of the main drivers of population dynamics help us develop an efficient control strategy? To answer these questions, the fate of individual plants was followed in permanent plots in the Czech Republic and Germany. Although *H. mantegazzianum* has been the object of many studies in the past (for review see Tiley *et al.*, 1996), there is little information on its population biology and life history. Moreover, this information was based on a limited number of plants (Otte and Franke, 1998), on observations under artificial conditions (Stewart and Grace, 1984) or was anecdotal (Tiley *et al.*, 1996). Another source of information is short notes on occasional observations by botanists and land managers (Morton, 1978; Brondegaard, 1990) or studies on control (for review see Pyšek *et al.*, Chapter 7, this volume). Finally, the studies on *H. mantegazzianum* and closely related species as fodder plants do not record the fate of individual plants or information on population biology, but focus on standing crop, biomass productivity and suitability for livestock (Satsyperova, 1984).

Population density of *H. mantegazzianum* is highly variable, ranging from isolated plants in sparse and small populations to dense and large populations covering several hectares. *H. mantegazzianum* occurs in many habitats, particularly those affected by former or present human activity: along transportation corridors (rivers, roads and railways), and abandoned meadows or forest edges (Pyšek and Pyšek, 1995; Thiele, 2006; Thiele *et al.*, Chapter 8, this volume). Thus, populations were studied in a range of environmental conditions in both the Czech Republic and Germany. In Germany the populations were classified as open and dense stands on the basis of the density of mature plants and extent of *H. mantegazzianum* cover (see later for details). In 2002, permanent plots were established in both regions. In Germany, 2–8 replicates of 1×2.5 m permanent plots were established at five sites and monitored every year for 3 years. This produced demographic data for two annual transitions. In the Czech Republic, eight 1×10 m plots were established and sampled at the beginning of summer and at the end of the vegetation season. These plots were monitored for 4 years yielding data for three transition periods.

When referring to 'population density', it should be noted that, for practical reasons, only plants above a certain threshold size (plants with leaves at least 8 cm long) were considered. This omits the majority of current year seedlings, which are unlikely to survive until the following spring. The stands in the Czech Republic had a mean density across sites and years of 5.4 plants/m² (min. 0.4; max. 20.2). In Germany, the dense and open stands harboured on average 7.7 (min. 1.3; max. 31.2) and 2.0 (0.3; 7.0) plants/m², respectively. Changes in density over time for particular localities in the Czech Republic are shown in Fig. 6.3. The decreasing trend in population densities, particularly in the overcrowded populations, agrees with projections of matrix models (see below). However, because the duration of the study was only 4 years, these results must be interpreted with care (Nehrbass *et al.*, 2006).

The proportion of plants that flowered varied considerably between years in the Czech populations and is difficult to interpret. The mean density, pooled across sites and years, is 0.7 flowering plants/m². In the dense stands in

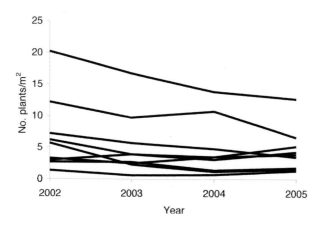

Fig. 6.3. Changes in population density of *H. mantegazzianum* over 4 years at eight sites studied in the Czech Republic. Each line represents one site. Number of plants was counted on permanent plots, 1×10 m in size. Only plants with leaves longer than 8 cm were counted.

Germany it is 0.8 (min. 0.0; max. 2.1) and in the open stands only 0.3, averaged across sites and years (min. 0.05; max. 0.8). Published reports correspond to our observations: one flowering plant per 0.5–1.0 m^2 (Tiley *et al.*, 1996) or 4–7 flowering individuals/m^2 in established stands (Gibson *et al.*, 1995, cited by Tiley *et al.*, 1996). Although the density of flowering plants is highly variable, the total number of seeds in the seed bank is related to the number of flowering plants (Krinke *et al.*, 2005).

To determine the effect of the size of a plant on its survival, the number of leaves and the length of the longest leaf were used as proxies of plant size in logistic regressions. For populations in the Czech Republic, both factors significantly and positively affected the probability of surviving to the following year. The survival was not dependent on the distance of the tested plant from the nearest neighbour or to the size of its Thiessen's polygon (J. Pergl *et al.*, unpublished). This indicates that survival is similar in the range of habitats studied and does not depend on local conditions.

The survival of *H. mantegazzianum* individuals in summer and winter was compared for the same stage classes as used in matrix models (see below). Survival of newly emerged seedlings (with the lamina of the largest leaf longer than 8 cm) was on average 22% during summer and 50% during winter. It was higher for larger plants (with 2–4 leaves) and varied between 60% and 67% for newly recorded plants at a given census and between 81% and 85% for those that were recorded previously. Large vegetative plants (with the longest leaf larger than 140 cm or with more than four leaves) have a slightly higher probability of surviving over summer (93%) than over winter (90%).

Similar to survival, flowering in *H. mantegazzianum* appears to be size dependent, which in turn is closely linked to the age of a plant and the time required to accumulate the necessary resources (Pergl *et al.*, 2006; Perglová *et al.*, Chapter 4, this volume). The results of studies in Germany (Hüls, 2005) (Fig. 6.4) and the Czech Republic (J. Pergl *et al.*, unpublished) suggest that the trigger for flowering is the size of the plant the 'previous' year and that the majority (90–100%) of plants that reach the minimal size flower.

Timing of flowering is crucial for monocarpic plants, so Pergl *et al.* (2006) used annual rings in the roots (Fig. 6.5) to determine the age structure of *H. mantegazzianum* populations in its native (Caucasus) and invaded (Czech Republic) distribution ranges. This study revealed that flowering occurred later in the native distribution range (Caucasus) and managed habitats (pastures) than in unmanaged sites in the Czech Republic. The later time of flowering in the native distribution range seems to be due to the higher altitude there, hence shorter period of growth compared to the sites in the invaded range. Grazing significantly prolonged the time needed for accumulating the resources necessary for flowering. The age structures of the populations in the different habitats in the Caucasus and Czech Republic are shown in Fig. 6.6. The number of 1–year-old plants is underestimated, as only plants with leaves at least 8 cm long are included. Interestingly, although plants from unmanaged sites in the Czech Republic flowered on average significantly earlier than those from other habitats, the oldest flowering plant was found in an extremely dry

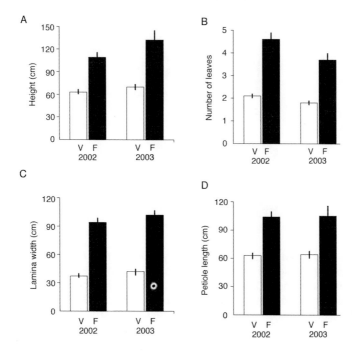

Fig. 6.4. Average height (A), number of rosette leaves (B), width of largest rosette leaf (C) and petiole length of largest rosette leaf (D) of plants in 2002 and 2003 that did not flower (V, white bars) and flowered (F, black bars) the following year. Error bars denote 95% confidence intervals. Differences between all the groups were statistically significant (logistic regressions: $\chi^2 > 34$, $df = 1$, $P < 0.001$). Data from Hüls (2005).

locality in the Czech Republic. These results suggest that the species is very tolerant and plastic in its response to environmental conditions and is able to postpone flowering for many years (up to 12 years) (for details see Pergl *et al.*, 2006; Perglová *et al.*, Chapter 4, this volume).

Matrix Model Approach: Life Tables from the Czech Republic and Germany

The populations and their dynamics based on data from Germany and the Czech Republic were compared by analysing the outputs from matrix models. The matrix models are based on transition probabilities between categories that were defined in terms of size categories rather than age. Despite slight differences in the sampling procedure used in these countries, plants were assigned to one of four size categories: seedlings (in the Czech Republic)/small vegetative plants (in Germany); juveniles/medium-sized vegetative plants; adult non-flowering plants/large vegetative plants; flowering plants (in both countries). The matrix model with data obtained at the end of each growth season

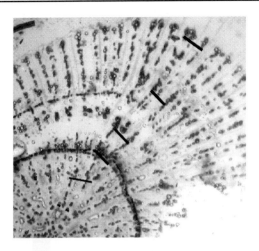

Fig. 6.5. Annual early wood rings in the secondary root xylem of *H. mantegazzianum* make it possible to determine the age of both vegetative and flowering plants. Age of plants was estimated by herb-chronology (Dietz and Ullmann, 1997, 1998; von Arx and Dietz, 2006). The figure shows annual rings in a cross-section of the root of a 7-year-old individual. Lines indicate the transitions between late wood of the previous year and early wood of the following year.

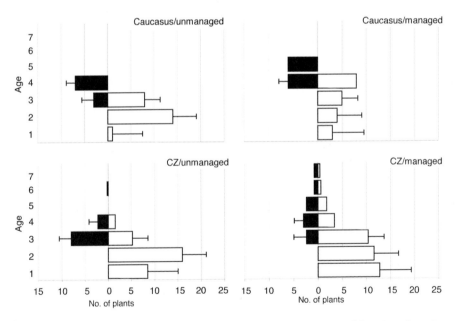

Fig. 6.6. Age structure of *H. mantegazzianum* populations at managed (pastures) and unmanaged sites in the native (Caucasus) and invaded (CZ – Czech Republic) distribution ranges. White bars represent mean number of non-flowering plants (lines = SD) per sample plot. Black bars represent flowering plants. For details of the sampling and method used to determine age see Pergl *et al.* (2006).

uses the year-to-year transitions for the years 2002–2005 in the Czech Republic and 2002–2004 in Germany.

As the data are for a limited number of transitions and sites, and the matrix models do not incorporate the spatial component of spread, the prediction of a decrease on population development in the future needs to be interpreted with caution (for a discussion of the differences between outputs from matrix and individual based models based on the same data, see Nehrbass *et al.*, 2006). The estimated population growth rates for individual sites, and the number of individuals used for each transition matrix within each plot, are summarized in Table 6.2. For seven out of the eight populations studied in the Czech Republic, it was possible to construct a pooled matrix across 2002–2005. For the majority of sites, however, the number of individuals within a plot was insufficient for constructing a robust transition matrix every year (Table 6.2). The values for the finite rate of population increase (λ) at particular sites pooled across years varies from 0.550 to 1.099. Those for particular sites and years are within the range 0.684–1.286 (Pergl *et al.*, unpublished). Pooled across years, values of λ are 1.15 and 1.16 for open and dense stands in Germany (Hüls, 2005). Populations in open stands have λ values of 0.76 and 1.24 for transitions 2002–2003 and 2003–2004 and in dense stands values of 0.75 and 1.38, respectively (Table 6.3). These population growth rates indicate stable or slightly decreasing local populations, which is to be expected as these populations invaded these sites a long time ago. Large-scale invasion dynamics depend on regional-scale processes such as seed dispersal, including long-distance dispersal and successful establishment of new populations. The populations that reach and remain at the carrying capacity act as sources for further invasions. It is clear that once a population

Table 6.2. Summary of finite rates of population increase (λ) based on matrix models for populations of *H. mantegazzianum* in the Czech Republic. For each locality and year, number of analysed plants (Plant no.) is shown as number of living/total number of individuals. Values marked with * are based on insufficient numbers of plants (missing data in diagonal or subdiagonal matrix elements); NA (not available). Numbers of localities correspond to those in the overview in Perglová *et al.*, Chapter 4, this volume.

Locality	Pooled λ	Pooled Plant no.	2002–2003 λ	2002–2003 Plant no.	2003–2004 λ	2003–2004 Plant no.	2004–2005 λ	2004–2005 Plant no.
3	0.749	59/93	0.93*	27/35	0.25*	14/34	1.052*	18/24
6	0.551	51/103	0.376*	23/59	NA	12/24	1.233*	16/20
8	1.099	112/151	0.924	25/45	NA	35/44	NA	52/62
9	0.994	111/157	1.08	39/51	0.684	33/56	1.286	39/50
12	NA	24/39	NA	6/14	NA	5/8	NA	13/17
13	0.878	433/600	0.945	167/234	0.867	140/195	0.849	126/171
15	0.83	113/170	0.721	39/66	0.842	31/49	1.012	43/55
16	0.796	140/200	NA	57/80	NA	48/65	NA	35/55
Pooled	0.84	1043/1513	0.935	383/584	0.813	318/475	0.953	342/454

Table 6.3. Finite rate of population increase (λ), bootstrap estimate of λ (λ_b) with lower and upper 95% confidence intervals in brackets (Dixon, 2001; Manly, 1993), expected numbers of replacements (R_0), stage distribution (ssd, stable stage distribution; osd, stage distribution observed during the second year of the transition interval), and Keyfitz's Δ (distance between observed and stable stage distribution) of dense and open stands of _H. mantegazzianum_ in 2002–2003 and 2003–2004. Abbreviations of life cycle stages: sv – small vegetative; mv – medium vegetative; lv – large vegetative; fl – flowering. Matrix analyses were based on paired permanent plots of 1 × 2.5 m established in dense and open stands at five study sites in Hesse, Germany (Hüls, 2005).

Interval	Stand type	λ	λ_b	R_0	Stage distribution sv/mv/lv/fl	Keyfitz's Δ
2002–2003	Dense	0.75	0.74 (0.56–0.87)	0.24	ssd: 0.29/0.32/0.24/0.15 osd: 0.24/0.29/0.27/0.20	0.07
	Open	0.76	0.73 (0.39–0.99)	0.45	ssd: 0.09/0.17/0.36/0.37 osd: 0.04/0.10/0.40/0.46	0.12
2003–2004	Dense	1.38	1.31 (0.70–1.63)	6.45	ssd: 0.38/0.34/0.24/0.04 osd: 0.64/0.23/0.10/0.02	0.25
	Open	1.24	1.07 (0.72–1.40)	3.94	ssd: 0.23/0.21/0.48/0.07 osd: 0.70/0.13/0.15/0.03	0.46

of _H. mantegazzianum_ reaches carrying capacity there is no potential for further growth. The variability in population growth rate at sites in Germany (Hüls, 2005) seems to be closely related to annual climatic variation and stochasticity. Extremely hot and dry conditions in summer 2003 strongly reduced primary productivity across Europe (Ciais _et al._, 2005), inducing dramatic changes in the population structure of _H. mantegazzianum_ due to increased mortality and low seedling establishment. This resulted in an increase in recruitment of new plants in gaps and high population growth rates between 2003 and 2004 (Table 6.3).

Matrix models make it possible to estimate a theoretical stable population stage structure, which can be compared with the observed stage of distribution. When the data are pooled across years, G-tests indicate no significant differences between the predicted and observed stage structures for each locality in the Czech Republic (Pergl _et al._, unpublished). However, when the test was performed using data for individual years, most differences were significant. This suggests that, although there is a significant year-to-year variation in stage structure, the populations show stable long-term dynamics. This is supported by the results from Germany. The perturbation caused by the extreme climatic conditions in 2003 resulted in large deviations between observed and expected stage structures in 2004 (measured as Keyfitz's delta; Table 6.3). Over this period, both open and dense stands showed similar stage structures and population dynamics. However, the differences between the observed and predicted stable stage distributions were smaller in 2002–2003, when populations experienced average weather. This was particularly true for dense stands, which were very close to equilibrium conditions, whereas open stands showed larger deviations from the expected stage structure.

The accuracy of matrix model projections can be verified by comparing expected and observed population age structure or age at reproduction if available (Cochran and Ellner, 1992). This was done by estimating the age at flowering (based on data from Czech Republic) using the program STAGECOACH (Cochran and Ellner, 1992) and the results of a study on the age structure of *H. mantegazzianum* (Pergl *et al.*, 2006). Results for a site in the Czech Republic indicated an estimated age at flowering of 4.36 ± 1.41 years (mean ± SD), while the observed median age was 3 years, with the oldest plant being 4 years old (Pergl, J. *et al.*, unpublished). This is corroborated by the results from the German study that indicate a generation time (age at flowering in monocarpic species), estimated according to Caswell (2001), of about 3 years for dense stands for the transition 2002–2003 (Hüls, 2005). The close match between the observed data and the results of independent matrix analyses for two regions indicate that, although there are some limitations in the use of simple, time-invariant, deterministic matrix models (Hoffmann and Poorter, 2002; Nehrbass *et al.*, 2006), they accurately describe the essential properties of the *H. mantegazzianum* populations studied. Elasticity analysis of these matrix models (Caswell, 2001) was used to identify transitions in the life cycle that have large effects on population growth rate, which might be used for developing management or control measures. Elasticity matrices for pooled populations in the Czech Republic and two stand types in Germany are shown in Table 6.4. The elasticity matrices averaged across years are similar in both regions and the highest elasticities are related to growth (transition to higher stage classes, i.e. sub-diagonal) and stasis (remaining within the same stage class, i.e. matrix diagonal). However, when analysed separately for each year-to-year transition, dense stands exhibit a high elasticity for stasis and open stands high elasticity for growth (Hüls, 2005). These results suggest that despite the enormous seed production, survival is crucial; the role of seed

Table 6.4. Elasticity matrices of *H. mantegazzianum* for pooled data from the Czech Republic, and for open and dense stands in Germany averaged across years and sites. Abbreviations: seedl – seedlings; juv – juveniles; ros – rosette plants; flow – flowering plants; sv – small vegetative; mv – medium vegetative, lv – large vegetative. Although the definition of stage classes varied slightly between regions, seedlings and small vegetative plants, juveniles and medium vegetative plants, and rosettes and large vegetative plants are considered to be equivalent developmental stages.

Czech Republic					Germany (open stands)				Germany (dense stands)			
	seedl	juv	ros	flow		sv	mv	lv	flow	sv	mv	lv	flow
seedl	0.07	0.03	0	0.11	sv	0.01	0	0	0.17	0.05	0.01	0.00	0.13
juv	0.11	0.13	0.02	0.07	mv	0.11	0.03	0	0.05	0.11	0.10	0	0.06
ros	0.04	0.13	0.11	0	lv	0.06	0.15	0.16	0.02	0.03	0.16	0.15	0.01
flow	0	0.04	0.14	0	flow	0	0	0.23	0	0	0	0.19	0

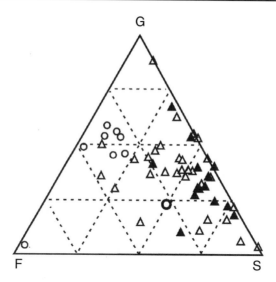

Fig. 6.7. Position of *H. mantegazzianum* in a rescaled elasticity space based on vital rates of survival (S), growth (G) and fecundity (F) using species from Franco and Silvertown (2004). Note that values of S, G and F are not simply sums of elasticity matrix elements; ● *H. mantegazzianum* (pooled across localities and years; data from the Czech Republic), ▲ iteroparous (polycarpic) forest herbs, △ iteroparous herbs from open habitats, ○ semelparous (monocarpic) herbs.

production is diminished by the poor establishment and high mortality of seedlings. This accords with the fact that within established populations there is little recruitment, while in the open the role of colonization is high. Compared to the other species, *H. mantegazzianum* fits near the group of polycarpic perennials in the fecundity, survival and growth 'elasticity space' (Hüls, 2005). Although the strictly monocarpic behaviour of *H. mantegazzianum* is confirmed and its average age at reproduction is between 3 and 5 years, this species seems to be rather isolated from the other short-lived monocarpic species analysed by Franco and Silvertown (2004) (Fig. 6.7).

Comparison of Open and Dense Stands of *Heracleum mantegazzianum*

Individual plants, linear stands, sparse, open populations and dense populations of *H. mantegazzianum* occur in nature (Thiele, 2006). Since the density of a population is not simply a function of time since its establishment (Müllerová *et al.*, 2005), understanding what determines the density of stands and if there are differences in the life cycles of plants in open and dense stands may provide guidelines for management and control. This is also important in the context of eradication of *H. mantegazzianum*, because control measures have so far focused mainly on dominant stands (Nielsen *et al.*, 2005). There is little information on whether open stands occur in suboptimal environmental conditions or represent initial population stages. An analysis of the variation in the life cycles of plants in dense and open stands using matrix population models was used to identify morphological traits and environmental conditions associated with this life-cycle variation. The data are from a field study of paired dense and open stands in five populations in Germany (Hüls, 2005).

In contrast to dense stands, where *H. mantegazzianum* cover approaches 100%, the cover of *H. mantegazzianum* in open stands is less than 10%. All individuals of *H. mantegazzianum* were marked in summer 2002, assigned to one of four life-cycle stages, i.e. small, medium, large vegetative and flowering plants, and revisited in the summer of 2003 and 2004. Soil samples, from areas adjacent to the permanent plots, were analysed for total nitrogen, available phosphorus, potassium, magnesium and pH. The height, number of rosette leaves and the length and width of the largest rosette leaf were recorded for each vegetative plant within a sample plot. The product of number of leaves × length × width served as a proxy for leaf area.

An analysis of stage-based (Lefkovitch) population matrices revealed that open stands of *H. mantegazzianum* did not have higher intrinsic population growth rates than dense stands (Table 6.4). Therefore, this could lead to the conclusion that open stands cannot be considered to be expanding populations or the front of an expanding population. Furthermore, there were considerable and biologically relevant differences in population dynamics, stage structure and elasticities between stand types (see above), at least during the transition 2002–2003, when populations developed under average climatic conditions. Nevertheless, these results must be interpreted with care; data used in this analysis are only for two transitions and established populations. However, there were no significant differences in abiotic environmental conditions, i.e. pH and soil nutrients, between the stand types. It is therefore possible that the differences result from the effect of a high density and biomass of *H. mantegazzianum*, which changes local abiotic conditions; the species acts as an 'ecosystem engineer' sensu Crooks (2002) (Hüls, 2005). In addition, a large survey of 202 stands in Germany (J. Thiele, unpublished) revealed no significant differences in environmental conditions between stand types.

The analysis of stand structure suggests that in open stands *H. mantegazzianum* exerts a strong competitive effect on the surrounding vegetation. By elongating its petiole the species is able to place its leaf area just above the resident vegetation and thus monopolize the light resource (Hüls, 2005). A considerable proportion of the young *H. mantegazzianum* plants are able to reach the leaf canopy in open stands. As a result they have a faster development and shorter generation time in open stands and outcompete the shade-intolerant grassland species (Kolbek *et al.*, 1994; Gibson *et al.*, 1995; Thiele, 2006). This successful competitive growth strategy is amplified by an early phenological development (Otte and Franke, 1998; Perglová *et al.*, Chapter 4, this volume). The structural and compositional changes, which occur in invaded communities, gradually result in species-poor or dense monospecific stands. In the course of this development, stand height and petiole length, but not leaf area, increase significantly (Hüls, 2005). The intensity of intraspecific competition increases and only a small fraction of the population reaches the closed leaf canopy. Consequently, the development of small individuals, growing in the shade of the canopy, is slowed down, which leads to protracted generation times, higher proportions of small- and medium-sized plants and a higher mortality of small individuals (Hüls, 2005).

Concluding Remarks and Future Perspective

This chapter summarizes the information on the population biology of *H. mantegazzianum* in various habitat types within its invaded distribution range in the Czech Republic and Germany. *H. mantegazzianum* is a monocarpic perennial (Pergl *et al.*, 2006; Perglová *et al.*, Chapter 4, this volume) with a poor seed bank (Moravcová *et al.*, Chapter 5, this volume). Although this species is invasive, once an area is colonized the population growth rate shows little variation between habitat types and fluctuates around $\lambda = 1$. As in many other plant species, environmental stochasticity, especially extreme climatic conditions, exerts a strong influence on population dynamics. In undisturbed conditions, a close match between the observed and stable stage distributions and a high elasticity for stasis suggest that dense monospecific stands have reached the carrying capacity of the habitat. In contrast, in open stands growth transitions have a large effect on the population growth rate. However, population dynamics in dense and open stand types respond similarly to disturbance. Gaps in the stand that result from drought, for example, are quickly filled by increased seedling establishment and recruitment. Although the time series of field observations is short, the predictions based on the data accord with reality. Invasive behaviour is associated with successful colonization of sites mostly resulting from human land use change and disturbance (Müllerová *et al.*, 2005; Thiele *et al.*, Chapter 8, this volume), especially increased abandonment (ruderalization) of landscapes. A similar invasive behaviour was observed at sites in the native range of *H. mantegazzianum* intensively used by humans, such as pastures.

The analyses presented here did not identify any special features responsible for the successful invasion of this species or a weak link in its life history on which control measures could focus. Future control measures need to consider every stage of the life cycle of *H. mantegazzianum*. Although a single *H. mantegazzianum* plant is able to produce thousands of seeds (see Perglová *et al.*, Chapter 4, this volume; Perglová *et al.*, 2006), their survival is site-dependent. For example, when competing with other species in managed meadows seedling survival is extremely low, particularly during the first year. *H. mantegazzianum* only reproduces by seed; such strictly monocarpic behaviour offers some possibilities for its eradication, e.g. by preventing seed release by depletion of mature plants at a site (Pyšek *et al.*, 2007 and Chapter 7, this volume).

Although the whole life cycle of *H. mantegazzianum* has been considered here, some issues remain to be studied. Namely the effect of various control measures on particular phases of the life cycle, employing long-term experiments and observations. Similarly, although regional landscape-scale dynamics over a historical time scale have been recently analysed (Müllerová *et al.*, 2005), metapopulation dynamics as well as the role of long-distance dispersal need to be addressed. Since some invasive species produce allelopathic substances (e.g. Hierro and Callaway, 2003), there is a need to determine whether *H. mantegazzianum* produces such substances and whether they are released from seed, decomposing litter or root exudates. Finally, a detailed study of seedling establishment and survival over a wider range of habitats and regions than considered here may improve our understanding of the population dynamics and invasion success of *H. mantegazzianum* in relation to land use.

Acknowledgements

We thank Hansjörg Dietz and Miguel Franco for comments on the manuscript, Zuzana Sixtová for technical assistance and land owners for permission to study *H. mantegazzianum* on their private land. Tony Dixon kindly improved our English. The study was supported by the project 'Giant Hogweed (*Heracleum mantegazzianum*) a pernicious invasive weed: developing a sustainable strategy for alien invasive plant management in Europe', funded within the 'Energy, Environment and Sustainable Development Programme' (grant no. EVK2-CT-2001-00128) of the European Union 5th Framework Programme, by institutional long-term research plans no. AV0Z60050516 from the Academy of Sciences of the Czech Republic and no. 0021620828 from Ministry of Education of the Czech Republic, and grant no. 206/05/0323 from the Grant Agency of the Czech Republic, and the Biodiversity Research Centre no. LC06073.

References

Brondegaard, V.J. (1990) Massenausbreitung des Bärenklaus. *Naturwissenschaftliche Rundschau* 43, 438–439.

Caffrey, J.M. (1999) Phenology and long-term control of *Heracleum mantegazzianum*. *Hydrobiologia* 415, 223–228.

Caffrey, J.M. (2001) The management of Giant Hogweed in an Irish river catchment. *Journal of Aquatic Plant Management* 39, 28–33.

Caswell, H. (2001) *Matrix Population Models. Construction, Analysis, and Interpretation*, 2nd edn. Sinauer, Sunderland, Massachusetts.

Ciais, P., Reichstein, M., Viovy, N., Granier, A., Ogée, J., Allard, V., Aubinet, M., Buchmann, N., Bernshofer, C., Carrara, A., Chevallier, F., De Noblet, N., Friend, A.D., Friedlingstein, P., Grünwald, T., Heinesch, B., Keronen, P., Knohl, A., Krinner, G., Loustau, D., Manca, G., Matteucci, G., Miglietta, F., Ourcival, J.M., Papale, D., Pilegaard, K., Rambal, S., Seufert, G., Soussana, J.F., Sanz, M.J., Schulze, E.D., Vesala, T. and Valentini, R. (2005) Europe-wide reduction in primary productivity caused by the heat and drought in 2003. *Nature* 437, 529–533.

Cochran, E.M. and Ellner, S. (1992) Simple methods for calculating age-based life history parameters for stage-structured populations. *Ecological Monographs* 62, 345–364.

Crawley, M.J. (1997) *Plant Ecology*. Blackwell, Oxford, UK.

Crooks, J.A. (2002) Characterizing ecosystem-level consequences of biological invasions: the role of ecosystem engineers. *Oikos* 97, 153–166.

Dietz, H. and Ullmann, I. (1997) Age-determination of dicotyledonous herbaceous perennials by means of annual rings: exception or rule? *Annals of Botany* 80, 377–379.

Dietz, H. and Ullmann, I. (1998) Ecological application of 'herbchronology': comparative stand age structure analyses of the invasive plant *Bunias orientalis* L. *Annals of Botany* 82, 471–480.

Dixon, P.M. (1993) The bootstrap and the jackknife: describing the precision of ecological indices. In: Scheiner, S.M. and Gurevitch, J. (eds) *Design and Analysis of Ecological Experiments*. Chapman & Hall, London, pp. 290–318.

Franco, M. and Silvertown, J. (2004) A comparative demography of plants based upon elasticities of vital rates. *Ecology* 85, 531–538.

Grime, J.P. and Hunt, R. (1975) Relative growth rate: its range and adaptive significance in a local flora. *Journal of Ecology* 63, 393–422.

Harper, J.L. (1977) *Population Biology of Plants*. Academic Press, London.

Hendry, G.A.F. and Grime, J.P. (1993) *Methods in Comparative Plant Ecology. A Laboratory Manual*. Chapman & Hall, London.

Hierro, J.L. and Callaway, R.M. (2003) Allelopathy and exotic plant invasion. *Plant Soil* 256, 25–39.

Hoffmann, W.A. and Poorter, H. (2002) Avoiding bias in calculations of relative growth rate. *Annals of Botany* 80, 37–42.

Hüls, J. (2005) Populationsbiologische Untersuchung von *Heracleum mantegazzianum* Somm. & Lev. in Subpopulationen unterschiedlicher Individuendichte. PhD Thesis, Justus-Liebig University Giessen, Germany.

Hunt, R., Causton, D.R., Shipley, B. and Askew, A.P. (2002) A modern tool for classical plant growth analysis. *Annals of Botany* 90, 485–488.

Kolbek, J., Lecjaksová, S. and Härtel, H. (1994) The integration of *Heracleum mantegazzianum* into the vegetation – an example from central Bohemia. *Biologia* 49, 41–51.

Krinke, L., Moravcová, L., Pyšek, P., Jarošík, V., Pergl, J. and Perglová, I. (2005) Seed bank in an invasive alien *Heracleum mantegazzianum* and its seasonal dynamics. *Seed Science Research* 15, 239–248.

Manly, B.F.J. (2001) *Randomization, Bootstrap and Monte Carlo Methods in Biology*, 2nd edn. Chapman & Hall/CRC, Boca Raton and London.

Moravcová, L., Perglová, I., Pyšek, P., Jarošík, V. and Pergl, J. (2005) Effects of fruit position on fruit mass and seed germination in the alien species *Heracleum mantegazzianum* (*Apiaceae*) and the implications for its invasion. *Acta Oecologica* 28, 1–10.

Morton, J.K. (1978) Distribution of giant cow parsnip (*Heracleum mantegazzianum*) in Canada. *Canadian Field Naturalist* 92, 182–185.

Müllerová, J., Pyšek, P., Jarošík, V. and Pergl, J. (2005) Aerial photographs as a tool for assessing the regional dynamics of the invasive plant species *Heracleum mantegazzianum*. *Journal of Applied Ecology* 42, 1042–1053.

Nehrbass, N., Winkler, E., Pergl, J., Perglová, I. and Pyšek, P. (2006) Empirical and virtual investigation of the population dynamics of an alien plant under the constraints of local carrying capacity: *Heracleum mantegazzianum* in the Czech Republic. *Perspectives in Plant Ecology, Evolution and Systematics* 7, 253–262.

Nielsen, C., Ravn, H.P., Nentwig, W. and Wade, M. (eds) (2005) *The Giant Hogweed Best Practice Manual. Guidelines for the Management and Control of Invasive Weeds in Europe*. Forest and Landscape, Hørsholm, Denmark.

Ochsmann, J. (1996) *Heracleum mantegazzianum* Sommier & Levier (*Apiaceae*) in Deutschland Untersuchen zur Biologie, Verbreitung, Morphologie und Taxonomie. *Feddes Repertorium* 107, 557–595.

Otte, A. and Franke, R. (1998) The ecology of the Caucasian herbaceous perennial *Heracleum mantegazzianum* Somm. et Lev. (Giant Hogweed) in cultural ecosystems of Central Europe. *Phytocoenologia* 28, 205–232.

Pergl, J., Perglová, I., Pyšek, P. and Dietz, H. (2006) Population age structure and reproductive behavior of the monocarpic perennial *Heracleum mantegazzianum* (*Apiaceae*) in its native and invaded distribution range. *American Journal of Botany* 93, 1018–1028.

Perglová, I., Pergl, J. and Pyšek, P. (2006) Flowering phenology and reproductive effort of the invasive alien plant *Heracleum mantegazzianum*. *Preslia* 78, 265–285.

Pyšek, P. (1991) *Heracleum mantegazzianum* in the Czech Republic – the dynamics of spreading from the historical perspective. *Folia Geobotanica & Phytotaxonomica* 26, 439–454.

Pyšek, P. and Prach, K. (1993) Plant invasions and the role of riparian habitats – a comparison of four species alien to central Europe. *Journal of Biogeography* 20, 413–420.

Pyšek, P. and Pyšek, A. (1995) Invasion by *Heracleum mantegazzianum* in different habitats in the Czech Republic. *Journal of Vegetation Science* 6, 711–718.

Pyšek, P., Kopecký, M., Jarošík, V. and Kotková, P. (1998) The role of human density and climate in the spread of *Heracleum mantegazzianum* in the Central European landscape. *Diversity and Distributions* 4, 9–16.

Pyšek, P., Krinke, L., Jarošík, V., Perglová, I., Pergl, J. and Moravcová, L. (2007) Timing and extent of tissue removal affect reproduction characteristics of an invasive species *Heracleum mantegazzianum*. *Biological Invasions* (in press doi 10, 107/s 10530-006-9038-0).

Satsyperova, I.F. (1984) *The Genus Heracleum of the Flora of the USSR – New Fodder Plants*. Nauka, Leningrad. [In Russian.]

Stewart, F. and Grace, J. (1984) An experimental study of hybridization between *Heracleum mantegazzianum* Sommier et Levier and *H. sphondylium* L. subsp. *sphondylium* (*Umbellifereae*). *Watsonia* 15, 73–83.

Thiele, J. (2006) Patterns and mechanisms of *Heracleum mantegazzianum* invasion into German cultural landscapes on the local, landscape and regional scale. PhD Thesis, Justus-Liebig University, Giessen, Germany.

Tiley, G.E.D. and Philp, B. (1994) *Heracleum mantegazzianum* (Giant hogweed) and its control in Scotland. In: de Waal, L.C., Child, L.E., Wade, P.M. and Brock, J.H. (eds) *Ecology and Management of Invasive Riverside Plants*. Wiley, Chichester, UK, pp. 101–109.

Tiley, G.E.D., Dodd, F.S. and Wade, P.M. (1996) *Heracleum mantegazzianum* Sommier et Levier. *Journal of Ecology* 84, 297–319.

von Arx, G. and Dietz, H. (2006) Growth rings in the roots of temperate forbs are robust annual markers. *Plant Biology* 8, 224–233.

7 Regeneration Ability of *Heracleum mantegazzianum* and Implications for Control

PETR PYŠEK,[1,2] IRENA PERGLOVÁ,[1] LUKÁŠ KRINKE,[3] VOJTĚCH JAROŠÍK,[1,2] JAN PERGL[1] AND LENKA MORAVCOVÁ[1]

[1]*Institute of Botany of the Academy of Sciences of the Czech Republic, Průhonice, Czech Republic;* [2]*Charles University, Praha, Czech Republic;* [3]*Museum of Local History, Kladno, Czech Republic*

> *Hurry now, we must protect ourselves and find some shelter*
> (Genesis, 1971)

Introduction

As populations of *Heracleum mantegazzianum* Sommier & Levier can only reproduce via seed (Tiley *et al.*, 1996; Krinke *et al.*, 2005; Moravcová *et al.*, 2005), control measures applied before flowering and fruit set will limit recruitment to subsequent generations (Nielsen *et al.*, 2005). The possibility of controlling this species by reducing fruit production has been repeatedly considered, assuming that if applied systematically over a number of years it would ultimately deplete the seed bank. Several studies have investigated the regeneration capacity of *H. mantegazzianum* (Pyšek *et al.*, 1995; Tiley and Philp, 1997, 2000; Otte and Franke, 1998; Caffrey, 1999, 2001; Nielsen, 2005; Pyšek *et al.*, 2007).

Unfortunately, the majority of the data on regeneration come from papers where this issue is only part of descriptive studies aimed at providing information on biomass production, fecundity and basic population parameters (Tiley and Philp, 1997; Otte and Franke, 1998). Four studies (Pyšek *et al.*, 1995, 2007; Tiley and Philp, 2000; Nielsen, 2005) focus explicitly on the response of *H. mantegazzianum* to the removal of tissues. The quality of these data may in some cases be compromised by: (i) regeneration being a secondary subject of the research; (ii) the need for a practical solution resulting in less rigid experimental design; and (iii) technical difficulties resulting from the size of the plants, which makes designing experiments and obtaining sufficient replicates difficult. Some papers dealing with these issues do not use statistical analysis (e.g. Tiley and Philp, 2000) or the response to the treatments is only described verbally (Otte and Franke, 1998). All but two (Caffrey, 1999; Pyšek *et al.*, 2007) were conducted at a single locality and do not consider the effect of site conditions, and all but one (Pyšek *et al.*, 2007) do not go beyond estimates of fruit number and size, i.e. they do not explore the germination ability of seeds produced after regeneration. Caffrey (1999) measure fruit length and width, and Tiley and Philp (1997) refer to seed from regenerated plants being viable, albeit without providing any data or evidence. Pyšek *et al.* (2007) is the only study we are aware of that explores in detail the quality of fruits and seeds produced by regeneration, in terms of fruit mass and germination characteristics.

In the present chapter the literature on regeneration in *H. mantegazzianum* is reviewed. In addition, primary data from two experiments are reported. The first experiment explored whether or not plants of *H. mantegazzianum*, in which flowering was initiated, could survive until the next year if damaged (a phenomenon previously reported and often cited, see Tiley *et al.*, 1996) and how they responded to repeated removal of tissues. The second experiment focused on the potential for fruit production of umbels removed from plants and left at a site, and considered not only fruit quality but also how it was affected by the time the umbels were removed.

As to the geographical coverage, data on species response to removal of organs come from Scotland, Ireland, Denmark, Czech Republic and Germany, and with one exception they state when and at what phenological stage the treatments were applied (Table 7.1). This allows the drawing of some general conclusions, valid for various regions, and the comparison of results from different parts of Europe.

Effect of Organ Removal on Mortality

From the management point of view, the ultimate aim is to kill the plant before it fruits. Tiley and Philp (1997) investigated the effect of cutting at different root depths and stem heights on regeneration and found that only cutting the tap root 15 cm below ground killed plants in the vegetative or reproductive stage so that none of them regenerated and produced flowers. Cutting the plants 5 cm below the soil surface or at ground level allowed regrowth of shoots from axillary buds below ground (Tiley and Philp, 1997).

No mortality was recorded among plants cut to ground level in Ireland (Caffrey, 1999), but treatment in this study was applied at an early phenological stage to plants that would have flowered that year (Table 7.1). Cutting plants at ground level at later phenological stages results in some mortality, but the pattern is fairly inconsistent among treatments (Table 7.2). Data from 2 subsequent years in the Slavkovský les region, Czech Republic, indicate that mortality may be affected by site conditions. Of ten plants treated in 2002 (one at each of 10 sites), only three set fruits, compared to 9–10 (from one site) that set fruits in 2003; interestingly, the average fruit number was the same in both years (Table 7.2).

Table 7.1. Overview of the studies on the regeneration capacity of *H. mantegazzianum* plants after removal of tissues. n.g. = not given.

Country (region)	Author	Number and type of treatments	Timing of treatments	Number of sites	Number of plants/ treatment	Characteristics assessed
Czech Republic (C Bohemia)	Pyšek *et al.* (1995)	3: Removal of umbels and/or leaves	1: Peak of flowering	1	8	Leaf area, fruit number, fruit mass
Scotland (Ayr River)	Tiley and Philp (1997)	6: Cut at different heights, incl. at ground level[1]	1: Flowering time	1	4	Mortality, weight of fruiting umbels
Germany (Lahn River)	Otte and Franke (1998)	2: Cut at two heights, incl. removal of plants at ground level[2]	2: Peak of flowering, cut off regenerated flowers	1	n.g.	Height, inflorescence size[3]
Ireland (Portmarnock and Mulkear Rivers)	Caffrey (1999)	1: Cut at ground level	2: Late March (vegetative stage of flowering plants), early May (beginning of flowering)[4]	2	15–30	Mortality, plant height, fruit number, fruit size[5]
Scotland (Ayr River)	Tiley and Philp (1997)	4: Cut at ground level; at 50 cm; umbels or leaves removed	1: Flowering time	1	4	Fruit mass, fruit number
Czech Republic (Slavkovský les region)	Pyšek *et al.* (2007)	9: Removed umbels, leaves or all above-ground organs	1: Peak of flowering	10	10	Mortality, fruit number, fruit mass, germination percentage, rate of germination
Czech Republic (Slavkovský les region)	Pyšek *et al.* (2007)	2: Cut at ground level or above rosette	3: Terminal umbel bud, peak of flowering, fruit formation	1	10	Mortality, fruit number, fruit mass, germination percentage
Czech Republic (Slavkovský les region)	This study	2: Removal of terminal umbels/flowering stems, and leaving cut parts at the site	3: Terminal umbel in receptive stigma/ post-receptive/ fruit development stage	1	4–6	Germination percentage
Czech Republic (Průhonice)	This study	2: Cut at ground level at two different times, then continuous removal of flowering shoots	First cut at early terminal bud stage/after it opened and inflorescence emerged, then continuously	1	10	Number of regenerating shoots and the date of their appearance, survival of the plants into next year

[1]Cut at six different levels from 15 cm below ground to below terminal inflorescence bud. [2]Cut at ground level and at the first node. [3]Not tested statistically, only verbally described. [4]Inferred from phenological data in the paper, not stated explicitly. [5]Fruit size (length and width) was only measured at one site.

Table 7.2. Regeneration of *H. mantegazzianum* after removal of all umbels. Treatment and phenological stage at which it was applied is indicated. Data come from Pyšek et al. (2007). Treatments 1–4 are different combinations of removal of leaves and flowers, treatments 5–10 are two treatments applied at three different times. Ten plants were subjected to the treatments, one at each of ten sites in the Slavkovský les region, Czech Republic (1–4) or ten plants at one locality (5–10) (see Table 7.1 for details of this study). Number of plants that regenerated and produced some fruit is indicated (*n*) and the values (mean ± SD) are based on those plants.

Treatment/phenological stage	n	Fruit number	% of control	Fruit mass	% of control	% germination	% of control
1. All flowers removed/peak of flowering	6	110.3 ± 250.6	1.1	10.6 ± 4.8	81.8	34.3 ± 35.5	40.0
2. All flowers and half leaf area removed/peak of flowering	5	52.6 ± 63.5	0.5	15.5 ± 8.9	119.7	57.0 ± 38.1	66.3
3. All flowers and all leaves removed/peak of flowering	4	32.5 ± 28.7	0.3	13.5 ± 5.9	104.3	65.7 ± 34.5	76.4
4. Cut at ground level/peak of flowering	3	268.9 ± 433.3	2.8	10.8 ± 1.8	83.3	81.3 ± 22.7	94.6
Control 1–4	10	9613.7 ± 9183.1	–	13.0 ± 3.6	–	85.9 ± 9.0	–
5. Cut at ground level/development of flowering stem	9	1631.8 ± 1247.3	16.5	14.4 ± 4.6	109.7	84.1 ± 14.6	108.3
6. Cut at ground level/peak of flowering	10	289.3 ± 395.4	2.9	9.4 ± 4.5	71.8	48.6 ± 43.1	62.6
7. Cut at ground level/formation of fruit	9	476.3 ± 611.1	4.8	11.9 ± 2.5	90.5	56.6 ± 22.7	72.9
8. Cut above rosette/development of flowering stem	9	4989.0 ± 4205.0	50.6	17.7 ± 7.1	135.0	64.0 ± 26.2	82.5
9. Cut above rosette/peak of flowering	9	652.8 ± 770.4	6.6	19.7 ± 7.4	150.2	61.3 ± 30.1	79.0
10. Cut above rosette/formation of fruit	7	407.9 ± 413.6	4.1	16.7 ± 5.3	127.5	64.0 ± 20.6	82.5
Control 5–10	10	9863.1 ± 5215.9	–	13.1 ± 2.6	–	77.6 ± 15.4	–

Flowering Plants of *Heracleum mantegazzianum* Do Not Survive into the Next Season

Although there is recent evidence that *H. mantegazzianum* is strictly monocarpic in both its invaded and native distribution ranges (Pergl *et al.*, 2006 and Chapter 6, this volume), it is reported to be polycarpic in Russia (Shumova in Tiley *et al.*, 1996). Moreover, Tiley *et al.* (1996) state that damaged flowering plants, which are not allowed to set fruit, may survive for one or more seasons.

To clarify this issue, an experiment to test the effect of removing regenerating shoots on the survival of plants was set up in the experimental garden of the Institute of Botany, Academy of Sciences of the Czech Republic, Průhonice. Ten plants growing in an experimental bed were cut at ground level shortly after the flowering stem appeared (7 June 2004; maximum height of plants was 60 cm), and another ten when the terminal bud opened and the inflorescence emerged from sheathed bracts (14–21 June 2004); then regenerating shoots were removed as they appeared during the rest of the growing season.

None of these plants survived until the following year. This, together with several years of observations at field sites in the Czech Republic, can be considered as evidence that *H. mantegazzianum* is strictly monocarpic in the invaded distribution range. As far as the native area is concerned, the same conclusion can be drawn from the estimates of age at flowering based on examining the roots of 473 plants (Pergl *et al.*, 2006). None of these plants exhibited any signs of repeated flowering.

Effect of Organ Removal on Vegetative Growth and Regeneration

The pattern of regeneration of flowering plants depends on the type of treatment. Plants cut to ground level regenerate from the stem base, while those with a stem or part of a stem left, mostly branch and produce new flowering shoots from leaf nodes between petioles and the stem (Tiley and Philp, 1997, 2000; Otte and Franke, 1998).

The only study that measured the ability of *H. mantegazzianum* to compensate for removal of leaves (Pyšek *et al.*, 1995) found that on average 12.4% of the leaf area removed at flowering time was regenerated by the end of the growing period (corresponding to 2752 cm^2/plant). In August, at the time of fruit ripening, plants that had their leaves removed in June had three times more leaf area than control plants, which at that time had lost most of their leaf area due to senescence (Pyšek *et al.*, 1995).

Plants in the experiment reported in the previous section did not survive to the next year but differed widely in the level of regeneration effort. Those cut when the flowering stem appeared produced an average of 7.6 ± 4.0 regenerating branches (mean ± SD, $n = 10$), while those cut after the terminal bud had opened produced 5.1 ± 3.6 (Fig. 7.1A). The difference was

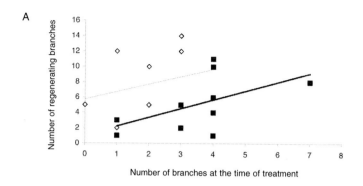

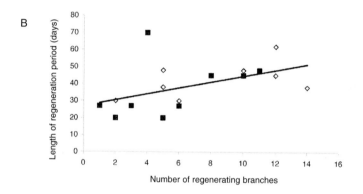

Fig. 7.1. (A) Vigorous plants are likely to regenerate better than weak plants. Ten plants were cut at ground level shortly after the flowering stem appeared (diamonds, dotted line), another ten when the terminal bud opened and the inflorescence emerged from sheathed bracts (solid square, solid line). The number of branches plants had at the time of the treatment is used as a proxy for plant vigour. Regeneration is measured as the number of branches that a plant produced under continuous branch removal. Branches were removed immediately as they developed. (B) Measures of regeneration effort are correlated. Plants that made many regeneration attempts over a long period regenerated more branches. This is indicated by a significant correlation between these measures ($r = 0.492$, $P < 0.05$). Note that the slopes are not significantly different and hence not distinguished in the plot. Symbols for the two treatments as in (A). Based on original data.

significant ($t = 2.36$; $df = 18$, $P = 0.03$), indicating that early treated plants have more resources that could be mobilized and allocated to regeneration. In addition, the regeneration effort in terms of the number of branches produced marginally significantly depended on plant vigour at the time of treatment ($t = 2.02$, $df = 18$, $P = 0.06$). Strong plants produced more regenerating branches than plants that were less vigorous at the time of tissue removal (Fig. 7.1A). The period over which regeneration occurred did not depend on the timing of the treatment ($t = 0.94$, $df = 18$, $P = 0.36$); plants produced new branches over 38.2 ± 14.0 days (mean $\pm$ SD, $n = 20$). Neither was the length of this period affected by plant vigour ($t = 0.51$, $df = 18$, $P = 0.62$). Nevertheless, both measures of regeneration effort are significantly correlated, which indicates that there might be an advantage of being large, manifested both in the period of time over which and extent to which regeneration occurs (Fig. 7.1B).

Effect of Organ Removal on Fecundity

All studies evaluated regeneration in terms of fecundity (the number of fruits produced) and in the majority umbels were removed at flowering (Tables 7.1–7.3). This in general led to marked reductions in fecundity, ranging from 2.8 to 9.9%, depending on whether or not the leaf rosette was also removed.

Data from Slavkovský les region, Czech Republic, suggest that cutting plants to ground level may be less effective in terms of reduction in fruit numbers than removing the umbels and leaving stems (Pyšek *et al.*, 2007). The former treatment yielded more than twice as much regeneration and the difference is even more marked when not only flowers but also leaves are removed (Table 7.2). Although these differences are not significant (Pyšek *et al.*, 2007), presumably because of low numbers of regenerating plants and high variation (Table 7.2), they should be taken into account when considering control measures. This result accords with the pattern of regeneration described above: plants regenerating from stem bases sometimes produce vigorous branches with umbels that are more fecund than those produced on branches, that result from the regrowth of the main stem.

Some studies combine removal of varying proportions of flowers and/or leaves (Table 7.1). Complete removal of leaves at the time of flowering always results in a reduced fruit set, but more vigorous individuals are able, after the loss of leaves, to produce more fruits than those that are weak at the time of treatment (Pyšek *et al.*, 2007).

Effect of Organ Removal on Seed Quality

Surprisingly, only recently have studies explored the germination of seed produced by regenerating plants. Because Tiley and Philp (1997) only note that such fruits are viable and do not give any figures or details on how this was assessed, the study by Pyšek *et al.* (2007) provides the first detailed information on this issue. In plants that rely exclusively on generative reproduction, not only the number of fruit produced is important but also their quality in terms of size and germination capability, which is even more crucial for population survival (Pyšek, 1997).

Removal of all leaves at the time of flowering (but not of flowers) reduces fruit mass (see also Tiley and Philp, 2000, not supported by a statistical analysis) but neither germination percentage nor the rate of germination are affected in a consistent way (Pyšek *et al.*, 2007). In general, removal of different proportions of the leaf canopy and flowers affected both final germination percentage and germination rate (assessed as the time needed for 50% of

Table 7.3. Comparison of the extent of regeneration reported in studies where plants of *H. mantegazzianum* were cut at ground level or above the basal rosette of leaves. Phenological stage at which the treatment was applied is indicated. Numbers are values recorded in particular treatments relative to control, and those reported as significant in the original studies are shown in bold. All plants, i.e. including those that did not regenerate, are included in the calculation of means. See Table 7.1 for details on particular studies.

Treatment	Application stage	Fruit number	Fruit size	Germination	Country	Source
Cut at ground level	Vegetative: early	**49.5**	**107.0**[1]		Ireland	Caffrey (1999)
Cut at ground level	Vegetative: development of flowering stem	**14.9**	98.7[2]	108.3	Czech Republic	Pyšek *et al.* (2007)
Cut at ground level	Flowering: beginning	**12.5**	**89.8**[1]		Ireland	Caffrey (1999)
Cut at ground level	Flowering: peak	9.9	63.8		Scotland	Tiley and Philp (2000)
Cut at ground level	Flowering: peak	4.3[3]			Scotland	Tiley and Philp (1997)
Cut at ground level	Flowering: peak	**2.8**	83.2[2]	94.6	Czech Republic	Pyšek *et al.* (2007)
Cut at ground level	Flowering: peak	**2.9**	71.8[2]	62.6	Czech Republic	Pyšek *et al.* (2007)
Cut at ground level	Flowering: peak	**4.3**	81.5[2]	72.9	Czech Republic	Pyšek *et al.* (2007)
Cut at ground level	Fruit formation	45.5	121.5[2]	82.5	Czech Republic	Pyšek *et al.* (2007)
Cut above rosette	Vegetative: development of flowering stem	16.3	48.6		Scotland	Tiley and Philp (1997)
Cut above rosette	Flowering: peak	15.8[3]			Scotland	Tiley and Philp (1997)
Cut above rosette	Flowering: peak	**6.0**	135.2[2]	79.0	Czech Republic	Pyšek *et al.* (2007)
Cut above rosette	Fruit formation	**2.9**	89.2[2]	82.5	Czech Republic	Pyšek *et al.* (2007)

[1] Expressed as fruit length.
[2] Expressed as fruit mass.
[3] Expressed as weight of fruiting umbels.

seed to germinate), but not significantly. This indicates that the reproductive characteristics of *H. mantegazzianum* are little affected by the loss of a large proportion of the leaves and flowers (Pyšek *et al.*, 2007). This accords well with previous findings that this plant is little affected by environmental conditions, which favours this species' invasion of Europe (Moravcová *et al.*, 2005 and Chapter 5, this volume; Müllerová *et al.*, 2005).

Timing of Control is Crucial

Of the studies reviewed, three explicitly consider the timing of the treatment as crucial for regeneration and manipulated this factor. However, Caffrey (1999) applied treatments at two different times, which are not explicitly related to a phenological stage; from the description of the population development in Ireland, it can be inferred that both treatments were applied early and with little effect, in terms of fruit production, if compared to other studies where the treatment was carried out later in the season (Table 7.3). Otte and Franke (1998) treated plants twice during the course of a growing season: new umbels produced by the post-treatment regrowth were also removed. Although this paper does not give quantitative assessment of regeneration, it provides important practical information – if umbels produced by regrowth were removed, no fruit was produced in this growing period. However, the results of the experiment reported in the section 'Effect of organ removal on vegetative growth and regeneration' (Fig. 7.1) indicate that the second cut is probably only effective if applied later to umbels with fruits already initiated (but no details of this are given by Otte and Franke, 1998). If the branches bearing regenerating flowering umbels are cut too early, regeneration continues (Fig. 7.1).

In spite of the different experimental designs and methods of assessment, some general conclusions can be drawn from a continental-wide comparison of studies that used the same sort of treatment, i.e. removal of all aboveground organs by cutting at ground level (Table 7.1). If regeneration is expressed as the number of fruits produced relative to the control (Fig. 7.2), there seem to be two qualitative thresholds. If plants are cut at an early vegetative stage (late March in Ireland from where the only data are available; Caffrey, 1999), fruit set is reduced by about 50%. Applying the treatment later, when the flowering stem has emerged and the flowers start to develop (phenological stages 2–3 on Fig. 7.2), results in a substantial reduction in fruit set and yields of 12.5–14.9% of the control. A further reduction is achieved if treatment is at peak flowering or beginning of fruit formation; the values of regeneration drop to 2.8–4.3% of the control for all but one data set (Fig. 7.2).

The effect of not removing leaf rosettes seems to disappear gradually as treatments are applied to phenologically more developed plants. The large difference observed in plants treated when the terminal bud forms (at this stage plants with rosettes produce three times more fruit than those with rosettes removed) is less obvious and insignificant when applied at the peak of flowering (Pyšek *et al.*, 2007) and indetectable when applied when fruits start to develop (Fig. 7.2).

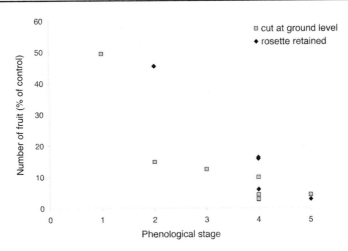

Fig. 7.2. Data from published studies allow exploration of the effect of phenological stage on the number of fruits produced by regenerating plants cut at ground level (squares) or with rosette leaves retained (diamonds). The later the cut is applied, the lower the number of fruits produced relative to the control. When both treatments were applied at the same phenological stage, diamonds positioned above squares indicate that up to stage 5 (fruit formation), leaving the basal rosette of leaves resulted in more regeneration. Phenological stages: 1: early vegetative; 2: vegetative (development of flowering stem); 3: beginning of flowering; 4: peak of flowering; 5: formation of fruits. Studies were carried out in the Czech Republic, Scotland and Ireland. See Table 7.1 for details of studies and Table 7.3 for data.

The study of Pyšek *et al.* (2007) makes it possible to assess the effect of timing of the treatment not only on the number of fruit produced by regenerating plants, but also their quality, in terms of size and germination. Mean fruit mass, although varying widely among treatments, both in absolute terms and relative to the control (Table 7.2), was not significantly affected by the treatment, and the same was true for the percentage of seed that germinated. This has important practical implications. Relative fecundity of treated plants is severely reduced by losing flower structures at some stage of development, but given the extraordinarily high number of fruits produced by plants of *H. mantegazzianum* (Perglová *et al.*, 2006 and Chapter 4, this volume) the absolute number of fruits that are available for recruitment of invading populations is very high (Table 7.2). Even plants subjected to the most effective of all treatments (the removal of all above-ground organs at ground level), which reduces the number of fruits down to 3–4% (Table 7.3), bear several hundred seeds (Table 7.2) that are viable and do not differ in size and germination characteristics from those produced by untreated plants (Pyšek *et al.*, 2007).

Removed Umbels Left at a Site Produce Viable Seeds

Two studies considered the possibility that cut-off umbels left at a site can produce viable fruits. The extent of this post-treatment fruit ripening is a warning. Pyšek *et al.* (2007) show that 85% of terminal umbels cut off at the beginning of fruit formation produce some fruits – less and of lower quality than the control (18.6% in terms of number and 43.8% in weight), but nevertheless producing an average of 1840 fruits per plant of which 24% germinated. That is, although cutting umbels off at the stage of fruit formation in the terminal umbel reduces fecundity of newly produced umbels to less than 5% of the controls (see Table 7.2), this may be ineffective if the umbels are not removed from the site. However, since there is no principle difference between treatments applied at the peak of flowering and early fruiting (Fig. 7.2; Pyšek *et al.*, 2007), the former seems to be the better management strategy, because umbels cut at flowering are less likely to give rise to fruits.

The above study, however, did not determine the effect of the time of removal on post-treatment fruit ripening in removed umbels. To explore this issue, an experiment was performed at the Slavkovský les research area in 2002 (for example, see Müllerová *et al.*, 2005 and Perglová *et al.*, Chapter 4, this volume, for details of the region). Umbels were removed at three phenological stages: (i) late stigma receptivity; (ii) post-receptive; and (iii) fruit development, with ovaries already of a flat shape but not with final fruit size. Two treatments were applied at each stage: (i) only the terminal umbel was removed and left lying on the ground until harvest; or (ii) the flowering stem was cut at ground level and left at the site; in the latter case, the umbels remained attached to a cut stem. The removed umbels were enclosed in a fine mesh to prevent loss of ripe fruits and left on the ground within the *H. mantegazzianum* stands until the end of August (fruit ripening time) to simulate mechanical control. Although many fruits decayed in the wet microclimate within the ground vegetation layer, some fruits matured. The umbels that remained attached to flowering stems produced seeds, many of which germinated (Fig. 7.3). The percentage germination was not significantly affected by the timing of the treatment and even flowers cut as early as the end of stigma receptivity produced viable seeds provided they were connected to a stem (Fig. 7.3).

The type of treatment had a significant effect. A significantly higher percentage of seeds from umbels that remained attached to a stem germinated: 9.0% (max. 20.0%) if removed at the earliest phenological stage, 19.0% (max. 30.0%) at the later stage and 15.0% (max. 50.0%) at the fruit development stage. Corresponding values for isolated umbels were 0.0%, 1.7% (max. 10.0%) and 3.3% (5.0%), respectively (Fig. 7.3). A probable explanation is that fruits attached to stems are supplied with resources for an extended period following the treatment. It is also possible that the rigid stem kept some umbels above the soil level and away from the wet conditions.

These results further emphasize that timing of treatment is critical. From a management point of view, there is a 'trade-off' between the risk of fruit developing and reduction in fecundity. This trade-off is affected by the phenology of the plant. Very early removal of umbels results in high levels of regeneration. Removal of umbels later in a season results in a marked reduction in fecundity but the removed fruits are likely to provide viable seeds. In addition, when manipulating umbels later in a season, it is difficult to avoid ripe fruits being released.

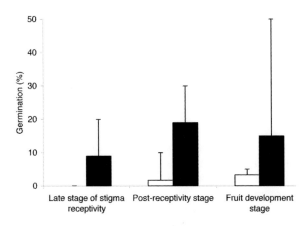

Fig. 7.3. Percentage of seeds that germinated of those maturing in umbels cut off at three phenological stages. Two treatments were adopted: either the whole flowering stem (dark columns) or only umbels (light columns) were cut and left at the site. Deletion tests revealed a significant difference between treatments ($F = 5.35$, $df = 1,12$, $P = 0.039$) but no effect of the timing of removal. Umbels attached to the cut-off stems produced seeds that germinated better than that those produced by cut-off umbels. Based on original data.

Conclusions: Guidelines for Control

Plants of *H. mantegazzianum* have a high regeneration capacity, which allows them to survive some control measures. Comparison of previous studies on this topic (Pyšek *et al.*, 1995, 2007; Tiley and Philp, 1997; Otte and Franke, 1998; Caffrey, 1999; Nielsen *et al.*, 2005) and the results of the experiments reported here identify certain principles that should be considered when designing a strategy for mechanical control.

1. The only treatment that immediately kills plants of *H. mantegazzianum* is cutting the tap root 15 cm below ground (Tiley and Philp, 1997). Any treatment that does not kill the plants, such as cutting at ground level, always results in a proportion of treated plants regenerating (Pyšek *et al.*, 1995, 2007; Caffrey, 1999).
2. Whatever the strategy for mechanical control, the life stage of the plants targeted for control is critical. Above-ground cutting of the vegetative (rosette) stage will not kill plants, but extends their life span by postponing the time of flowering. Vegetative plants can only be killed by cutting the root. In the case of flowering plants, it is not necessary to cut the root below the soil surface as once flowering is initiated these plants will not survive until the next year. Therefore, the best strategy is to kill plants at the rosette stage by cutting their roots and preventing those at the flowering stage from producing fruits. Alternatively, if a long-term programme is feasible, only flowering plants need to be targeted until the population is depleted (but see Pergl *et al.*, 2007 and Chapter 6, this volume, for data on life span).

3. Timing of the treatment is crucial. If carried out too early, plants will regenerate to a high level. Removal of umbels is effective if carried out at the peak of flowering or at the beginning of fruit formation. Subsequent cutting of regenerated flowering umbels, as they emerge, prevents plants from producing fruit. Removal of leaf rosette does not increase the effect of this treatment; there is some evidence that cutting flowering plants at ground level is less efficient than removing flowers and leaves from the stems (Pyšek *et al.*, 2007).

4. Umbels must be removed from the site. Even umbels cut at late flowering or early fruiting are able to produce viable seeds and thus should be collected and destroyed (burnt). Cutting whole flowering stems and leaving them at a site is not recommended.

5. Because of the extraordinary fecundity of *H. mantegazzianum* (see Perglová *et al.*, 2006 and Chapter 4, this volume), even a severe reduction in the number of fruits produced by regenerating plants, relative to the control, still results in large quantities of fruit in absolute numbers. More importantly, these seeds are generally of a good quality, in terms of size and viability (Pyšek *et al.*, 2007).

Acknowledgements

We thank Javier Puntieri and Lois Child for helpful comments on the manuscript, Tony Dixon for editing the English, and Zuzana Sixtová and Ivan Ostrý for technical assistance. The study was supported by the project 'Giant Hogweed (*Heracleum mantegazzianum*) a pernicious invasive weed: developing a sustainable strategy for alien invasive plant management in Europe', funded within the 'Energy, Environment and Sustainable Development Programme' (grant no. EVK2-CT-2001-00128) of the European Union 5th Framework Programme, and by institutional long-term research plan no. AV0Z60050516 of the Academy of Sciences of the Czech Republic, and no. 0021620828 of the Ministry of Education of the Czech Republic, and Biodiversity Research Centre no. LC06073.

References

Caffrey, J.M. (1999) Phenology and long-term control of *Heracleum mantegazzianum*. *Hydrobiologia* 415, 223–228.

Caffrey, J.M. (2001) The management of giant hogweed in an Irish river catchment. *Journal of Aquatic Plant Management* 39, 28–33.

Krinke, L., Moravcová, L., Pyšek, P., Jarošík, V., Pergl, J. and Perglová, I. (2005) Seed bank of an invasive alien, *Heracleum mantegazzianum*, and its seasonal dynamics. *Seed Science Research* 15, 239–248.

Moravcová, L., Perglová, I., Pyšek, P., Jarošík, V. and Pergl, J. (2005) Effects of fruit position on fruit mass and seed germination in the alien species *Heracleum mantegazzianum* (Apiaceaae) and the implications for its invasion. *Acta Oecologica* 28, 1–10.

Müllerová, J., Pyšek, P., Jarošík, V. and Pergl, J. (2005) Aerial photographs as a tool for assessing the regional dynamics of the invasive plant species *Heracleum mantegazzianum*. *Journal of Applied Ecology* 42, 1042–1053.

Nielsen, C., Ravn, H.P., Nentwig, W. and Wade, M. (eds) (2005) *The Giant Hogweed Best Practice Manual. Guidelines for the Management and Control of an Invasive Alien Weed in Europe*. Forest and Landscape Denmark, Hørsholm, Denmark.

Nielsen, S.T. (2005) The return of the Giant Hogweed: population ecology and management of *Heracleum mantegazzianum* Sommier & Levier. MSc thesis, Department of Biological Sciences, University of Aarhus, Denmark.

Otte, A. and Franke, R. (1998) The ecology of the Caucasian herbaceous perennial *Heracleum mantegazzianum* Somm. et Lev. (Giant Hogweed) in cultural ecosystems of Central Europe. *Phytocoenologia* 28, 205–232.

Pergl, J., Perglová, I., Pyšek, P. and Dietz, H. (2006) Population age structure and reproductive behavior of the monocarpic perennial *Heracleum mantegazzianum* (*Apiaceae*) in its native and invaded distribution range. *American Journal of Botany* 93, 1018–1028.

Perglová, I., Pergl, J. and Pyšek, P. (2006) Flowering phenology and reproductive effort of the invasive alien plant *Heracleum mantegazzianum*. *Preslia* 78, 265–285.

Pyšek, P. (1997) Clonality and plant invasions: can a trait make a difference? In: de Kroon, H. and van Groenendael, J. (eds) *The Ecology and Evolution of Clonal Plants*. Backhuys, Leiden, The Netherlands, pp. 405–427.

Pyšek, P., Kučera, T., Puntieri, J. and Mandák, B. (1995) Regeneration in *Heracleum mantegazzianum* – response to removal of vegetative and generative parts. *Preslia* 67, 161–171.

Pyšek, P., Krinke, L., Jarošík, V., Perglová, I., Pergl, J. and Moravcová, L. (2007) Timing and extent of tissue removal affect reproduction characteristics of an invasive species *Heracleum mantegazzianum*. *Biological Invasions* (in press doi 10.107/s 10530-006-9038-0).

Tiley, G.E.D. and Philp, B. (1997) Observations on flowering and seed production in *Heracleum mantegazzianum* in relation to control. In: Brock, J.H., Wade, M., Pyšek, P. and Green, D. (eds) *Plant Invasions: Studies from North America and Europe*. Backhuys Publishers, Leiden, The Netherlands, pp. 123–137.

Tiley, G.E.D. and Philp, B. (2000) Effects of cutting flowering stems of Giant Hogweed *Heracleum mantegazzianum* on reproductive performance. *Aspects of Applied Biology* 58, 77–80.

Tiley, G.E.D., Dodd, F.S. and Wade, P.M. (1996) Biological flora of the British Isles. 190. *Heracleum mantegazzianum* Sommier et Levier. *Journal of Ecology* 84, 297–319.

8 Ecological Needs, Habitat Preferences and Plant Communities Invaded by *Heracleum mantegazzianum*

JAN THIELE, ANNETTE OTTE AND R. LUTZ ECKSTEIN

University of Giessen, Giessen, Germany

> *They infiltrate each city with their thick dark warning odour*
> (Genesis, 1971)

Introduction

Heracleum mantegazzianum Sommier & Levier is a monocarpic, perennial tall forb of the *Apiaceae* family. In its native region in the Western Greater Caucasus, it is distributed over a wide altitudinal range from the foothills (50 m a.s.l.) to the subalpine zone (2200 m a.s.l.). The main native habitat types are alluvial softwood forests of the foothills and valley bottoms, forest clearings and abandoned grasslands in the montane zone, and subalpine tall-herb vegetation (see Otte *et al.*, Chapter 2, this volume). In the 19th century, *H. mantegazzianum* was introduced to Europe and cultivated in botanic and private gardens (Pyšek, 1991). Since its introduction the species has escaped from cultivation repeatedly and invaded a variety of habitats (Ochsmann, 1996).

Several studies have investigated habitat preferences of *H. mantegazzianum* in its invaded range in Europe. The extent of invasion in European landscapes was studied by Pyšek and Pyšek (1995) in the Czech Republic, and by Schepker (1998) and Thiele (2006) in Germany. These studies found that stands of *H. mantegazzianum* can sometimes cover a few hectares but are mostly constrained to small patches. Some information on invaded habitat types has been provided recently by Pyšek (1994), Pyšek and Pyšek (1995), Ochsmann (1996), Tiley *et al.* (1996), Wade *et al.* (1997) and Thiele (2006). Ecological needs of *H. mantegazzianum* have been studied by Pyšek and Pyšek (1995) and Ochsmann (1996) in a number of sites based on Ellenberg indicator values of the invaded vegetation. Exact measurements of soil nutrients and other soil parameters are given by Clegg and Grace (1974), Neiland *et al.* (1986), Tiley *et al.* (1996), Otte and Franke (1998) and Thiele and Otte (2006). Plant communities that include *H. mantegazzianum* have been described by Weber (1976), Dierschke (1984), Klauck (1988), Kolbeck *et al.* (1994), Otte and Franke (1998), Sauerwein (2004) and Thiele (2006).

This chapter describes the habitats, plant communities and ecological needs of *H. mantegazzianum* in its invaded range in Europe, based on literature and our field research (Thiele, 2006; Thiele and Otte, 2006). If not otherwise indicated, results presented in this chapter are based on our field studies carried out in 20 study areas in Germany in 2002–2003. Study areas were landscape sections of 1 km^2 chosen to represent the most heavily invaded areas in Germany. Mostly, the study areas were situated in the natural geographic region 'Western low mountain ranges' and secondly in the region 'Foothills of the Alps' (Thiele, 2006). For all stands of *H. mantegazzianum* in the study areas, we recorded the habitat type and stand area, and estimated the abundance of *H. mantegazzianum*. Vegetation relevés and site conditions were sampled from all stands larger than 25 m^2 (hereafter referred to as 'extensive stands'), unless the stands were destroyed by management measures such as cutting or rotovating. The total sample size was 202. Each study area contained between three and 20 extensive stands of *H. mantegazzianum*.

Habitats Invaded by *Heracleum mantegazzianum*

H. mantegazzianum occurs in a variety of habitat types in its invaded range in Europe. Among the most common habitat types of *H. mantegazzianum* are linear structures along traffic routes (roadsides, railway margins) and flowing waters (Fig. 8.1; Pyšek and Pyšek, 1995; Thiele, 2006). These habitats can be completely open or partly shaded by tree lines, single trees or shrubs. Furthermore, *H. mantegazzianum* can often be found at fringes and margins of woodlands and grasslands. In terms of area covered by stands of *H. mantegazzianum*, abandoned grasslands are the most represented habitat type, followed by tall-herb stands which can be found at long-abandoned former grassland sites or at other disused sites. Moreover, ruderal places, i.e. sites that recently have been severely disturbed by human activities, are suitable habitats (Neiland *et al.*, 1986; Pyšek, 1994; Ochsmann, 1996). Examples are sand pits, building sites and rubbish dumps. Closed tree canopies prevent invasion and growth of *H. mantegazzianum*, but it can occur beneath sparse canopies or in light gaps. Managed grasslands are a marginal habitat type for *H. mantegazzianum* in which the species sometimes can establish if there is a high pressure of *H. mantegazzianum* seeds from adjacent stands. However, regular mowing or grazing with adequate intensity (e.g. mowing twice a year) adversely affects the performance of *H. mantegazzianum* (lower height, prevention or reduction of fruit set, slower life cycle). As a result, occurrences of *H. mantegazzianum* in managed grasslands are not invasive as long as regular management is applied. In addition to rural habitat types, *H. mantegazzianum* also occurs in urban areas, gardens and parks (Pyšek, 1994; Pyšek and Pyšek, 1995).

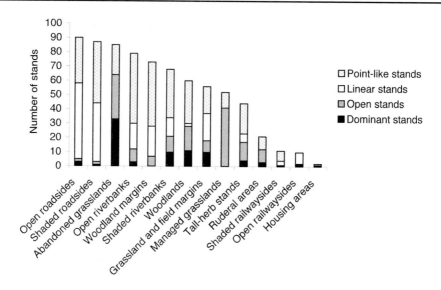

Fig. 8.1. Frequencies of habitat types invaded by *H. mantegazzianum* in 20 study areas in Germany. Number of stands recorded in each habitat, based on 738 records from Germany, is shown and representation of different stand types within habitat types indicated. Stands exceeding 25 m² in size were classified into dominant (with more than 50% cover of *H. mantegazzianum*) and open (with less than 50% cover). Other categories distinguished are point-like (smaller than 25 m²) and linear (narrower than 1 m) stands. Data from Thiele (2006). 'Open' habitat types are virtually treeless and shrubless, while 'shaded' habitat types had more than 10% tree or shrub cover. 'Abandoned grasslands' are former agricultural grasslands which are currently disused. 'Woodland margins' are the outer 5(–10) m of copses, forests and shrubland. 'Woodlands' refers to the interior of copses and shrubland but not to forest interiors. 'Grassland and field margins' are marginal parts of grassland and arable fields, which are not used agriculturally. 'Tall-forb stands' can be found in long-abandoned sites of former agricultural grasslands or other long disused sites. 'Ruderal areas' are recently disturbed sites, such as sand pits, building ground and rubbish dumps, which are in an early stage of secondary succession.

     H. mantegazzianum develops stands of varying extent and density. In our field studies, we recognized point-like stands (smaller than 25 m²), linear stands (narrower than 1 m) and extensive stands (larger than 25 m²). Extensive stands were classified into dominant (with more than 50% cover of *H. mantegazzianum*) and open (with less than 50% cover). Of all extensive stands found during our field studies in Germany, 36% were dominant. The highest proportions of dominant stands (among extensive stands) were found in open roadsides where inadequate maintenance was applied, in abandoned grasslands, and in neglected grassland and field margins (Fig. 8.1). Particularly extensive invasions by *H. mantegazzianum* in abandoned pastures and former settlements are currently present in the Slavkovský les region, Czech Republic (see Pyšek *et al.*, Chapter 3, this volume; Pyšek and Pyšek, 1995; Müllerová *et al.*, 2005) and by *H. sosnowskyi* Manden. in abandoned agricultural land in Latvia (see Ravn *et al.*, Chapter 17, this volume).

Plant Communities Invaded by *Heracleum mantegazzianum*

An alternative and more detailed perspective on habitats of *H. mantegazzianum* is provided by phytosociological analysis of vegetation relevés. We made 202 vegetation relevés in the 20 study areas in Germany during 2002 and 2003. The size of relevé plots was 25 m². We recorded all vascular plant species in the plots and estimated their abundance based on the modified Braun-Blanquet scale (Braun-Blanquet, 1965; Wilmanns, 1989). The relevés were classified according to the Central European system of plant communities (Ellenberg, 1988; Oberdorfer 1993). Species names follow Wisskirchen and Haeupler (1998). The floristic composition, basic parameters of vegetation structure and species numbers of plant communities with *H. mantegazzianum* are presented in Table 8.1.

Table 8.1. Composition of plant communities with *H. mantegazzianum* based on 202 vegetation relevés from 20 study areas in Germany. Constancy of plant species (% of the number of vegetation relevés in which they were recorded) is presented for the types of communities distinguished. Cover (%) and height of the vegetation layers were assessed separately for *H. mantegazzianum* and the remaining resident herbaceous plant species. Height refers to the main leaf canopy of the vegetation layers. FQ = total frequency of the species in the data set. Species are arranged according to their affinity to particular community types; those which did not exceed a constancy of 10% in at least one community type were omitted. Species names follow Wisskirchen and Haeupler (1998).

	Managed grasslands	Ruderal grasslands	Tall-herb communities	Woodlands	Other open vegetation	FQ
Number of relevés	36	53	78	19	16	
Average cover (%)						
Total	93.5	93.3	87.8	83.7	72.6	
Heracleum mantegazzianum	16.5	44.2	48.1	23.2	16.7	
Herb layer	88.5	73.1	52	39.6	64.3	
Tree layer	–	–	–	66.1	–	
Average height (m)						
Heracleum mantegazzianum	0.6	1.0	1.2	1.0	0.7	
Herb layer	0.4	0.6	0.7	0.5	0.6	
Tree layer	–	–	–	15.1	–	
Species number	24.5	22.1	14.3	16.8	22.1	
Heracleum mantegazzianum	100	100	100	100	100	202
Grasslands						
Dactylis glomerata	92	87	36	58	37	124
Holcus lanatus	78	68	12	16	32	81
Arrhenatherum elatius	58	75	22		19	81
Alopecurus pratensis	86	49	22		13	76
Ranunculus repens	75	47	19	21	43	78
Galium mollugo agg.	47	51	9	11	24	57
Taraxacum officinale agg.	69	30	8	5	19	51
Festuca rubra agg.	31	40	3	5	24	39
Rumex acetosa	42	32	3		24	38

Continued

Table 8.1. *Continued.*

	Managed grasslands	Ruderal grasslands	Tall-herb communities	Woodlands	Other open vegetation	FQ
Anthriscus sylvestris	36	36	13	11	6	45
Phleum pratense	47	23	4	5	33	
Veronica chamaedrys s.l.	22	36	1	5	29	
Heracleum sphondylium	39	19	4		6	28
Achillea millefolium agg.	39	19	3		6	27
Agrostis stolonifera	33	21	8	16	24	36
Lathyrus pratensis	33	19	6		24	31
Festuca pratensis	47	8				21
Poa pratensis s.str.	25	21	3	5		23
Trifolium repens	36	11		5		20
Angelica sylvestris	17	23	3		43	27
Bistorta officinalis	39	8	4		6	22
Plantago lanceolata	33	9	1			18
Cirsium palustre	14	21	5		32	25
Agrostis capillaris	33	8	8	5	37	29
Vicia cracca	14	19	3		19	20
Cerastium holosteoides	28	9				15
Lolium perenne	36	2	1		6	16
Leucanthemum vulgare	11	17			13	15
Trisetum flavescens s.str.	19	11	3		6	16
Trifolium pratense	25	6			6	13
Myosotis nemorosa	19	6		16	19	16
Prunella vulgaris	14	9	1		6	12
Lotus pedunculatus	11	11			32	15
Ranunculus acris agg.	17	6				9
Anthoxanthum odoratum	22	2				9
Centaurea jacea	19	4				9
Tall herb communities						
Urtica dioica dioica	22	60	97	84	32	137
Galium aparine agg.	8	60	74	53	19	106
Poa trivialis	83	75	69	63	32	141
Aegopodium podagraria	39	19	51	37	6	72
Glechoma hederacea	22	30	42	42		65
Impatiens glandulifera	3	2	40	32		39
Symphytum officinale	11	2	37	16	6	38
Stellaria nemorum	6	4	32	21		33
Alliaria petiolata			31	11		26
Calystegia sepium	6	17	26	5	13	34
Galeopsis tetrahit	3	42	22	32	13	48
Geum urbanum	6	26	21	47	6	42
Elymus repens	47	45	17		19	57
Solidago gigantea		6	17	5		17
Carduus crispus		4	15			14
Filipendula ulmaria	22	21	14	5	13	33
Petasites hybridus	11	8	14			19
Stachys sylvatica	3	19	12	11		22

Table 8.1. *Continued*

	Managed grasslands	Ruderal grasslands	Tall-herb communities	Woodlands	Other open vegetation	FQ
Cirsium oleraceum	14	6	12			17
Humulus lupulus		2	12		13	12
Phalaris arundinacea	3	11	10	26	13	22
Poa nemoralis		6	10	47		20
Moehringia trinervia		2	10	11	6	12
Lamium album	3	4	10			11
Woodlands						
Acer pseudoplatanus	3		6	32	19	15
Fraxinus excelsior	6	6	10	26		18
Stellaria holostea	3	11	6	21	6	17
Alnus glutinosa			5	21		8
Festuca gigantea	6		9	16		12
Elymus caninus		4	10	11		12
Salix fragilis			8	11		8
Salix eleagnos			1	11		3
Alnus incana				11		2
Populus nigra				11		2
Companion species						
Cirsium arvense	58	42	12		43	59
Rubus fruticosus agg.		25	18	37	32	39
Deschampsia cespitosa	17	19	9	26	43	35
Hypericum perforatum	19	28	3	5	32	30
Rubus idaeus	3	17	12	11	56	30
Vicia sepium	14	36	4	5	6	29

Plant communities of *H. mantegazzianum* in its Central European invaded range predominantly belong to the vegetation classes of semi-natural grasslands and nitrophilous tall-herb communities (Kolbek *et al.*, 1994; Ochsmann, 1996; Otte and Franke, 1998; Sauerwein, 2004; Thiele and Otte, 2006b). Woodlands with *H. mantegazzianum* belong partly to more or less natural alluvial forests (Kolbek *et al.*, 1994; Sauerwein, 2004; Thiele and Otte, 2006b), but anthropogenic woodlands often do not match with the system of plant communities and, therefore, are not described in detail here. The same applies to some pioneer plant communities of *H. mantegazzianum* at severely disturbed sites, e.g. sand pits, and other anthropogenic vegetation types (Ochsmann, 1996; Thiele and Otte, 2006). Klauck (1988) classified the stands with dominant *H. mantegazzianum* as a separate association Urtico-Heracleetum mantegazzianii, but this proposal has been rejected by the majority of authors studying these plant communities (Schwabe and Kratochwil, 1991; Ochsmann, 1996; Otte and Franke, 1998; Sauerwein, 2004; Thiele and Otte, 2006b).

Within the class of semi-natural grasslands (Molinio-Arrhenatheretea) *H. mantegazzianum* is confined to communities of nutrient rich, mesic to moist sites. These belong to the order Arrhenatheretalia, which comprises oat grass meadows (Arrhenatherion) and rye grass–white clover pastures (Cynosurion). Principally, these communities are characterized by regular agricultural land use (mowing, pasturing). However, *H. mantegazzianum* is more prevalent in variants without regular land-use regimes. These are, on the one hand, ruderal grasslands maintained by rather irregular mowing or removal of shrubs, such as neglected road verges and grassland margins, and on the other hand, abandoned former agricultural grasslands.

Nitrophilous tall-herb communities of the class Galio-Urticetea which feature *H. mantegazzianum* are mainly terrestrial ground elder (*Aegopodium podagraria* L.) communities at mesic to moist sites (Aegopodion) and riparian tall-herb communities (Calystegion). In addition to these, *H. mantegazzianum* occasionally occurs in the alliance Alliarion, which comprises communities of shady fringes. Within the class Galio-Urticetea, *H. mantegazzianum* has the highest affinity to plant species that are especially typical or characteristic of Aegopodion communities. Apart from the name-giving species, ground elder, these are most frequently *Urtica dioica* L., *Galium aparine* L. and *Glechoma hederacea* L. In riparian tall-herb communities, *H. mantegazzianum* is usually confined to zones that are only inundated during floods but otherwise offer aerated top soils (Ochsmann, 1996; Thiele and Otte, 2006b). Therefore, from the phytosociological point of view, *H. mantegazzianum* is a species of terrestrial tall herb communities of the alliance Aegopodion (Sauerwein, 2004; Thiele and Otte, 2006b).

According to Sauerwein (2004), *H. mantegazzianum* can also establish in ruderal annual communities of the alliance Sisymbrion (class Stellarietea mediae) in urban areas. However, a precondition for occurrences in annual communities is that frequent disturbances, essential for such vegetation types, have stopped.

Alluvial woodlands with *H. mantegazzianum* can partly be classified as alder–ash–gallery forests (Alnenion glutinoso-incanae, class Querco-Fagetea) along colline (100–200 m a.s.l.) to montane (500–900 m a.s.l.) rivers, or grey willow scrub (Salicion elaeagni, class Salicetea purpureae) along montane to subalpine (up to 2000 m a.s.l.) rivers. These are the most natural vegetation types of *H. mantegazzianum* described from Central Europe. Other woodlands in which *H. mantegazzianum* occurs include young forestry plantings in river valleys and pioneer forests of *Populus tremula* L. or *Salix caprea* L., for example (Sauerwein, 2004; Thiele and Otte, 2006b). It is noteworthy that almost all woodlands with *H. mantegazzianum* that were found during our field studies had developed from abandoned grasslands during the last 50 years (see later sections). Generally, *H. mantegazzianum* in woodlands is restricted to sparse canopies, gaps or margins where the species can benefit from increased light supply compared with closed canopies of forest interiors.

The frequencies of plant communities with *H. mantegazzianum* in the sites that we studied are presented in Fig. 8.2. *H. mantegazzianum* occurs with about the same frequency in grasslands and tall-herb communities, which accounted for most of the relevés, while other open vegetation types and woodlands are less represented.

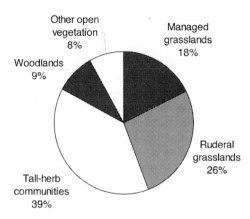

Fig. 8.2. Relative frequencies of plant communities with *H. mantegazzianum* found at 202 sites in 20 study areas in Germany. Simplified from Thiele and Otte (2006b).

Types of Disturbance and Soil Conditions in Sites Invaded by *Heracleum mantegazzianum*

Generally, sites invaded by *H. mantegazzianum* are not subject to regular land use except for marginal occurrences in managed grasslands. In our data set, 71% of relevé plots were disused and 17% were maintained by rather irregular mowing or removal of shrubs and trees, while only 12% of sites were under agricultural land use. The difference between the percentages of agricultural land use (12%) and the plant community type 'managed grasslands' (18%, Fig. 8.2) are due to some recently abandoned grasslands that were classified as 'managed grasslands' from a phytosociological perspective.

Disturbances apart from agricultural land use or maintenance mowing were detectable in 57% of extensive stands of *H. mantegazzianum*. Human-caused disturbances included clearing of trees or shrubs (10%), mechanical disturbing of the sward (7%), deposition of organic and inorganic waste (5%) and mining (5%). Furthermore, 34% of sites were disturbed by periodical flooding of rivers, which sometimes overlapped with human-caused disturbances.

The history of relevé plots during the last 50 years was reconstructed from longitudinal series of aerial photographs (see also Pyšek *et al.*, Chapter 3, this volume; Thiele and Otte, 2006b). Dates of the time series were the 1950s, 1970s and approximately 2000. The prevalent change was abandonment of grassland management and arable land use (54% of plots). Another trend was severe disturbance or destruction of sites (18%) by clearing of woodlands or open-cast mining (sand pit, rock quarry), for example. The remaining sites were either under persistent management (agricultural land use, maintenance) or derelict over the whole period. In general, sites preferred by *H. mantegazzianum* (i.e. sites without regular land use) are characterized by considerable habitat changes during recent decades.

Table 8.2. Chemical characteristics of topsoil samples (five cores per plot, 25 cm depth) and available field capacity of the effective root zone (AFC) from 202 plots with *H. mantegazzianum* in 20 study areas in Germany. Medians and the 10 and 90 percentiles are presented. The AFC gives the maximum amount of plant-available water in the soil in the effective root zone (i.e. up to approx. 1 m depth depending on soil density and texture). P and K were extracted with calcium-acetate-lactate solution using the CAL method (Schüller, 1969), while Mg was extracted with calcium chloride solution (Schachtschabel, 1954). N and C were analysed with a CN analyser. In 2002, the pH values of topsoil samples were measured in H_2O with a laboratory pH meter and additional drillings were conducted up to 1 m depth, if possible, to assess the AFC ($n = 118$). Data from Thiele and Otte (2006b).

Parameter	n	Median	10%	90%
P_{CAL} (mg/100g)	202	1.7	0.2	8.1
K_{CAL} (mg/100g)	202	8.3	4.2	21.6
N_t (% SDM)	202	0.3	0.2	0.4
Mg_{CaCl_2} (mg/100g)	202	14.3	7.0	27.0
C_{org} (% SDM)	192	2.8	1.6	5.2
C/N ratio	192	9.8	8.2	16.3
pH_{H_2O}	118	5.6	4.9	6.4
$AFC_{root\ zone}$ (mm)	118	168	140	220

Concerning soil texture, soils were mostly fairly deep (> 0.5 m) and had a medium or high capacity for soil moisture (available field capacity usually between 140 and 220 mm, Table 8.2). With respect to nutrient content (N, P, K, Mg, Ca), soils of *H. mantegazzianum* sites are variable but usually fair or rich, and the range of C/N ratios is markedly narrow (Table 8.2; Neiland *et al.*, 1986; Tiley *et al.*, 1996; Thiele and Otte, 2006b). These site characteristics indicate that *H. mantegazzianum* needs fairly high nutrient and moisture levels for optimal growth. With respect to pH values, *H. mantegazzianum* was found in a wide range of conditions from acidic to alkaline (Table 8.2). Extreme pH values were 4.0 (Thiele and Otte, 2006b) and 8.5 (Clegg and Grace, 1974).

In our field study, the light supply of relevé plots was estimated on a five-step ordinal scale ranging from deep shade to full light. Most plots were found at open sites with full light supply (78%), which shows that *H. mantegazzianum* needs high light levels. Average Ellenberg values of vegetation relevés indicate light to semi-shade situations (Ochsmann, 1996; Thiele and Otte, 2006b). Growth of the species in semi-shade is fairly good, but it cannot grow in full shade.

Population Characteristics of *Heracleum mantegazzianum* in Relation to Plant Communities and Site Conditions

Cover of *H. mantegazzianum* in the vegetation sampled during our field study varied between 1% and 95%. Low cover percentages were common, and

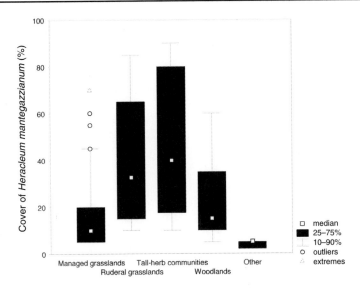

Fig. 8.3. Cover of *H. mantegazzianum* (%) in different vegetation types. High cover percentages of *H. mantegazzianum* in the community type 'managed grasslands' are due to freshly abandoned grasslands classified as 'managed grasslands' from a phytosociological perspective. 'Other' community types mostly included severely disturbed sites, which generally had low plant cover and, therefore, also *H. mantegazzianum* cover was low. Data from Thiele and Otte (2006b). Outliers are further from the upper box level than 1.5 inter-quartile range. For extremes the coefficient is 3. Sample sizes: managed grasslands = 36; ruderal grasslands = 53; tall herb communities = 78; woodlands = 19; other = 16.

about half of sampled plots had *H. mantegazzianum* cover of less than 20%. Nevertheless, in 31% of sampled plots (*n* = 202) and 36% of all encountered extensive stands (*n* = 233), *H. mantegazzianum* was dominant, with cover exceeding 50%.

There was no statistically significant relationship between soil characteristics and cover of *H. mantegazzianum*. However, in a few sites with low nutrient status and impeded drainage the cover and height of *H. mantegazzianum* was conspicuously low, which suggests that poor nutrient supply and/or wetness constrain the species. With respect to plant communities, *H. mantegazzianum* cover could take on virtually any value in ruderal grasslands and tall-herb communities, whereas cover was constrained to moderate or low percentages in managed grasslands and woodlands due to regular land use and shading, respectively (Fig. 8.3).

Furthermore, regular land use and shading significantly reduced the abundance of flowering individuals of *H. mantegazzianum* in managed grasslands (median/maximum per 25 m²: 1.5/26, *n* = 36) and woodlands (0.0/5, *n* = 19) compared to ruderal grasslands (4.5/37, *n* = 53) and tall-herb communities (7.0/54, *n* = 78). In managed grasslands fruit set was also strongly reduced because of mowing or grazing of primary stems, whereas fruit set was

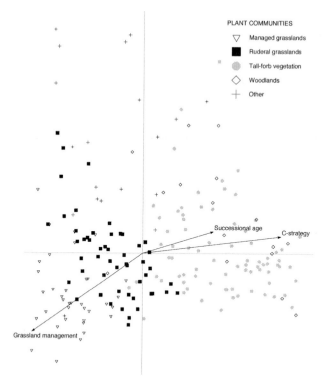

Fig. 8.4. Correspondence analysis (CA) of 202 vegetation relevés with *H. mantegazzianum* from 20 study areas in Germany. The first (*x*) and second (*y*) axes of the CA are presented. The main gradient (axis 1) represented secondary successional seres after abandonment of grassland management or severe disturbance. For definition of plant communities see text. Arrows indicate trends in the environmental variables. 'Successional age' was derived from multi-temporal series of aerial photographs and represents the time since abandonment of land use or severe disturbance events on a binary scale (before 1970s, after 1970s). 'C-strategy' refers to the classification of plant strategies according to Grime *et al.* (1988) and was calculated as the sum of proportions of C-strategy of plant species in the relevés. Data from Thiele and Otte (2006b).

generally abundant in any other vegetation type, even in woodlands when flowering occurred.

Heracleum mantegazzianum in Secondary Successional Seres

A gradient analysis (correspondence analysis) of our vegetation relevés with *H. mantegazzianum* revealed a main sequence of vegetation types from managed grassland to ruderal grasslands to tall-herb communities and, finally, to woodlands. Subordinately, a parallel sequence from severely disturbed sites to tall-herb stands to woodlands was found (Fig. 8.4; for details see Thiele and Otte, 2006b). These two parallel sequences represented the main gradient (first axis of the correspondence analysis) in the vegetation data. Towards the upper end of this gradient the age of abandoned or disturbed sites and the proportion of C-strategy (Grime *et al.*, 1988; Grime, 2001) among resident plant species increased, whereas grassland management declined. Generally, these sequences of vegetation types with *H. mantegazzianum* can be interpreted as successional seres. Although some sites were actually in a stable state because of regular land use or maintenance, the majority of sites with *H. mantegazzianum* were in the process of secondary succession following abandonment or severe disturbances. These secondary successions mostly started from grassland swards, but some were from bare ground after severe disturbance (e.g. sand pits, mining or clearing of forests). Both seres will, ultimately, result in forests (Kahmen, 2004) unless land use is resumed or severe disturbances recur.

Along these successional seres, the cover of *H. mantegazzianum* showed a unimodal response (Fig. 8.5). In managed grasslands the cover of

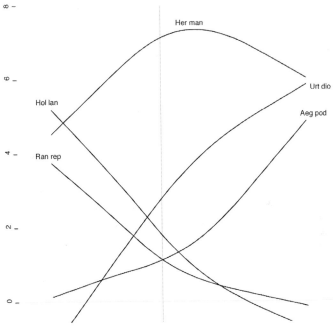

Fig. 8.5. Response curves of *H. mantegazzianum* and selected resident species along the main gradient in 202 vegetation relevés from 20 study areas in Germany. Gradients were analysed by correspondence analysis (CA). Response curves were calculated from cover-abundance estimates on the modified Braun-Blanquet scale by Generalized Additive Models in CANOCO. The *x*-axis depicts the first CA axis representing a successional gradient from managed grasslands to tall-herb stands and woodlands. Along this gradient grassland species declined and tall herbs increased. The *y*-axis depicts predicted cover-abundances classes of the species. The nine classes of the modified Braun-Blanquet scale were coded numerically (1–9). The maximum predicted cover-abundance class of *H. mantegazzianum* (Her man) of '7' corresponds to 25–50% cover. Abbreviations of species names: Aeg pod = *Aegopodium podagraria* L., Hol lan = *Holcus lanatus* L., Ran rep = *Ranunculus repens* L., Urt dio = *Urtica dioica* L. Modified from Thiele and Otte (2006b).

H. mantegazzianum was constrained by land use and in successional seres starting on bare ground plant cover was generally low due to recent disturbance. The highest cover of *H. mantegazzianum* was found in young stages of succession or in sites where succession was blocked by permanent ongoing low-intensity maintenance, such as neglected road verges and grassland margins. With increasing successional age, cover of *H. mantegazzianum* declined again, whereas that of native tall herbs steadily increased. Finally, when woody components took over, cover of *H. mantegazzianum* was more and more constrained by increasing shade.

Hence, it appears that declining cover of *H. mantegazzianum* with increasing successional age is attributable to interspecific competition with other tall-herbs and woody species. However, it should be borne in mind that the data were single records from different sites in different successional stages and not multi-temporal observations of the same sites. Therefore, we do not have direct evidence for native species reducing *H. mantegazzianum* cover during succession at a particular site. The relationship between *H. mantegazzianum* cover and successional age could, on the one hand, be attributed to less successful invasion into old successional stages or, on the other hand, to successful invasion into young successional stages followed by declining *H. mantegazzianum* cover due to competition increasing with successional age.

Land-use Changes as Drivers of Invasion

Comparison of current and historical aerial photographs of the 20 study areas in Germany indicated that landscapes with *H. mantegazzianum* have undergone considerable changes during the last 50 years (1950s to approx. 2000). The area covered by agricultural land (arable fields, managed grasslands) decreased dramatically over this period, while that covered by forests increased (Fig. 8.6). Therefore, the predominant trend of land-cover changes in the study areas has been abandonment of agricultural land and development of woodlands, which were partly planted but mostly developed through natural colonization and succession. Succession on former agricultural land determined the dynamics of habitats available for invasion by *H. mantegazzianum*.

To determine the long-term dynamics, habitats were ranked according to their vulnerability to invasion. The affinity of *H. mantegazzianum* to particular habitat types was evaluated using the electivity index E, used by Ivlev (1961) as a measure of selectivity in prey selection. The application of this formula (see Table 8.3) gives a range of values from $+1.0$ for a very high degree of selection to -1.0 for a complete avoidance. Thus, high values in our data indicate that *H. mantegazzianum* exhibits a strong affinity to a given habitat type, while low values indicate that it is rarely present in that habitat. The ranking of habitat types according to the electivity index showed that *H. mantegazzianum* has a high affinity to completely open (i.e. tree- and shrub-less) habitat types, such as abandoned grasslands and ruderal sites, but little affinity with woody habitat types (>10% cover of trees or shrubs).

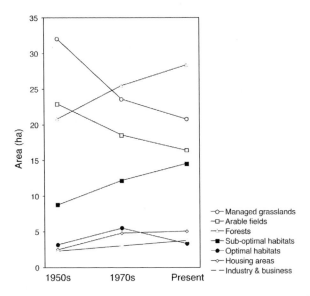

Fig. 8.6. Area covered by land-cover types and habitats of *H. mantegazzianum* averaged over 20 study areas of 100 ha in Germany for three dates: 1950s, 1970s and approx. 2000. For classification of optimal and sub-optimal habitats see Table 8.3.

Based on a conspicuous 'gap' in the electivity index values between completely open and woody habitats, we classified habitats with *H. mantegazzianum* into optimal and sub-optimal (Table 8.3). Dynamics of these two groups of habitats over the last 50 years were different. Sub-optimal habitat types accounted for the majority of the total area of habitats with *H. mantegazzianum* and exhibited a steady increase over the whole period, while optimal habitats showed a unimodal trend. From the 1950s to the 1970s, the area of optimal habitats of *H. mantegazzianum* rose by 73.4%, but since then it has dropped back to approximately the same level as in the 1950s. In total, the cover of habitat types suitable for *H. mantegazzianum* in the study areas has been increasing, resulting in a 1.5-fold increase (Fig. 8.6).

A large proportion of optimal habitats of *H. mantegazzianum* in the 1970s (61%) and a substantial proportion of sub-optimal habitats (27%) originated from former agricultural land (Fig. 8.7A, B). From the 1970s to the present, 35% of optimal habitats have become sub-optimal due to the establishment of woody plants in the course of succession (Fig. 8.7C). Further, 38% of sub-optimal habitats have developed into forests (Fig. 8.7D). These results show that the increase of optimal habitats of *H. mantegazzianum* from the 1950s to the 1970s is mainly due to abandonment of land, whereas the subsequent decrease of optimal habitats is largely attributable to secondary succession. Decreasing land use has been a post-war trend in isolated and poorer areas of Europe, especially in mountainous areas (Baldock *et al.*, 1996; MacDonald *et al.*, 2000). Significantly, the study areas in Germany were mostly situated in mountainous areas where

Table 8.3. Habitat types for *H. mantegazzianum* mapped from aerial photographs from 20 study areas in Germany. Affinity to particular habitat types was assessed using the electivity index $E = (r - p)/(r + p)$ (Ivlev, 1961), where r is the proportion of the habitat area covered by *H. mantegazzianum* and p is the proportion of the total area of the landscape sampled that is covered by that habitat. High values of E indicate that *H. mantegazzianum* exhibits a strong affinity to a given habitat type, while low values indicate that it tends to avoid that habitat. The value of zero indicates that *H. mantegazzianum* is present in a given habitat in the same proportion as the habitat is represented in the landscape. Based on E, habitats were ranked and classified into optimal and sub-optimal based on a conspicuous gap between completely open habitats and habitats containing woody components (>10% tree or shrub cover).

Habitat type	Electivity index
Optimal habitats:	
Overhead and buried cable routes	0.96
Abandoned grasslands, neglected grassland and field margins, and tall-herb stands	0.94
Open railway margins	0.87
Open riverbanks	0.82
Ruderal areas	0.82
Open roadsides	0.69
Sub-optimal habitats:	
Afforestations	0.39
Copses	0.33
Shaded railway margins	0.33
Shaded riverbanks	0.32
Tree fallow	0.32
Shaded roadsides	−0.17
Partly shaded railway margins	−0.52
Partly shaded roadsides	−0.61
Forest margins	−0.74
Partly shaded riverbanks	−0.76

Bethe and Bolsius (1995) have recognized that agriculture is likely or very likely to be reduced.

Conclusions

H. mantegazzianum invades a variety of habitat types in Europe. While road-sides, riverbanks, ruderal places and woodland fringes have previously been documented as preferred habitats, results from our field study in Germany emphasize the importance of abandoned grasslands as habitats of *H. mantegazzianum* in European landscapes.

H. mantegazzianum can occur under a wide spectrum of environmental conditions. However, preferred habitats are rather similar, characterized by rich resource supply and disturbance but lack of regular management. Primary environmental factors constraining invasion of *H. mantegazzianum* are

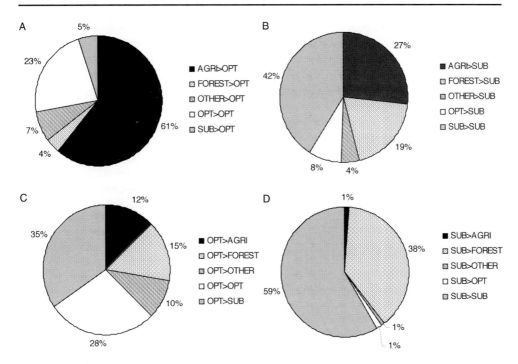

Fig. 8.7. Origin and fate of optimal and sub-optimal habitats of *H. mantegazzianum* that existed in the 1970s in 20 study areas in Germany. Habitats were mapped from multi-temporal time series of study areas for three dates: 1950s, 1970s and approx. 2000. Habitats were classified into optimal and sub-optimal based on electivity indices (see Table 8.3). Optimal habitats are open habitat types, whereas sub-optimal habitats contain woody components (>10% tree or shrub cover). A and B: origin of 1970s habitats, i.e. land-cover types from which they had developed since the 1950s. C and D: fate of 1970s habitats, i.e. land-cover types into which they have developed until approximately 2000. Abbreviations of land-cover and habitat types: OPT = optimal habitats; SUB = sub-optimal habitats; AGRI = agricultural land (arable fields, managed grasslands); FOREST; OTHER = all other land-cover types which are not habitats of *H. mantegazzianum*.

regular land use and shading by trees. Furthermore, it appears that low soil nutrient status and/or wetness constrain invasion.

From a phytosociological perspective, abandoned or neglected semi-natural grasslands are a major plant community type of *H. mantegazzianum*, although the sociological centre of the species is on nitrophilous tall-herb communities. Plant communities with *H. mantegazzianum* form a successional gradient from grasslands, or subordinately severely disturbed sites, to tall-herb communities and woodlands. The successional age of the sites affects the cover of *H. mantegazzianum*. In old successional stages, cover is constrained by interspecific competition from other tall herbs and woody species.

Habitats of *H. mantegazzianum* in German study areas have been very dynamic during the last 50 years. On the whole, the area available for invasion has increased considerably, which is mainly due to abandonment of agricultural land use. However, optimal (i.e. open) habitats are characterized by a substantial turnover. Abandonment creates new habitat patches, while secondary succession resulting in forests eliminates habitats. Altogether it appears that abandonment of land use is the primary driver of habitat dynamics and may well have enhanced invasion of *H. mantegazzianum* during recent decades.

Acknowledgements

We thank M.J.W. Cock for language revision. The study was supported by the project 'Giant Hogweed (*Heracleum mantegazzianum*) a pernicious invasive weed: developing a sustainable strategy for alien invasive plant management in Europe', funded within the 'Energy, Environment and Sustainable Development Programme' (grant no. EVK2-CT-2001-00128) of the European Union 5th Framework Programme.

References

Baldock, D., Beaufoy, G., Brouwer, F. and Godeschalk, F. (1996) *Farming at the Margins. Abandonment or Redeployment of Agricultural Land in Europe.* Institute for European Environmental Policy, London.

Bethe, F. and Bolsius, E.C.A. (1995) *Marginalisation of Agricultural Land in the Netherlands, Denmark and Germany.* National Spatial Planning Agency, The Hague, Netherlands.

Braun-Blanquet, J. (1965) *Plant Sociology: the Study of Plant Communities*, 1st edn. Hafner, London.

Clegg, L.M. and Grace, J. (1974) The distribution of *Heracleum mantegazzianum* (Somm. et Lev.) near Edinburgh. *Transactions of the Botanical Society Edinburgh* 42, 223–229.

Dierschke, H. (1984) Ein *Heracleum mantegazzianum*-Bestand im NSG 'Heiliger Hain' bei Gifhorn (Nordwestdeutschland). *Tuexenia* 4, 251–254.

Ellenberg, H. (1988) *Vegetation Ecology of Central Europe*, 4th edn. Cambridge University Press, Cambridge.

Grime, J.P. (2001) *Plant Strategies, Vegetation Processes, and Ecosystem Properties*, 2nd edn. Wiley, Chichester, UK.

Grime, J.P., Hodgson, J.G. and Hunt, R. (1988) *Comparative Plant Ecology: a Functional Approach to Common British Species.* Unwin Hyman, London.

Ivlev, V.S. (1961) *Experimental Ecology of the Feeding of Fishes.* Yale University Press, New Haven, Connecticut.

Kahmen, S. (2004) Plant trait responses to grassland management and succession. *Dissertationes Botanicae* 382, 1–123.

Klauck, E.J. (1988) Das Urtico-Heracleetum mantegazzianii. Eine neue Pflanzengesellschaft der nitrophytischen Stauden- und Saumgesellschaften. *Tuexenia* 8, 263–267.

Kolbek, J., Lecjaksová, S. and Härtel, H. (1994) The integration of *Heracleum mantegazzianum* into the vegetation – an example from central Bohemia. *Biologia* 49, 41–51.

MacDonald, D., Crabtree, J.R., Wiesinger, G., Dax, T., Stamou, N., Fleury, P., Gutierrez Lazpita, J. and Gibon, A. (2000) Agricultural abandonment in mountain areas of Europe: environmental consequences and policy response. *Journal of Environmental Management* 59, 47–69.

Müllerová, J., Pyšek, P., Jarošík, V. and Pergl, J. (2005) Aerial photographs as a tool for assessing the regional dynamics of the invasive plant species *Heracleum mantegazzianum*. *Journal of Applied Ecology* 42, 1042–1053.

Neiland, M.R.M., Proctor, J. and Sexton, R. (1986) Giant hogweed (*Heracleum mantegazzianum* Somm. & Lev.) by the River Allan and part of the River Forth. *Forth Naturalist and Historian* 9, 51–56.

Oberdorfer, E. (1993) *Süddeutsche Pflanzengesellschaften III. Wirtschaftswiesen und Unkrautgesellschaften*, 3rd edn. Fischer, Jena.

Ochsmann, J. (1996) *Heracleum mantegazzianum* Sommier & Levier (*Apiaceae*) in Deutschland – Untersuchungen zur Biologie, Verbreitung, Morphologie und Taxonomie. *Feddes Repertorium* 107, 557–595.

Otte, A. and Franke, R. (1998) The ecology of the Caucasian herbaceous perennial *Heracleum mantegazzianum* Somm. et Lev. (Giant Hogweed) in cultural ecosystems of Central Europe. *Phytocoenologia* 28, 205–232.

Pyšek, P. (1991) *Heracleum mantegazzianum* in the Czech Republic: dynamics of spreading from the historical perspective. *Folia Geobotanica & Phytotaxonomica* 26, 439–454.

Pyšek, P. (1994) Ecological aspects of invasion by *Heracleum mantegazzianum* in the Czech Republic. In: de Waal, L.C., Child, L.E., Wade, P.M. and Brock, J.H. (eds) *Ecology and Management of Invasive Riverside Plants*. Wiley, Chichester, UK, pp. 45–54.

Pyšek, P. and Pyšek, A. (1995) Invasion by *Heracleum mantegazzianum* in different habitats in the Czech Republic. *Journal of Vegetation Science* 6, 711–718.

Sauerwein, B. (2004) *Heracleum mantegazzianum* Somm. et Lev., eine auffällige *Apiaceae* bracher Säume und Versaumungen. *Philippia* 11, 281–319.

Schachtschabel, P. (1954) Das pflanzenverfügbare Magnesium im Boden und seine Bestimmung. *Zeitschrift für Pflanzenernährung, Düngung, Bodenkunde* 67, 9–23.

Schepker, H. (1998) *Wahrnehmung, Ausbreitung und Bewertung von Neophyten – Eine Analyse der problematischen nichteinheimischen Pflanzenarten in Niedersachsen*. Ibidem, Stuttgart, Germany.

Schüller, H. (1969) Die CAL-Methode, eine neue Methode zur Bestimmung des pflanzenverfügbaren Phosphates in Böden. *Zeitschrift für Pflanzenernährung und Bodenkunde* 123, 48–63.

Schwabe, A. and Kratochwil, A. (1991) Gewässer-begleitende Neophyten und ihre Beurteilung aus Naturschutz-Sicht unter besonderer Berücksichtigung Südwestdeutschlands. *NNA Berichte* 4, 14–27.

Thiele, J. (2006) Patterns and mechanisms of *Heracleum mantegazzianum* invasion into German cultural landscapes on the local, landscape and regional scale. PhD Thesis, Justus-Liebig University, Giessen, Germany.

Thiele, J. and Otte, A. (2006b) Analysis of habitats and communities invaded by *Heracleum mantegazzianum* Somm. et Lev. (Giant Hogweed) in Germany. *Phytocoenologia* 36, 280–312.

Tiley, G.E.D., Dodd, F.S. and Wade, P.M. (1996) *Heracleum mantegazzianum* Sommier & Levier. *Journal of Ecology* 84, 297–319.

Wade, M., Darby, E.J., Courtney, A.D. and Caffrey, J.M. (1997) *Heracleum mantegazzianum*: a problem for river managers in the Republic of Ireland and the United Kingdom. In: Brock, J.H., Wade, M., Pyšek, P. and Green, D. (eds) *Plant Invasions: Studies from North America and Europe*. Backhuys, Leiden, The Netherlands, pp. 139–151.

Weber, R. (1976) Zum Vorkommen von *Heracleum mantegazzianum* Somm. et Levier im Elstergebirge und den angrenzenden Gebieten. *Mitteilungen zur floristischen Kartierung (Halle)* 2, 51–57.

Wilmanns, O. (1989) *Ökologische Pflanzensoziologie*, 4th edn. Quelle and Meyer, Heidelberg, Germany.

Wisskirchen, R. and Haeupler, H. (1998) *Standardliste der Farn- und Blütenpflanzen Deutschlands*. Ulmer, Stuttgart, Germany.

9 Impact of *Heracleum mantegazzianum* on Invaded Vegetation and Human Activities

JAN THIELE AND ANNETTE OTTE

University of Giessen, Giessen, Germany

They are approaching! Mighty Hogweed is avenged. Human bodies soon will know our anger

(Genesis, 1971)

Introduction

Heracleum mantegazzianum Sommier & Levier is a monocarpic perennial tall forb of the *Apiaceae* family. Originating from the Western Greater Caucasus, the species was introduced to European botanic gardens in the 19th century and subsequently dispersed widely as an ornamental plant (see Jahodová *et al.*, Chapter 1, this volume; Pyšek, 1991). Since its introduction, *H. mantegazzianum* has escaped cultivation repeatedly and during the second half of the 20th century its incidence increased greatly in several European countries (Pyšek 1991, 1994; Ochsmann, 1996; Tiley *et al.*, 1996; Wade *et al.*, 1997). Stands of *H. mantegazzianum* can become dominant rapidly in particularly suitable habitats (see Thiele *et al.*, Chapter 8, this volume). Therefore, *H. mantegazzianum* is commonly regarded as a hazardous invasive species (see Pyšek *et al.*, Chapter 19, this volume).

Our results presented in this chapter are based on field studies in 20 study areas (1 km² landscape sections) in Germany. Data records from study areas included inventories of *H. mantegazzianum* stands in the field and mapping of all suitable habitats, whether invaded by *H. mantegazzianum* or not, based on aerial photographs.

Which Habitat Patches Are Most Likely to Become Invaded?

A trivial prerequisite for invasion is that habitats are suitable for the invasive species. However, even in suitable habitats the probability of invasion of par-

Table 9.1. Significant predictors of the probability of occurrence of *H. mantegazzianum* in suitable habitat patches (*n* = 1555). Significance was tested by logistic regression. *P*-levels: * < 0.01; ** < 0.001; *** < 0.0001. The shape index was calculated as patch perimeter divided by the square root of patch area. Habitat connectivity was assessed by the area-informed proximity index of McGarigal and Marks (1995) with a search radius of 100 m.

Predictor	Effect	*P*-level
Cover percentage of *H. mantegazzianum* in adjacent patches	+	***
Distance from rivers	−	***
Distance from agricultural roads	−	**
Shape index	+	***
Habitat connectivity	+	**
Woody habitat (> 10% woody cover)	−	*
Patch area	+	*

Wald test of overall model significance: *P* < 0.001

ticular habitat patches is not equal. A logistic regression used to analyse which predictors determine whether *H. mantegazzianum* occurs in a habitat patch or not showed that the probability of invasion is determined by both local and landscape factors (Table 9.1).

First, occurrences of *H. mantegazzianum* are spatially auto-correlated, i.e. habitat patches adjacent to invaded sites have an increased probability of invasion due to local seed dispersal. Furthermore, the vegetation structure of habitats affects the probability of invasion. Specifically, woody habitats (i.e. > 10% tree or shrub cover) have a lower probability of being invaded than completely herbaceous ones. On the one hand, this effect may be directly caused by shading, which reduces the amount of suitable habitat for recruitment and growth but, on the other hand, it may also be attributable to low disturbance intensities in these woody habitats, accompanied by increased competition from other tall herbs. Furthermore, large areas and elongated shapes of habitat patches increase the chance of *H. mantegazzianum* seeds being spread to them and, thus, the probability of invasion.

At the landscape scale, distances from transport corridors for *H. mantegazzianum* seeds (rivers, roads) and the connectivity of habitat patches have a marked effect on the probability of invasion. Probability of invasion decreases with increasing distance from rivers and road corridors. With regard to road corridors, only agricultural roads (including dirt tracks) had a significant effect in our logistic regression model, while the effect of large roads (including highways) was not significant. This might be due to greater maintenance effort (e.g. mowing twice a year) on verges of large roads compared to agricultural roads. Finally, the probability of invasion increases with habitat connectivity, which we assessed using the area-informed proximity index of McGarigal and Marks (1995) with a search radius of 100 m.

Impact on Local Plant Communities

As described by Thiele *et al.* (Chapter 8, this volume), *H. mantegazzianum* stands have cover varying from 1 to almost 100%. While it can be assumed that a low cover percentage of *H. mantegazzianum* will not affect native plant communities substantially, a high cover percentage can be expected to lead to far-reaching alterations. These alterations concern the vegetation structure, abundance of resident species and floristic composition. If *H. mantegazzianum* becomes established in vegetation types with a relatively low-growing herb layer, such as grasslands or ruderal pioneer vegetation, it introduces a new vegetation layer (tall-herb layer) above the resident herbs and grasses. This leads to shading of the resident herb layer, which increases with increasing cover of *H. mantegazzianum* or other colonizing tall herbs. When *H. mantegazzianum* attains high cover percentages, the cover of other co-occurring plant species is constrained to low percentages. This affects low-growing herb species and grasses, e.g. *Ranunculus repens* L. and *Holcus lanatus* L., as well as tall-herb species, such as *Urtica dioica* L. and *Aegopodium podagraria* L. (see Thiele *et al.*, Chapter 8, this volume).

High cover percentages of *H. mantegazzianum* can also decrease the number of vascular plant species per unit area. An analysis of 202 sampling plots (25 m^2) with *H. mantegazzianum* from 20 study areas in Germany revealed a negative relationship between *H. mantegazzianum* cover percentage and number of vascular plant species (Fig. 9.1). Regression of species numbers on *H. mantegazzianum* cover percentage yielded a significant negative slope esti-

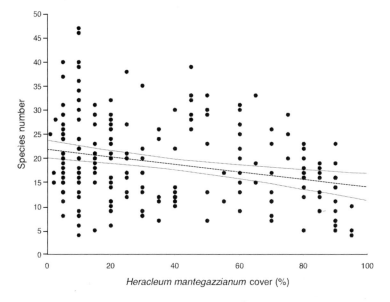

Fig. 9.1. Numbers of vascular plant species in plots 25 m^2 in size plotted against the cover of *H. mantegazzianum*. The regression slope of –0.083 differed significantly from zero ($P <$ 0.001; $R^2 = 0.07$). Dashed lines indicate 95% confidence interval of estimates. Data are from 202 plots in 20 study areas in Germany.

Table 9.2. Generalized Linear Model of numbers of vascular plant species in sampling plots (25 m^2) on *H. mantegazzianum* cover percentage and plant community type.

Effect	Estimate	*P*-level
Intercept[2]	21.360	< 0.001
H. mantegazzianum cover percentage[1]	−0.047	0.019
Plant community type[1]		< 0.001
Managed grasslands[2]	3.913	< 0.001
Ruderal grasslands[2]	2.759	0.009
Tall-herb communities[2]	−4.781	< 0.001
Woodlands[2]	−3.436	0.020

[1] Significance tested by Type III likelihood ratio tests.
[2] Significance tested by Wald tests.

mate of $b = -0.083$ ($P < 0.001$). This suggests that increasing *H. mantegazzianum* cover generally reduces resident species richness. However, significant regression slopes do not unambiguously imply a unilateral causal relationship between species numbers and *H. mantegazzianum* cover. In order to thoroughly assess impacts of *H. mantegazzianum* on plant species diversity it is necessary to distinguish different community types and to consider mechanisms of impact.

Species numbers differed between community types (ANOVA: $P < 0.001$). In particular, tall-herb communities (14.3 ± 6.1; $n = 78$) had lower species numbers than the other open community types, managed grasslands (24.5 ± 8.5; $n = 36$), ruderal grasslands (22.1 ± 9.1; $n = 53$) and 'other' open vegetation types (22.1 ± 9.1; $n = 16$; Tukey HSD tests: $P < 0.005$). In order to evaluate the relative importance of community type and *H. mantegazzianum* cover percentage for species numbers, we applied a Generalized Linear Model (Table 9.2). The model revealed a highly significant effect of community type ($P < 0.001$), while cover percentage of *H. mantegazzianum* had less significance ($P = 0.02$). Parameter estimates predicted that the number of species was reduced by 4.8 in tall-herb communities compared to the average and that an increase in *H. mantegazzianum* cover by 50 percentage points decreased the number of species by 2.4.

Separate regression analyses (Table 9.3) showed almost no significant relationships between *H. mantegazzianum* cover percentage and species numbers within particular community types. Marginal *P*-values were found for ruderal grasslands ($P = 0.06$) and 'other' community types ($P = 0.04$), which mostly comprised ruderal pioneer vegetation on disturbed ground (for definitions of community types see Chapter 8). Similarly, comparing species numbers of open and dominant stands (> 50% cover) of *H. mantegazzianum* (Fig. 9.2) we found a marginally significant lower species richness in dominant stands for ruderal grasslands (Mann-Whitney U-test: $P < 0.05$) while no significant differences could be found for the remaining community types (Fig. 9.3). Thus, negative trends in species numbers due to *H. mantegazzianum* cover were confined to ruderal grasslands and, presumably, 'other' open community types.

The main mechanism by which *H. mantegazzianum* can outcompete other plant species is shading out lower-growing species. *H. mantegazzianum*

Table 9.3. Results of simple regressions of the number of vascular plant species in sampling plots (25 m^2) on the cover (%) of *H. mantegazzianum*, conducted for different plant community types. Species numbers were standardized within communities prior to the analyses. Additionally, means and ranges of cover percentages of *H. mantegazzianum* within the community types are presented.

Plant community type	Slope	R^2	P	n	*H. mantegazzianum* cover (%) Mean	Min	Max
Managed grasslands	0.0002	0.00	0.47	36	16.5	1	70
Ruderal grasslands	−0.0092	0.07	0.06	53	44.2	5	90
Tall-herb communities	−0.0026	0.01	0.47	78	48.1	5	95
Woodlands	−0.0148	0.09	0.21	19	23.2	5	70
Other open vegetation	−0.0291	0.27	0.04	16	16.7	2	70

efficiently places its leaves above resident herbs and grasses by extending its leaf stalks just as far as necessary (the emergent strategy; Hüls, 2005). Therefore, it is plausible that high cover percentages of *H. mantegazzianum* suppress or exclude light-demanding grasses and herbs in irregularly maintained or abandoned (i.e. ruderal) grasslands and in pioneer communities on disturbed grounds (community type 'other'). In contrast, *H. mantegazzianum* does not significantly reduce species numbers in managed grasslands, tall-herb communities and woodlands, i.e. where it is constrained by management or shading or where other tall-herbs are present.

Pyšek and Pyšek (1995) found significant differences in number of vascular plant species between uninvaded vegetation and dominant stands of *H. mantegazzianum*. This is in agreement with results from our plots in Germany, where species numbers of tall-herb communities, regardless of *H. mantegazzianum* cover, were considerably lower than in other community types. Altogether, it can be stated that community type corresponding to different successional stages (see Thiele *et al.*, Chapter 8, this volume) is the key factor for species numbers, while *H. mantegazzianum* cover percentage affects species numbers only in comparatively early successional stages of vegetation (ruderal grasslands, ruderal pioneer vegetation). A similar reduction in species numbers of herbaceous vegetation stands in the course of succession after abandonment of land use or large-scale disturbance has also been observed in other studies of uncontrolled secondary successions in native vegetation (Schmidt, 1981; Neuhäusl and Neuhäuslová-Novotná, 1985; Meiners *et al.*, 2001).

In conclusion, the negative relationship between *H. mantegazzianum* cover percentage and number of vascular plant species per unit area is attributable to generally decreasing species numbers in the course of succession from low-growing and light-demanding vegetation types towards tall-herb stands and, finally, woodlands. This reduction in species numbers is mediated by native tall herbs as well as *H. mantegazzianum* or other neophytes. Thus, loss of plant species diversity in such cases is a general symptom of successional changes rather than a particular effect of invasive species.

Fig. 9.2. (A) Open stand and (B) dominant stand of *H. mantegazzianum* at an abandoned grassland site near the city of Kassel, Germany. Photo: J. Hüls.

Impact on Regional Flora

The assumed impacts of invasive plant species that attain high cover in indigenous vegetation are suppression and, possibly, local exclusion of native plant species. On the regional scale, a dominant plant invader could cause a decline of regional populations of native species. To make a native species endangered, in the sense of a high risk of regional extinction, would require that the invasive species dominates a large proportion of the habitat area of a particular indigenous population. Thus, since fairly common species normally co-

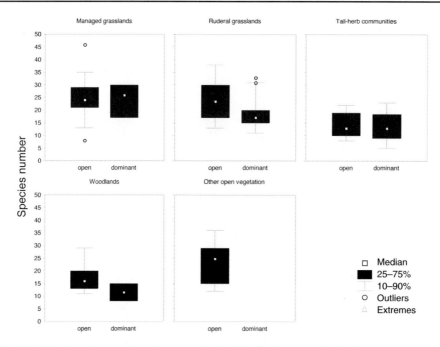

Fig. 9.3. Number of vascular plant species in 25 m² plots sampled in five vegetation types. Plots are classified according to cover percentage of *H. mantegazzianum* into open (< 50% ground cover) and dominant stands (> 50% ground cover). The only significant difference in species numbers between open and dominant stands was found for ruderal grasslands (Mann-Whitney U-test: $P < 0.05$). Dominant stands in managed grasslands ($n = 3$) are recently abandoned grassland sites that were phytosociologically classified as 'managed grasslands'. Data are from 202 plots in 20 study areas in Germany (see Chapter 8 for details). Sample sizes for community types: managed grasslands = 36, ruderal grasslands = 53, tall-herb communities = 78, woodlands = 19, other open vegetation = 16.

occur with *H. mantegazzianum* (see Thiele *et al.*, Chapter 8, this volume), they would only be regionally endangered if the invader: (i) attains high rates of habitat occupancy (i.e. percentage of suitable habitat patches invaded); (ii) builds up extensive stands; and (iii) commonly attains dominance which would, altogether, result in: (iv) a high habitat saturation, i.e. percentage of total habitat area covered by the invader (Pyšek and Pyšek, 1995).

To assess the impact of *H. mantegazzianum* on regional flora, the current stage of invasion was surveyed in 20 study areas representing the most heavily invaded landscapes in Germany (Thiele, 2006). For this purpose, all habitat types in which *H. mantegazzianum* was found during our field surveys were considered to be suitable habitats. After the field surveys, we mapped patches of suitable habitats from aerial photographs and calculated area sums for each habitat type in a GIS. *H. mantegazzianum* was present in 15.9% of suitable habitat patches, while extensive stands (> 25 m²) occurred in 11.8% of all suitable habitat patches. Dominant stands (with > 50% cover of *H. mantegazzianum*) represented 36% of these extensive stands.

The resulting habitat saturation and invasion extent of *H. mantegaz-zianum* stands are presented in Table 9.4 for each habitat type separately. Habitat saturation is the percentage of total suitable habitat area covered by *H. mantegazzianum* plants, while invasion extent is the percentage of habitat area invaded by *H. mantegazzianum* stands. The area invaded by *H. mantegazzianum* stands may be considerably larger than the area covered by individual plants, depending on the cover percentage of *H. mantegazzianum* stands. The highest value of habitat saturation was 8.7%, found in abandoned grasslands, neglected grassland and field margins and tall-herb stands, followed by open (i.e. treeless and shrubless) railway margins (4.2%), open riverbanks (2.8%), ruderal areas (2.8%) and open roadsides (1.6%; Table 9.4). The ranking of habitat types was quite similar for invasion extent. The highest value was 18.5%, again found in abandoned grasslands, neglected margins and tall-herb stands. By far the largest proportions of habitat saturation and invasion extent were attributable to extensive stands of *H. mantegazzianum* (> 25 m^2), whereas the contribution of point-like (< 25 m^2) and linear (narrower than 1 m) stands was generally negligible except for open roadsides, where about 40% of the invaded area was attributable to these stand types.

Table 9.4. Invasion extent and saturation of habitats by *H. mantegazzianum* recorded in 20 study areas (1 km^2 landscape sections) in Germany. Invasion extent is the percentage of the available habitat area that is invaded by *H. mantegazzianum* stands (the area sum of stands divided by the total available habitat area). Habitat saturation is the percentage of the available habitat area that is covered by *H. mantegazzianum* plants. Habitat areas were obtained from aerial photographs using GIS methods. Grassland margins were marginal areas of grassland parcels no longer used agriculturally and were, therefore, combined with abandoned grasslands. Tall-herb stands had to be combined with abandoned grasslands because it was not possible to discern them with certainty in aerial photographs.

	Total available habitat area (m^2)	Invasion extent		Habitat saturation	
		Area invaded (m^2)	Invasion extent (%)	Area covered (m^2)	Habitat saturation (%)
Abandoned grasslands, grassland margins and tall-herb stands	427,804	78,117	18.5	37,214	8.7
Open railway sides	19,647	1,594	9.7	830	4.2
Open riverbanks	65,747	8,614	13.8	1,855	2.8
Ruderal areas	79,259	4,513	5.8	2,189	2.8
Open roadsides	67,001	1,364	3.3	1,085	1.6
(Partly) shaded riverbanks	219,569	4,271	2.0	2,108	0.7
Woodlands (including tree fallow, forestry plantings)	1,284,723	21,733	1.8	5,760	0.7
(Partly) shaded railway margins	172,833	809	0.6	706	0.4
(Partly) shaded roadsides	212,431	1173	0.8	339	0.2
Forest margins	1,115,017	2028	0.2	393	0.04

Altogether, the current invasion extent and habitat saturation of *H. man-tegazzianum* in the most heavily invaded landscapes of Germany were moderate for most habitat types. In its most preferred habitats, i.e. abandoned grasslands, neglected grassland and field margins, and tall-herb stands, *H. mantegazzianum* covers roughly 10% of the available habitat area. Impacts include local alteration of the vegetation structure, exclusion of light-demanding herbs from ruderal grasslands and pioneer communities on disturbed grounds, and reduction of abundance of native tall-herb species. As species that co-occur with *H. mantegazzianum* are generally widespread and abundant, it appears that regional populations of associated plant species have not been endangered at the current level of invasion. However, the invasion pattern of *H. mantegazzianum* in the study areas is merely a snapshot and does not provide a way to predict future development.

Other Environmental Impacts

In addition to impact on plant communities and populations, dense stands of *H. mantegazzianum* can lead to riverbank erosion. This is mediated through the suppression or exclusion of native species, which play an important role in riverbank stabilization (Caffrey, 1999). When *H. mantegazzianum* plants in dense stands die off in winter they leave behind bare soil that can be eroded by rainfall or winter floods (Williamson and Forbes, 1982; Tiley and Philp, 1994). Deposition of eroded silt can alter substrate characteristics in rivers and, for example, render gravel substrates unsuitable for salmonid spawning (Caffrey, 1999).

Potential for Conflicts with Nature Conservation

In a Germany-wide questionnaire survey addressed to district nature conservation authorities (Thiele, 2006), *H. mantegazzianum* was quite frequently reported to occur in nature reserves and even plant communities of conservation concern. However, no rare habitats, communities or co-occurring plant species were found associated with *H. mantegazzianum* during our field studies in Germany. Furthermore, analysis of preferred site conditions showed that *H. mantegazzianum* is barely capable of invading sites offering suitable conditions (drought, wetness, poor nutrient status) for rare species and communities and, if so, *H. mantegazzianum* would be constrained to low abundances (Thiele and Otte, 2006b). Therefore, it seems that *H. mantegazzianum* cannot endanger plant communities and plant species of concern for nature conservation.

Explanations for reports of *H. mantegazzianum* in protected plant communities may be found in the details of spatial arrangement alongside environmental gradients. Given steep gradients (e.g. soil moisture), *H. mantegazzianum* might occur in the vicinity of protected communities, which might be misinterpreted as impending invasion. An example could be observed by comparing a report by nature conservation authorities with a case study of a nature reserve (Schepker, 1998). The authorities stated that *H. mantegazzianum* occurred within protected habitat types (calcareous marsh, acidic marsh and salt meadows), whereas the case study showed that *H. mantegazzianum* was growing close to these plant communities but not within them.

However, in a few cases *H. mantegazzianum* was found at sites that formerly featured protected community types (e.g. nutrient-poor chalk grassland), but these sites had degenerated due to abandonment of appropriate management, eutrophication or other reasons. In such situations *H. mantegazzianum* is not the cause, but is rather a symptom of human-caused habitat deterioration. In conclusion, we would argue that *H. mantegazzianum* does not seriously conflict with the aims of nature conservation.

Impact on Human Health and Activities

A particular impact on human health is photodermatitis elicited by furanocoumarins in the sap of *H. mantegazzianum* (e.g. Drever and Hunter, 1970; Lagey *et al.*, 1995; Jaspersen-Schib *et al.*, 1996; Wade *et al.*, 1997; Hattendorf *et al.*, Chapter 13, this volume). Photodermatitis occurs 24–48 h after physical contact of skin with the sap of *H. mantegazzianum* when the skin is exposed to sunlight (UV light). The sap remains phototoxic for some hours after plants have been cut (Wade *et al.*, 1997). Symptoms of *H. mantegazzianum* dermatitis include mild to severe erythematous reactions (red colouring of the skin) with or without painful blisters, depending on the quantities of sap and UV light received by the skin. Hyper-pigmentation of burned parts of the skin occurs within 3–5 days of contact and may persist for months or years.

To prevent photodermatitis it is advisable to avoid any contact with *H. mantegazzianum*. In cases where this is not possible, people should wear appropriate clothing, which does not leave skin exposed. During management measures (cutting etc.), long-sleeved shirts, trousers, closed footwear, gloves and face-protecting devices should be used. If the skin comes in contact with the sap it should be thoroughly rinsed with water immediately. The sap will also remain toxic on exposed clothes for some hours, so these should be handled with caution.

H. mantegazzianum can also have implications for recreational and economic human interests by restricting public access to sites such as riverbanks, amenity areas and trails or paths (Williamson and Forbes, 1982; Lundström, 1984; Tiley and Philp, 1994) and thus can affect various interest groups. For example, obstruction of lake shores and riverbanks by stands of *H. mantegazzianum* affects anglers, water sports enthusiasts, swimmers, birdwatchers, hikers and those working along river systems (Wade *et al.*, 1997).

Economic Impact

In a pilot study on economic implications of invasive species in Europe, Reinhardt *et al.* (2003) estimated annual costs of *H. mantegazzianum* in Germany based on several surveys that were extrapolated to the whole country. According to their estimates, accumulated annual costs of *H. mantegazzianum* amounted to 12,313,000 € which was distributed among the health system (1,050,000 €), nature reserves (1,170,000 €), road management (2,340,000 €), municipal management (2,100,000 €) and district management (5,600,000 €).

Conclusions

Local impacts of *H. mantegazzianum* concern the vegetation structure, cover of resident plant species and species composition of invaded plant communities. Impacts are especially marked in ruderal grasslands and ruderal pioneer vegetation. *H. mantegazzianum* excludes typical plant species of these community types and reduces the local species richness in the course of succession towards tall-herb communities. In tall-herb communities, high cover of *H. mantegazzianum* restricts the abundance of native species. However, a further reduction of species richness was not observed for tall-herb communities.

Impacts on plant communities and local plant species richness are largely driven by successional changes following abandonment of land use or after large-scale disturbances. In the course of succession, competitive native tall herbs, such as *Urtica dioica*, have similar impacts on resident vegetation. Therefore, these impacts could be seen as symptoms of human-driven changes rather than a particular effect of *H. mantegazzianum*.

Impacts on local plant communities will affect regional populations of native plant species through the reduction or restriction of local cover percentages or local displacement of sub-populations. Although *H. mantegazzianum* affects up to 10% of the area of suitable habitat types in our study areas, it appears that regional populations of native plant species have not been endangered until now as these co-occurring species are very common.

Conflicts with nature conservation are unlikely, as habitats of current conservation concern are not suitable habitats for *H. mantegazzianum*.

Impacts on humans, such as human health and recreational activities, are a major problem arising from *H. mantegazzianum*. Where conflicts with humans are imminent it is advisable to apply appropriate monitoring and management (Chapters 14–18, this volume).

Acknowledgements

We thank M.J.W. Cock for language revision. The study was supported by the project 'Giant Hogweed (*Heracleum mantegazzianum*) a pernicious invasive weed: developing a sustainable strategy for alien invasive plant management in Europe', funded within the 'Energy, Environment and Sustainable Development Programme' (grant no. EVK2-CT-2001-00128) of the European Union 5th Framework Programme.

References

Caffrey, J.M. (1999) Phenology and long-term control of *Heracleum mantegazzianum*. *Hydrobiologia* 415, 223–228.

Drever, J.C. and Hunter, J.A. (1970) Giant hogweed dermatitis. *Scottish Medical Journal* 15, 315–319.

Hüls, J. (2005) Untersuchungen zur Populationsbiologie an *Heracleum mantegazzianum* Somm. & Lev. in Subpopulationen unterschiedlicher Individuendichte. PhD thesis, University of Giessen, Germany.

Jaspersen-Schib, R., Theus, L., Guirguis-Oeschger, M., Gossweiler, B. and Meier-Abt, P.J. (1996) Acute poisonings with toxic giants in Switzerland between 1966 and 1994. *Schweizerische Medizinische Wochenschrift* 126, 1085–1098.

Lagey, K., Duinslaeger, L. and Vanderkelen, A. (1995) Burns induced by plants. *Burns* 21, 542–543.

Lundström, H. (1984) Giant hogweed, *Heracleum mantegazzianum*, a threat to the Swedish countryside. In: *Weeds and Weed Control, 25th Swedish Weed Conference*. Uppsala, Sweden, pp. 191–200.

McGarigal, K. and Marks, B.J. (1995) *FRAGSTATS: Spatial Pattern Analysis Program for Quantifying Landscape Structure*. USDA Forest Service Gen. Technical Report PNW-351.

Meiners, S.J., Pickett, S.T.A. and Cadenasso, M.L. (2001) Effects of plant invasions on the species richness of abandoned agricultural land. *Ecography* 24, 633–644.

Neuhäusl, R. and Neuhäuslová-Novotná, Z. (1985) Verstaudung von aufgelassenen Rasen am Beispiel von Arrhenatherion-Gesellschaften. *Tuexenia* 5, 249–258.

Ochsmann, J. (1996) *Heracleum mantegazzianum* Sommier & Levier (*Apiaceae*) in Deutschland: Untersuchungen zur Biologie, Verbreitung, Morphologie und Taxonomie. *Feddes Repertorium* 107, 557–595.

Pyšek, P. (1991) *Heracleum mantegazzianum* in the Czech Republic: dynamics of spreading from the historical perspective. *Folia Geobotanica & Phytotaxonomica* 26, 439–454.

Pyšek, P. (1994) Ecological aspects of invasion by *Heracleum mantegazzianum* in the Czech Republic. In: de Waal, L.C., Child, L.E., Wade, P.M. and Brock, J.H. (eds) *Ecology and Management of Invasive Riverside Plants*. Wiley, Chichester, UK, pp. 45–54.

Pyšek, P. and Pyšek, A. (1995) Invasion by *Heracleum mantegazzianum* in different habitats in the Czech Republic. *Journal of Vegetation Science* 6, 711–718.

Reinhardt, F., Herle, M., Bastiansen, F. and Streit, B. (2003) *Ökologische Folgen der Ausbreitung von Neobiota*. Forschungsbericht 201 86 211, Umweltbundesamt, Germany.

Schepker, H. (1998) Wahrnehmung, Ausbreitung und Bewertung von Neophyten: Eine Analyse der problematischen nichteinheimischen Pflanzenarten in Niedersachsen. PhD thesis, University of Hannover, Germany.

Schmidt, W. (1981) Über das Konkurrenzverhalten von *Solidago canadensis* und *Urtica dioica*. *Verhandlungen der Gesellschaft für Ökologie* 9, 173–188.

Thiele, J. (2006) Patterns and mechanisms of *Heracleum mantegazzianum* invasion into German cultural landscapes on the local, landscape and regional scale. PhD Thesis, Justus-Liebig University, Giessen, Germany.

Thiele, J. and Otte, A. (2006) Analysis of habitats and communities invaded by *Heracleum mantegazzianum* Somm. et Lev. (Giant Hogweed) in Germany. *Phytocoenologia* 36, 281–320.

Tiley, G.E.D. and Philp, B. (1994) *Heracleum mantegazzianum* (Giant Hogweed) and its control in Scotland. In: de Waal, L.C., Child, L., Wade, P.M. and Brock, J.H. (eds) *Ecology and Management of Invasive Riverside Plants*. Wiley, Chichester, UK, pp. 19–26.

Tiley, G.E.D., Dodd, F.S. and Wade, P.M. (1996) *Heracleum mantegazzianum* Sommier & Levier. *Journal of Ecology* 84, 297–319.

Wade, M., Darby, E.J., Courtney, A.D. and Caffrey, J.M. (1997) *Heracleum mantegazzianum*: a problem for river managers in the Republic of Ireland and the United Kingdom. In: Brock, J.H., Wade, M., Pyšek, P. and Green, D. (eds) *Plant Invasions: Studies from North America and Europe*. Backhuys, Leiden, The Netherlands, pp. 139–151.

Williamson, J.A. and Forbes, J.C. (1982) Giant Hogweed (*Heracleum mantegazzianum*): its spread and control with glyphosate in amenity areas. In: *Proceedings of the British Crop Protection Conference – Weeds*, pp. 967–972.

10 Seed Ecology of *Heracleum mantegazzianum* and *H. sosnowskyi*, Two Invasive Species with Different Distributions in Europe

LENKA MORAVCOVÁ,[1] ZIGMANTAS GUDŽINSKAS,[2] PETR PYŠEK,[1,3] JAN PERGL[1] AND IRENA PERGLOVÁ[1]

[1]Institute of Botany of the Academy of Sciences of the Czech Republic, Průhonice, Czech Republic; [2]Institute of Botany, Vilnius, Lithuania; [3]Charles University, Praha, Czech Republic

Waste no time! They are approaching!

(Genesis, 1971)

Introduction

Heracleum mantegazzianum Sommier & Levier and *H. sosnowskyi* Manden. (*Apiaceae*) belong to the group of 'large' or 'giant' hogweeds in the section *Pubescentia* and both species are among the worst invasive aliens in Europe. The term 'giant' reflects their size – the flowering plants of *H. mantegazzianum* grow up to 5 m and those of *H. sosnowskyi* up to 4 m (Nielsen *et al.*, 2005). Although there is another species of this group, *H. persicum* (see Jahodová *et al.*, Chapter 1, this volume), which is now invasive in Scandinavia, this chapter only compares the germination ecology of *H. mantegazzianum* and *H. sosnowskyi*.

Both species are natives of the Caucasus Mountains, where they grow in medium- to high-altitude meadows (see Jahodová *et al.*, Chapter 1, this volume; Satsyperova, 1984). The invaded distribution range of *H. mantegazzianum* is central and western Europe, while that of *H. sosnowskyi* is mainly in Poland and some countries of the former USSR (for details of the distribution in Europe see Jahodová *et al.*, Chapter 1, this volume), where it was widely used as a fodder plant.

Both species are monocarpic perennials (Tkachenko, 1989); after several years as a vegetative rosette, they flower once and die. *H. sosnowskyi* is reported to live for up to 6 years when planted in pastures and fields for biomass and silage production (Satsyperova, 1984). However, only the life span of *H. mantegazzianum* is well studied. Under natural conditions, this species usually flowers between the third and fifth year, with a recorded maximum of 12 years, and is strictly monocarpic (see Perglová *et al.*, Chapter 4, this volume; Pergl *et al.*, 2006).

Both species reproduce exclusively by seeds and are very prolific. *H. mantegazzianum* produces on average 10,000–20,000 fruits in Europe, with maxima occasionally reaching around 50,000 fruits (see Perglová *et al.*, Chapter 4, this volume). An average plant of *H. sosnowskyi* is reported to have produced 8836 fruits in the Leningrad area, Russia (Tkachenko, 1989).

Fruits of both species are broadly winged mericarps, which are connected in pairs by a carpophore and split when mature (Holub, 1997). Each mericarp contains one seed. For simplicity the mericarp is termed a 'fruit' throughout this chapter and the term 'seed' is used when referring to germination. The fruits of *H. sosnowskyi* are oval to elliptical; fruits collected in Lithuania (from terminal umbels) and used for experiments in the present study were 10.5–16.5 mm long (mean 13.4 ± 1.3 mm, $n = 60$) and 5.3–8.7 mm wide (mean 7.2 ± 0.7 mm). Mandenova (1950) gives the ranges of the length and width of mericarps as 7–9 and 4–6 mm, respectively, without specifying to which umbel order these sizes relate. The fruits of *H. mantegazzianum* are oval-elliptical; Tiley *et al.* (1996) give the ranges of the length and width as 6–18 and 4–10 mm, respectively, also without specifying which umbel. The fruits of *H. mantegazzianum*, originating from terminal umbels, used for the experiments in the present study, were 8.8–14.6 mm long (mean 11.7 ± 1.2 mm, $n = 200$) and 5.3–9.2 mm wide (mean 7.4 ± 0.8 mm).

Seeds of both species have a morphophysiological dormancy (Nikolaeva *et al.*, 1985; Baskin and Baskin, 1998), which is broken by the cold and wet conditions of autumn and winter stratification in the field and in the laboratory by temperatures within the range of 1–6°C. Seeds of both species germinate early in spring but not during summer (see Moravcová *et al.*, Chapter 5, this volume; Z. Gudžinskas, unpublished data). Autumn germination is possible in *H. sosnowskyi* (but was not recorded). The reproductive traits of *H. mantegazzianum* are well studied (Krinke *et al.*, 2005; Moravcová *et al.*, 2005, 2006; Chapters 4 and 5, this volume), but poorly so for *H. sosnowskyi*; so far the research on this species has focused mainly on biomass production (Satsyperova, 1984), seed production (Tkachenko, 1989) and content of furanocoumarins (Tkachenko and Zenkevich, 1987). The present study, based on both field research in the invaded region of Lithuania and common garden experiments, is the first published information on its germination, stratification and dormancy requirements, and seed bank dynamics.

There are differences in germination between invasive species and their less invasive congeners (Baker, 1965; Lambrinos, 2002; Mihulka *et al.*, 2003) or invasive species and their native congeners or related taxa (Dreyer *et al.*, 1987; Callaway and Josselyn, 1992; Vilà and D'Antonio, 1998; Radford and Cousens, 2000; Van Clef and Stiles, 2001). Similarly, some successful invasive aliens have larger (Richardson *et al.*, 1987; Honig *et al.*, 1992; Radford and Cousens,

2000) and/or longer persisting seed banks (Pyke, 1990; Van Clef and Stiles, 2001) than native or less invasive congeners. In this chapter the reproductive traits related to seed ecology of the two *Heracleum* species are compared to determine whether the differences can explain their varying success as invaders.

Differences in Soil Seed Bank of *Heracleum mantegazzianum* and *H. sosnowskyi*

Dynamics of the seed banks of both species were studied in the field by sampling three times a year and in a common garden burial experiment, where sampling was carried out repeatedly in the course of the year. Because the results for *H. mantegazzianum* have been published in detail elsewhere (Krinke *et al.*, 2005), the following account focuses on *H. sosnowskyi*, and is based on primary data and a comparison of both species.

Seed bank dynamics and composition

To obtain data on seed bank composition and dynamics comparable to those that are available for *H. mantegazzianum* (Krinke *et al.*, 2005), a similar study, using the same methods, was carried out at sites dominated by *H. sosnowskyi* in its invaded range (Table 10.1). Soil samples were taken in spring before seed germination (April), summer before seed release (July) and autumn after seed release (October). *H. mantegazzianum* was studied at seven sites in the Czech Republic (Krinke *et al.*, 2005), *H. sosnowskyi* at three sites in Lithuania. The geographical location, altitude and characteristics of the *H. sosnowskyi* populations are given in Table 10.1.

The vertical distribution of seeds in the soil seed bank is similar for both species. In the spring sample of *H. sosnowskyi*, 98.2% of the total seed, including dead seeds, are in the upper soil layer of 0–5 cm, with little in the deeper layers of 6–10 cm (1.5%) and 11–15 cm (0.3%) (Fig. 10.1). Nevertheless, no living seeds were found in the deepest soil layer (11–15 cm). The vertical distribution of living and dead seeds also varied significantly within

Table 10.1. Geographical location, altitude and characteristics of populations of *H. sosnowskyi* at three sites in Lithuania.

	Latitude N	Longitude E	Altitude	Population size (m^2)	Year of invasion	Density of flowering plants/m^2 (2003)	Mean plant height (m)
Santariškės	54°44′55.7″	25°16′39.9″	191	4560	1987	1.1	3.50
Bajorai	54°45′14.6″	25°15′25.0″	182	1452	1990	0.4	3.21
Visoriai	54°45′07.9″	25°16′06.8″	183	9640	1989	0.9	2.57

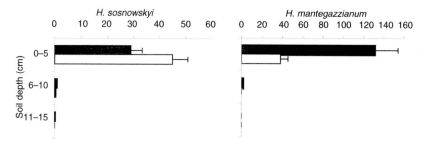

Fig. 10.1. Vertical distribution in the soil in spring of living (empty bars) and dead (black bars) seeds of *H. sosnowskyi*. Deletion tests (Crawley, 2002) on square-root + 0.5 numbers of seeds indicate that the seeds occur mainly in the upper soil layer (0–5 cm) (*P* < 0.001), and the numbers in the 6–10 cm and 11–15 cm soil layers do not differ (*P* < 0.001). Numbers of seeds per core sample (49.8 cm²) are shown; horizontal lines are standard errors of means. Corresponding figures for *H. mantegazzianum* are based on data from Krinke *et al.* (2005). Note different scales are used for the two species.

the individual sites (Table 10.2). The results for *H. mantegazzianum* show the same trend, with 95% of seeds in the upper soil layer (Fig. 10.1) and significant differences within sites (Krinke *et al.*, 2005, Table 6).

As the vast majority of the seeds are located in the upper soil layer, the study of the seasonal dynamics of the seed bank was based only on samples taken from the 0–5 cm layer. The variation among study sites of *H. sosnowskyi* was significant for all seed groups (dormant, living and total, i.e. their sum). For the total seed bank of *H. sosnowskyi*, averaged across spring, summer and autumn samples, 31.7% of the variation was linked to among sites and 68.3% to within sites (Table 10.3). If compared with results for *H. mantegazzianum* (77.9% of variation attributed among and 22.1% within sites; Krinke *et al.*, 2005), these figures indicate that the Lithuanian sites in which *H. sosnowskyi* was studied were less heterogeneous than the *H. mantegazzianum* sites sampled in the Slavkovský les, Czech Republic.

The composition of the *H. sosnowskyi* seed bank in the course of the season, expressed as the numbers of non-dormant, living and total seeds,

Table 10.2. Nested ANOVA of the vertical distribution of *H. sosnowskyi* seeds in the soil (layers: 0–5, 6–10, 11–15 cm) in spring. Data were transformed to square root numbers + 0.5 of living and dead seeds. Layer is evaluated as a fixed effect. *** *P* < 0.001, NS – not significant.

Source of variation	Living			Dead		
	df	MS	F	df	MS	F
Layer	2	323.12	51.402***	2	175.23	31.409***
Sites within layers	6	6.286	6.257***	6	5.579	5.788***
Replicates within sites	81	1.005		81	0.964	

Table 10.3. ANOVA of the soil seed bank of *H. sosnowskyi* among sites and within sites. Data are log transformed numbers plus 0.5 of the dormant, living and the total seeds, averaged for autumn, spring and summer samples. Sites are evaluated as random effects and variance is expressed in percentages. *** *P* < 0.001, * *P* < 0.05.

Source of variation	Dormant				Living				Total			
	df	MS	*F*	Variance	*df*	MS	*F*	Variance	*df*	MS	*F*	Variance
Among sites	2	2.450	9.301***	40.8	2	0.715	5.895*	30.4	2	0.783	6.263*	31.7
Within sites	27	0.263		59.2	27	0.121		69.6	27	0.125		68.3

showed significant differences among the spring, summer and autumn samples and varied significantly within individual sites (Table 10.4).

Average numbers of non-dormant, living and total seeds in the seed bank of *H. sosnowskyi* were highest in spring and lowest in summer (Fig. 10.2).

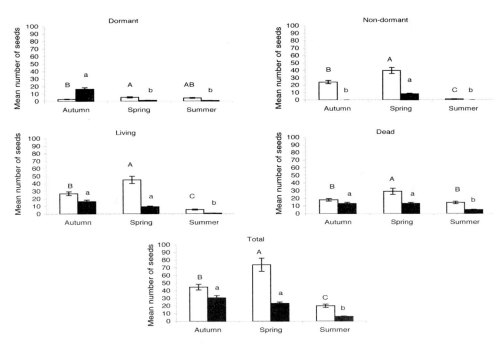

Fig. 10.2. Seasonal dynamics of the *H. sosnowskyi* seed bank (empty bars), inferred from autumn, spring and summer samples. Bars are mean numbers of dormant, non-dormant, living, dead and total seeds. Each value is pooled across three localities and ten replicates within each locality. Vertical lines are standard errors of the means. Bars with the same letters did not differ significantly (*P* < 0.05) in deletion tests (Crawley, 2002); capital letters refer to *H. sosnowskyi* and lower case letters to *H. mantegazzianum*. Corresponding values for *H. mantegazzianum* (black bars) are from Krinke *et al.* (2005). Germinated seeds were considered as non-dormant; non-germinated seeds were tested for viability by staining with tetrazolium; viable seeds were considered as dormant.

Table 10.4. Nested ANOVAS of the variation of the soil seed bank of *H. sosnowskyi* among seasons. Data are square rooted numbers + 0.5 of the dormant, non-dormant, living, dead and the total of seeds. Season is evaluated as a fixed effect. *** $P < 0.001$, * $P < 0.05$.

Source of variation	Dormant			Non-dormant			Living			Dead			Total		
	df	MS	F	df	MS	F	df	MS	F	df	MS	F	df	MS	F
Season	2	1.163	0.269 n.s.	2	193.50	33.292***	2	134.92	15.213*	2	16.045	2.135 n.s.	2	12.610	7.481*
Sites within season	6	4.317	7.796***	6	5.812	4.274***	6	8.869	5.245***	6	7.515	4.192***	6	1.133	
Replicates within sites	81	0.554		81	1.360		81	1.691		81	1.792		81	0.265	

Table 10.5. Number of seeds of *H. sosnowskyi* per m^2 in the soil seed bank at the three localities studied. Each value is the mean ± SD of ten replicates. Values per m^2 were extrapolated from the original data, which were used in statistical analyses.

Locality	Spring				Summer				Autumn			
	Non-dormant	Dormant	Dead	Total	Non-dormant	Dormant	Dead	Total	Non-dormant	Dormant	Dead	Total
1	11,477 ± 6,744	2,312 ± 1,663	9,407 ± 5,937	23,195 ± 13,859	462 ± 403	1,286 ± 753	3,055 ± 2,175	4,804 ± 2,835	4,502 ± 1,254	704 ± 437	4,040 ± 2,241	9,246 ± 3,501
2	6,512 ± 3,532	503 ± 977	4,201 ± 3,543	11,216 ± 7,032	161 ± 208	643 ± 388	1,809 ± 957	2,613 ± 1,180	3,437 ± 1,771	402 ± 268	2,854 ± 2,348	6,693 ± 4,120
3	5,970 ± 1,700	362 ± 247	3,899 ± 1,283	10,231 ± 2,702	141 ± 269	663 ± 519	3,819 ± 2,127	4,623 ± 2,720	6,613 ± 4,582	543 ± 391	3,980 ± 1,805	11,135 ± 6,363
Total	7,986 ± 3,035	1,059 ± 1,087	5,836 ± 3,097	14,881 ± 7,217	255 ± 180	864 ± 366	2,894 ± 1,014	4,013 ± 1,216	4,851 ± 1,616	550 ± 151	3,625 ± 668	9,025 ± 2,229

This seemingly contradicts the pattern found for *H. mantegazzianum*, where most are found in autumn, and then the number of dormant, living, dead and total seeds decreases (Fig. 10.2). This discrepancy seems to be a result of differences in the sampling regime. The seed bank in Lithuania was sampled over a single season (from spring to autumn), while in the Czech Republic it was over two subsequent seasons (from autumn to summer the following year). Thus, the pattern in total numbers of *H. sosnowskyi* seeds reflects between-year fluctuations in the number and density of flowering plants. This explains why more seeds were found in spring than after seed set in autumn (Fig. 10.2); the seeds present in spring and summer were produced in the previous year.

To avoid the bias caused by the different sequence of sampling times, the percentages of non-dormant, dormant and dead seeds were compared (Fig. 10.3A). The percentage of living seeds of *H. sosnowskyi* in the total seed bank did not change from autumn (59.8%) to spring (60.8%), but decreased to 22.8% in summer. The percentage of non-dormant seeds among living seeds was similar in autumn (89.8%) and spring (88.3%), but decreased to 27.9% in summer (Fig. 10.3B).

The main difference between the species is in the autumn seed bank, when almost all the seeds of *H. mantegazzianum* were dormant and nearly 90% of *H. sosnowskyi* seeds were non-dormant (Fig. 10.3B). It needs to be noted that due to unusual climatic conditions in Lithuania in the year of sampling, ripe seeds were covered by early snow. Thus, the autumn sample was taken after the snow had melted. This short period of wet and cold conditions might have been enough to stratify the seeds and break their dormancy. An easy breaking of seed dormancy in autumn accords well with the laboratory finding that seeds of *H. sosnowskyi* require a shorter period of cold stratification for breaking dormancy than those of *H. mantegazzianum* (see below).

The average density of *H. sosnowskyi* seeds, expressed per m^2 and pooled across localities, was 9025 ± 2229 (mean ± SD) in autumn, 14,881 ± 7217 in spring and 4013 ± 1216 in summer for total seeds, and 5400 ± 3281, 9045 ± 6411 and 1119 ± 889, respectively, for living seeds (Table 10.5). For *H. mantegazzianum*, it was 6719 ± 4119 total seeds in autumn, 4907 ± 2278 in spring and 1301 ± 1036 in summer (Krinke *et al.*, 2005).

Seed bank depletion

Field data on changes in the seasonal dynamics of the seed bank provide important information on the strategy of alien species in terms of population regeneration and competition with native taxa (Van Clef and Stiles, 2001). The results, however, can be biased by factors beyond an investigator's control, such as the seasonal variation in the weather and the fact that the amount of seeds entering the soil is not known precisely. Burial of controlled numbers of seeds and the monitoring of their germination on a fine temporal scale can provide more reliable information on the temporal pattern of seed bank depletion. Moravcová *et al.* (2006) record the fate of buried seeds of

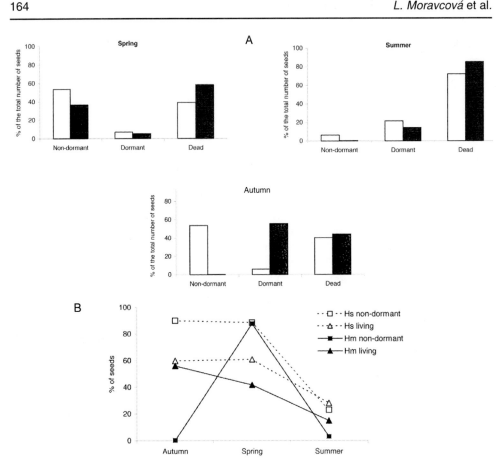

Fig. 10.3. (A) Changes in the percentage of dormant, non-dormant and dead seeds in the seed bank of *H. sosnowskyi* (empty bars) during the course of a year. Samples were taken in autumn (after seed release), spring (before germination) and summer (before new seeds are shed). Mean values shown are pooled across three localities in Lithuania. (B) Changes in the percentage of living seeds that are not dormant, and of the total seeds that are living for *H. sosnowskyi* (empty symbols). The percentage of non-dormant seeds is the same in autumn and spring, and decreases to a low value in summer, after the vast majority of non-dormant seed germinated in spring. The percentage of living seed is highest in spring and lowest in summer. Germinated seeds were considered as non-dormant; non-germinated seeds were tested for viability by staining with tetrazolium; viable seeds were considered as dormant. Corresponding data on *H. mantegazzianum* (Hm; black symbols and bars, respectively) are taken from Krinke *et al.* (2005).

both species in the experimental garden of the Institute of Botany, Průhonice, Czech Republic (50° 00′ 03.9″ N, 14° 33′ 31.7″ E). Seeds of *H. sosnowskyi* were placed in bags made of a fine mesh and buried to a depth of 5–7 cm in the autumn of 2004, those of *H. mantegazzianum* 2 years earlier, and removed at monthly intervals (except during the winter months when the soil was frozen). Living seeds were classified as non-dormant if they germinated

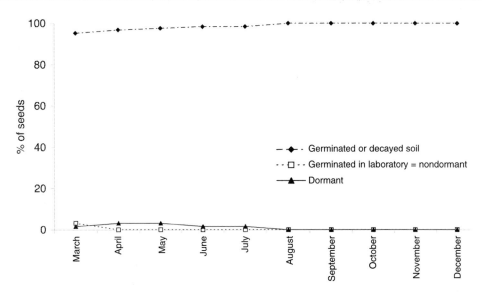

Fig. 10.4. Seasonal pattern in the depletion of the *H. sosnowskyi* seed bank during the course of one season. Percentages of the seeds buried in the experimental garden of the Institute of Botany, Průhonice, Czech Republic that were dormant and non-dormant are shown. Each percentage is based on five replicates. Seeds collected earlier in 2004 were buried at a depth of 5–8 cm at the end of October and followed until December 2005. They were taken from the soil every month, except when the soil was frozen, and those that germinated were recorded; those that did not were tested for dormancy by germinating them in the laboratory at 10/5°C and for viability using tetrazolium.

within 1 month following transfer to a climate chamber, or dormant after testing for viability using tetrazolium staining (Baskin and Baskin, 1998).

Of seeds of *H. sosnowskyi* in the first spring sample (March 2005), 95.2% had already germinated or decayed in the soil. This sample contained a very small proportion of non-dormant and dormant seed, which was ascertained in the laboratory. By May and July, only 3.2% and 1.6%, respectively, of the seeds in the soil had not germinated and all were dormant. From August onwards the soil samples did not contain any living seeds (Fig. 10.4; cf. *H. mantegazzianum*, see Fig. 5.5). These results suggest that seeds of *H. sosnowskyi* are unable to survive for more than one season; the seed bank was very quickly depleted by rapid germination in spring and later on by the rapid decay of dormant seeds.

In contrast, at least a small amount of *H. mantegazzianum* seeds remained viable for considerably longer, a minimum of 3 years (1.2%, see Moravcová *et al.*, Chapter 5, this volume). This may be linked to the fact that a higher percentage of *H. mantegazzianum* seeds is located in soil layers deeper than 5 cm. The difference is rather small (5% compared to 2% in *H. sosnowskyi*), but given the fecundity of both species and the fact that the percentage of seeds that survive is generally very low, it may be important. The more seeds that occur in the lower soil layers, the higher the probability of a persistent seed bank (sensu Thompson *et al.*, 1997).

Further indication that the dynamics of both species' seed banks differ is the easier breaking of dormancy in *H. sosnowskyi*, which enables this species to germinate in autumn when climatic conditions are favourable; however, the survival of seedlings that emerge in autumn and their role in the population dynamics and renewal are unclear; no seedlings were found in the field in Lithuania (Z. Gudžinskas, unpublished). The seed bank of *H. mantegazzianum* is classified as 'short-term persistent' (see Moravcová *et al.*, Chapter 5, this volume; Krinke *et al.*, 2005), but that of *H. sosnowskyi*, based on the results of this study, must be considered to be 'transient' (in sense of Thompson *et al.*, 1997). This is suggested despite the relatively high percentage of dormant seeds in Lithuania in summer, which is a feature of short-term persistent seed banks. Nevertheless, other field results (very high percentage of the seeds in the upper soil layer) together with those obtained in the common garden experiment (seeds lost dormancy very rapidly and did not survive more than one season) and germination studies (seeds germinate even more readily than those of *H. mantegazzianum*) indicate that *H. sosnowskyi* has a transient soil seed bank. However, to verify this, seed bank experiments in those regions of Europe where the species is invasive (see Jahodová *et al.*, Chapter 1, this volume; Nielsen *et al.*, 2005) and the climate is different from Central Europe are needed.

Differences in the Germination Characteristics of *Heracleum mantegazzianum* and *H. sosnowskyi*

To compare the germination of both species, seeds of *H. mantegazzianum* (Moravcová *et al.*, 2006 and Chapter 5, this volume) and *H. sosnowskyi* were germinated at different temperature regimes. Seeds of both species were first stratified for 2 months at temperatures of 4–6°C to break the dormancy and then germinated at different temperatures: 2, 6, 10/5, 20/5, 15/10, 25/10 and 22°C (see Fig. 10.5).

The majority (71–94%) of the seeds of *H. sosnowskyi* germinated almost regardless of the temperature regime. The lowest germination percentages were recorded at 22°C (Fig. 10.5). However, 10% germinated during the stratification process, before setting the temperature for germination. These results suggest that the majority of the seeds of *H. sosnowkyi* break dormancy at almost the same time, following a stratification period as short as or less than 2 months, and immediately germinate, independently of the temperature. This accords with the early spring massive germination of *H. sosnowskyi* seeds, observed in the field in Lithuania as well as in the common garden burial experiment conducted in the Czech Republic. This experiment also showed that the seeds of *H. sosnowskyi* only germinated before March.

This is very different from *H. mantegazzianum*, where dormancy is broken gradually and germination can extend over several years (3 years minimally; cf. Figs 5.6 and 5.7). A study of the germination requirements, using the same design as described here for *H. sosnowskyi*, provided different

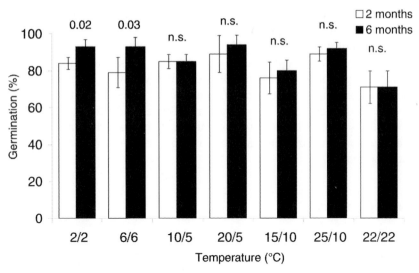

Fig. 10.5. Percentage of *H. sosnowskyi* seeds (mean ± SE) that germinated when subjected to various temperature regimes and under two germination periods. The seeds were cold-stratified for 2 months at 4–6°C prior to the experiment. Seven temperature regimes were used: 2, 6, 10/5, 20/5, 15/10, 25/10 and 22°C (with alternating day/night temperatures; the day and night each lasted 12 hours). Percentage of seeds that germinated was ascertained after 2 and 6 months. Differences between times within the temperature regime are shown above bars (t-test for paired comparisons).

results for *H. mantegazzianum*. The stratification period for seeds of *H. mantegazzianum* needed for breaking dormancy lasts at least 2 months. The highest percentage and fastest germination occurred at alternating temperatures of 20/5°C, with slower but comparably high percentages of germination at 6°C, following stratification at the same temperature. In *H. mantegazzianum*, all germination at the higher temperatures occurred in the first 2 months and then stopped. However, germination at low temperatures of 2°C and 6°C continued (Moravcová *et al.*, 2006 and Chapter 5, this volume). This indicates that the higher temperatures represent a constraint on the breaking of dormancy in *H. mantegazzianum* and seeds remain dormant until the next cold period.

The above results can be interpreted with regard to the distribution limits of both species in Europe. The European distribution of these two *Heracleum* species is distinct, with *H. sosnowskyi* confined to the northern and eastern parts of the continent and *H. mantegazzianum* having the centre of the invaded distribution range in central and western Europe (see Jahodová *et al.*, Chapter 1, this volume). An attempt to elucidate climatic factors that might have played a role in shaping this distribution was made using *H. mantegazzianum* as a model species. Pyšek *et al.* (1998) show that the distribution of this species in the Czech Republic is significantly affected by the temperature in January; the number of reported localities decreases with increasing temperature, but the distribution is also determined by human population density, which reflects possibilities for dispersal.

Although the transient seed bank and immediate germination not extending over a period longer than 1 year may be a disadvantage in terms of the long-term population dynamics (Pyke, 1990; Van Clef and Stiles, 2001), the germination characteristics of *H. sosnowskyi* indicate that invasion by this species is unlikely to be limited by temperature. The current distribution of *H. sosnowskyi* seems to be driven mainly by human activities and the history of introductions, namely the massive planting in eastern Europe, than by ecological constraints.

Acknowledgements

Thanks to Zuzana Sixtová and Vendula Čejková for technical assistance. Tony Dixon kindly improved our English. The study was supported by the project 'Giant Hogweed (*Heracleum mantegazzianum*) a pernicious invasive weed: developing a sustainable strategy for alien invasive plant management in Europe', funded within the 'Energy, Environment and Sustainable Development Programme' (grant no. EVK2-CT-2001-00128) of the European Union 5th Framework Programme, by institutional long-term research plans no. AV0Z60050516 from the Academy of Sciences of the Czech Republic and no. 0021620828 from Ministry of Education of the Czech Republic, and grant no. 206/05/0323 from the Grant Agency of the Czech Republic.

References

Baker, H.G. (1965) Characteristics and modes of origin of weeds. In: Baker, H.G. and Stebbins, G.L. (eds) *The Genetics of Colonizing Species*. Academic Press, New York, pp. 147–172.

Baskin, C.C. and Baskin, J.M. (1998) *Seeds. Ecology, Biogeography and Evolution of Dormancy and Germination*. Academic Press, San Diego, California.

Callaway, J.C. and Josselyn, M.N. (1992) The introduction and spread of smooth cordgrass (*Spartina alterniflora*) in South San Francisco Bay. *Estuaries* 15, 218–225.

Crawley, M.J. (2002) *Statistical Computing: an Introduction to Data Analysis Using S-plus*. Wiley, Chichester, UK.

Dreyer, G.D., Baird, L.M. and Fickler, C. (1987) *Celastrus scandens* and *Celastrus orbiculatus* comparisons of reproductive potential between a native and an introduced woody vine. *Bulletin of the Torrey Botanical Club* 114, 260–264.

Holub, J. (1997) *Heracleum* – hogweed. In: Slavík, B., Chrtek Jr, J. and Tomšovic, P. (eds) *Flora of the Czech Republic 5*. Academia, Praha, pp. 386–395. [In Czech.]

Honig, M.A., Cowling, R.M. and Richardson, D.M. (1992) The invasive potential of Australian banksias in South African fynbos: a comparison of the reproductive potential of *Banksia ericifolia* and *Leucadendron laureolum*. *Australian Journal of Ecology* 17, 305–314.

Krinke, L., Moravcová, L., Pyšek, P., Jarošík, V., Pergl, J. and Perglová, I. (2005) Seed bank of an invasive alien, *Heracleum mantegazzianum*, and its seasonal dynamics. *Seed Science Research* 15, 239–248.

Lambrinos, J.G. (2002) The variable invasive success of *Cortaderia* species in a complex landscape. *Ecology* 83, 518–529.

Mandenova, I.P. (1950) *Caucasian Species of the Genus* Heracleum. Georgian Academy of Sciences, Tbilisi. [In Russian.]

Mihulka, S., Pyšek, P. and Martínková, J. (2003) Invasiveness of *Oenothera* congeners in Europe related to their seed characteristics. In: Child, L.E., Brock, J.H., Brundu, G., Prach K., Pyšek, P., Wade, M. and Williamson, M. (eds) *Plant Invasions: Ecological Threats and Management Solutions*. Backhuys, Leiden, The Netherlands, pp. 213–225.

Moravcová, L., Perglová, I., Pyšek, P., Jarošík, V. and Pergl, J. (2005) Effects of fruit position on fruit mass and seed germination in the alien species *Heracleum mantegazzianum* (*Apiaceae*) and the implication for its invasion. *Acta Oecologica* 28, 1–10.

Moravcová, L., Pyšek, P., Pergl, J., Perglová, I. and Jarošík, V. (2006) Seasonal pattern of germination and seed longevity in the invasive species *Heracleum mantegazzianum*. *Preslia* 78, 287–301.

Nielsen, Ch., Ravn, H.P., Netwig, W. and Wade, M. (eds) (2005) *The Giant Hogweed Best Practice Manual. Guidelines for the Management and Control of Invasive Weeds in Europe*. Forest and Landscape, Hørsholm, Denmark.

Nikolaeva, M.G., Rasumova, M.V. and Gladkova, V.N. (1985) *Reference Book on Dormant Seed Germination*. Nauka, Leningrad. [In Russian.]

Pergl, J., Perglová, I., Pyšek, P. and Dietz, H. (2006) Population age structure and reproductive behavior of the monocarpic perennial *Heracleum mantegazzianum* (*Apiaceae*) in its native and invaded distribution ranges. *American Journal of Botany* 93, 1018–1028.

Pyke, D.A. (1990) Comparative demography of co-occurring introduced and native tussock grasses: persistence and potential expansion. *Oecologia* 82, 537–543.

Pyšek, P., Kopecký, M., Jarošík, V. and Kotková, P. (1998) The role of human density and climate in the spread of *Heracleum mantegazzianum* in the Central European landscape. *Diversity and Distributions* 4, 9–16.

Radford, I.J. and Cousens, R.D. (2000) Invasiveness and comparative life history traits of exotic and indigenous *Senecio* species in Australia. *Oecologia* 125, 531–542.

Richardson, D.M., Van Wilgen, B.W. and Mitchell, D.T. (1987) Aspects of the reproductive ecology of four Australian *Hakea* species (*Proteaceae*) in South Africa. *Oecologia* 71, 345–354.

Satsyperova, I.F. (1984) *Hogweeds in the Flora of the USSR: New Forage Plants*. Nauka, Leningrad. [In Russian.]

Thompson, K., Bakker, J.P. and Bekker, R.M. (1997) *The Soil Seed Bank of North West Europe: Methodology, Density and Longevity*. Cambridge University Press, Cambridge.

Tiley, G.E.D., Dodd, F.S. and Wade, P.M. (1996) Biological flora of the British Isles. 190. *Heracleum mantegazzianum* Sommier et Levier. *Journal of Ecology* 84, 297–319.

Tkachenko, K.G. (1989) Peculiarities and seed productivity in some *Heracleum* species grown in Leningrad area. *Rastitelnye Resursy* 1, 52–61. [In Russian.]

Tkachenko, K.G. and Zenkevich, I.G. (1987) The composition of essential oils from fruits of some species of the genus *Heracleum* L. *Rastitelnye Resursy* 1, 87–91. [In Russian.]

Van Clef, M. and Stiles, E.W. (2001) Seed longevity in three pairs of native and non-native congeners: assessing invasive potential. *Northeastern Naturalist* 8, 301–310.

Vilà, M. and D'Antonio, C.M. (1998) Fruit choice and seed dispersal of invasive vs. noninvasive *Carpobrotus* (*Aizoaceae*) in coastal California. *Ecology* 79, 1053–1060.

11 Herbivorous Arthropods on *Heracleum mantegazzianum* in its Native and Invaded Distribution Range

STEEN O. HANSEN,[1] JAN HATTENDORF,[1] CHARLOTTE NIELSEN,[2] RUEDIGER WITTENBERG[3] AND WOLFGANG NENTWIG[1]

[1]University of Bern, Bern, Switzerland; [2]Royal Veterinary and Agricultural University, Hørsholm, Denmark; [3]CABI Switzerland Centre, Delémont, Switzerland

Nothing can stop them!

(Genesis, 1971)

Introduction

Heracleum mantegazzianum Sommier & Levier (*Apiaceae*) is native to the Western Greater Caucasus, where it occurs in the alpine and subalpine belt, mainly in meadows, clearings and forest margins (Mandenova, 1950). Spatially separated by regions with dryer climates, the plant only reached Western Europe after being introduced into botanical gardens in the 19th century. Meanwhile it has naturalized along waterways and roads and on fallow and disturbed land all over Europe. Its good competitive ability and high seed production make it an aggressive invasive species, thus causing problems for many European authorities, especially in regions where the land use is changing. It is a typical representative of the competitive/ruderal strategy type (Otte and Franke, 1998). There are two main reasons for stopping this weed from spreading further in Europe: (i) the plant affects the structure and function of ecosystems by reducing the biodiversity of communities and landscapes; and (ii) the plant is a toxic public nuisance because its sap causes a serious UV-induced photodermatitis (see Hattendorf *et al.*, Chapter 13, this volume).

More than 350 species of insects have been found on different *Heracleum* species (Hansen *et al.*, 2006a). The majority of these are merely visitors, using this tall plant as a temporary resting place, and do not interact with the plant directly. However, some of these insects do feed on

H. mantegazzianum, but only a few are intimately associated with this plant. Few reports are available on the specialization and impact of insects associated with *H. mantegazzianum* (e.g. Sampson, 1990; Seier *et al.*, 2003) and so the potential of herbivores for biological control of *H. mantegazzianum* is still largely unknown.

In this chapter we analyse the host specificity of these herbivorous insects and test the hypothesis that proportionally more specialist herbivores are found in the native range of *H. mantegazzianum*. The increased competitive ability of non-indigenous plant species is often attributed to the absence of their specialized natural enemies in the invaded range (enemy release hypothesis, Torchin *et al.*, 2001; Mitchell and Power, 2003). Additionally, this analysis will provide a list of associated herbivores and their host range which can be taken into consideration in developing a classical biological control programme (see Cock and Seier, Chapter 16, this volume). We also examine if certain herbivore insect orders have a higher representation in their native region as compared to the invaded region.

It has been tested for single invasive species but rarely for several species, as to whether larger proportions of specialists occur in their native region as compared to the invaded region. Mitchell and Power (2003) demonstrated that the invasiveness of some species is correlated with their release from pathogens. Memmott *et al.* (2000) demonstrated that the biomass of specialist herbivores of *Cytisus scoparius* (*Fabaceae*) is higher in the native region. Wolfe (2002) also confirmed a higher level of attack by herbivores and pathogens on *Silene latifolia* Poiret (*Caryophyllaceae*) in its native range.

Plants continuously develop specific anti-feeding defence systems, such as secondary metabolites, trichomes, thick epidermis etc. Some plant defence systems are particularly effective against some insect guilds. Therefore in this study we also investigated whether insects from certain orders feed specifically on *H. mantegazzianum*. Since the insect orders are not equally represented, it is necessary to compare the number of species on *H. mantegazzianum* with the average (expected) frequency of the herbivore guild. Therefore we compare our results with the worldwide distribution of herbivores and the distribution in the former USSR (for details see Hansen *et al.*, 2006a).

The Herbivorous Guilds on *Heracleum mantegazzianum*

Field surveys on *H. mantegazzianum* were carried out in Belgium, the Czech Republic, Denmark, Germany, the Netherlands, Latvia, Switzerland as well as in the Caucasian areas of Georgia and Russia. From early May to September in 2002 and from May to mid-August in 2003, data from 37 different locations were acquired. Of these, 21 were located in Europe and the other 16 in the Western Greater Caucasus up to 2050 m a.s.l., which is considered to be the native region of *H. mantegazzianum* (Mandenova, 1950; Otte and Franke, 1998; see Jahodová *et al.*, Chapter 1, this volume). Approximately the same amount of time was spent collecting insects in both regions. More detailed descriptions of the methods and results of this chapter can be found in Hansen (2005) and Hattendorf (2005). Data from previous large-scale investigations of herbivores on *H. mantegazzianum* in England (Sampson, 1990), Switzerland and Slovakia (Bürki and Nentwig, 1997) supplemented our data and are included in this study. Additionally, we gathered the scattered information from 161 publications and various insect keys containing species information from Europe and the Caucasus for all *Heracleum* species.

The terms monophagous and oligophagous are not uniformly defined in the literature (e.g. Memmot *et al.*, 2000; Imura, 2003). We are attempting to avoid any such confusion here by using the terms 'species-specific', 'genus-specific' or 'family-specific'. Polyphagous herbivores feed on different families. We have classified the herbivores into these groups according to records from the literature.

Overall we gathered information on 358 insect species occurring on 16 different *Heracleum* species. Of these, 265 were herbivores and were used for analysis. There were 162 herbivore species on *H. mantegazzianum*, of which 123 were polyphagous or had unknown specificity. Some of the genus-specific and family-specific herbivore species are presented in Table 11.1.

We found that the proportion of specialists was significantly dependent on whether sampling was done in the invaded or native range of the host species (Fig. 11.1) but also that the proportion differed significantly among the four invasive plant species, *H. mantegazzianum, Solanum carolinense, Solidago altissima* and *Cytisus scoparius* for both the native and invaded range ($P < 0.001$, $df = 7$). The insects feeding on *S. altissima, S. carolinense* and *C. scoparius* were divided into generalists and specialists by Jobin *et al.* (1996, specialists defined as feeding within genus, $n = 276$), Imura (2003, specialists within family, $n = 57$) and Memmott *et al.* (2000, specialists within tribe Genistea, $n = 42$) respectively. The enemy release hypothesis (Torchin *et al.*, 2001; Mitchell and Power, 2003) predicts that a larger proportion of specialists and/or a higher density or biomass of herbivores should be found in the native Caucasus area, where they would inflict more damage to the host plant. Hattendorf (2005) has also demonstrated that the defence systems (furanocoumarins and trichomes) of the giant hogweed are different in the native region as compared to the invaded regions (see Hattendorf *et al.*, Chapter 13, this volume). Indirectly, this again indicates that the composition of the herbivore species or herbivore biomass on *H. mantegazzianum* differs in the two regions. Of the four mentioned species, this difference between the native and invaded ranges is by far the least with *H. mantegazzianum*. This could be related to the fact that our sampling was the most detailed or that it could be family-specific.

The chewers feeding on different plant organs of *H. mantegazzianum* belong mainly to the Coleoptera and constitute a large feeding guild (41%) (Fig. 11.2). The root borers are also coleopterans. The gall-forming insects belong predominantly to the Cecidomyiidae (Diptera) and leaf miners chiefly involve the Agromyzidae (Diptera) (Fig. 11.2).

The proportions of species in each insect order found on *H. mantegazzianum* in the native area of Caucasus as compared with the invaded part of Europe were not significantly different ($P > 0.3$, $df = 5$). The over-representation of sapsucking hemipteran species on *H. mantegazzianum* in the

Table 11.1. Selected herbivore species found on *H. mantegazzianum* in Europe and in the Caucasus. Systematics and nomenclature are according to *Fauna Europaea* (2005).

Order Family	Species	Feeding specificity	Stage collected[a]	Way of feeding[b]	Plant organ[c]	Locality	Source
Hemiptera							
Pentatomidae	*Graphosoma lineatum* L.	Family	L/A	Sap sucker	Umbel	EU/CAU	Wagner (1966); Jakob *et al.* (1998); Hansen and Hattendorf[e]
Miridae	*Orthops basalis* Costa	Family	A	Sap sucker	Stem, umbel, leaves	EU[oc]	Sampson (1990); Nielsen and Ravn[e]
Aphididae	*Anuraphis subterranea* Walker	Family	L/A	Sap sucker	Leaf envelope	EU/CAU	Bürki and Nentwig (1997); Hansen and Hattendorf[e]
	Cavariella theobaldi Gillette and Bragg	Family	L/A	Disease transmitter, sap sucker	Stem, umbel, leaves	EU/CAU	Sampson (1990); Nielsen and Ravn[e]; Hansen and Hattendorf[e]
	Dysaphis lauberti Börner	Family	L/A	Sap sucker	Stem, umbel, leaves	EU/CAU	Hansen and Hattendorf[e]
	Aphis fabae Scopoli	Polyphagous	L/A	Sap sucker	Umbel, leaves	EU/CAU	Hansen and Hattendorf[e]
	Paramyzus heraclei Börner	Genus	L/A	Disease transmitter, sap sucker	Leaves	EU/CAU	Sampson (1990); Heie (1994); Hansen and Hattendorf[e]
Coleoptera							
Cerambycidae	*Phytoecia boeberi* Ganglbauer		A	Root borer, leaf chewer	Stem, leaves	CAU	Hansen and Hattendorf[e]
Curculionidae	*Liophloeus tessulatus* Müller	Family	L/A	Root borer, leaf chewer	Stem, leaves, root	EU[oc]	Bürki and Nentwig (1997); Hansen and Hattendorf[e]
	Lixus iridis Olivier	Family	L/A	Stem borer, leaf chewer	Stem, leaves	EU/CAU	Hansen and Hattendorf[e]
	Otiorhynchus tatarchani Reitter		A	Root borer, leaf chewer	Root, leaves	CAU	Hansen and Hattendorf[e]
	Nastus fausti Reitter		L/A	Root borer, leaf chewer	Root, stem, leaves	CAU	Hansen and Hattendorf[e]

Table 11.1. *Continued*

Order Family	Species	Feeding specificity	Stage collected[a]	Way of feeding[b]	Plant organ[c]	Locality	Source
Lepidoptera							
Epermeniidae	*Epermenia chaerophyllella* Goeze	Family	E/L/P	Leaf miner	Leaves	EU[OC]	Sampson (1990); Emmet (1996)
Noctuidae	*Dasypolia templi* Thunberg	Family	L	Chewing	Root, stem, umbel, leaves	CAU[OE]	Seppänen (1970); Hansen and Hattendorf[e]
Depressariidae	*Depressaria radiella* Goeze	Family	L/P	Chewing	Umbel	EU/CAU	Sampson (1990); Bürki and Nentwig (1997); Hansen and Hattendorf[e]
	Agonopterix heracleana L.	Family	L	Leaf roller, umbel chewing	Leaves	EU/CAU	Emmet (1979); Sampson (1990); Hansen and Hattendorf[e]; Karsholt *et al.* (2006)
	Agonopterix caucasiella Karsholt		L/P	Umbel chewing			
Tortricidae	*Pammene (Cydia) gallicana* Guenée	Family	L	Chewing	Umbel	EU[OC]	Emmet (1979); Sampson (1990)
	Pammene aurana F.	Family	L	Chewing	Umbel	EU	Sampson (1990); Harper *et al.* (2002)
Diptera							
Tephritidae	*Euleia heraclei* L.	Family	L	Leaf miner, chewing	Leaves	EU[OC]	Sampson (1990); Hansen and Hattendorf[e]
Agromyzidae	*Melanagromyza angeliciphaga* Spencer	Family	L/P/A	Stem borer	Stem	EU/CAU	Spencer (1972); Bürki and Nentwig (1997); Jakob *et al.* (1998); Hansen and Hattendorf[e]
	Melanagromyza heracleana Zlobin	Family	L/P/A	Stem borer	Stem	CAU	Hansen and Hattendorf[e]

| | *Phytomyza spondylii* Goureau | Family | L/P | Leaf miner | Leaves | EU[OC] | Ashwood-Smith *et al.* (1984); Bürki and Nentwig (1997); Sampson (1990); Nielsen and Ravn[e] |
| Psilidae | *Psila rosae* F. | Family | L | Root borer | Root | EU[OC] | Hardman and Ellis (1982); Nielsen and Ravn[e]; Hansen and Hattendorf[e] |

[a] Stages collected: E = eggs, L = larvae, P = pupae, A = adults.
[b] Genus = feeds only on *Heracleum* spp., family = feeds on *Apiaceae*, polyphagous = feeds on several plant families.
[c] Plant organ: umbel = feeding on seeds and flower stalks but not on pollen and nectar.
[d] The locality, where the species is found, is noted as Europe = EU and Caucasus = CAU; EU/CAU = found in both regions. EU does not mean that this species is not occurring in the Caucasus, but just that it had not been found so far on *H. mantegazzianum* in the Caucasus.
OC = Occurs in Caucasus, but so far not found on *H. mantegazzianum*.
OE = Occurs in Europe, but so far not found on *H. mantegazzianum*.
[e] Collected during field trips in 2002 in the Caucasus, unpublished.

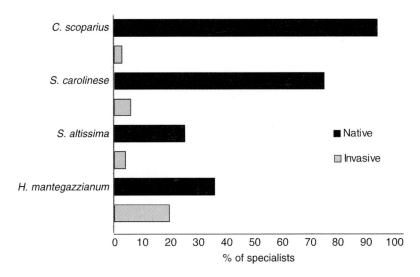

Fig. 11.1. Specialist herbivore species on invasive plants make up a significantly higher proportion of the herbivores in the native areas compared to the invaded areas ($P < 0.001$, $df = 7$) and significantly different for each of the four invasive plant species ($P < 0.001$, $df = 7$). Data for *Solidago altissima* L. were obtained from Jobin *et al.* (1996, $n = 276$), for *Solanum carolinense* L. from Imura (2003, $n = 57$), and for *Cytisus scoparius* L. from Memmott *et al.* (2000, $n = 42$).

invaded region (Fig. 11.3) and the under-representation of Lepidoptera and Diptera has also been found in comparable studies of other invasive plants (Imura, 2003; Simberloff, 2003). There are various ways in which over- and under-representations of some insect orders could have developed. Some insect groups are known to contain more generalist feeders, which could then be responsible for the observed differences. Dipteran and hymenopteran herbivores generally are comprised of a larger proportion of specialists than Orthoptera and Coleoptera, for example. Another explanation could be that certain plant defence systems (e.g. furanocoumarins) are particularly active against certain insect orders on *H. mantegazzianum*. A third explanation is that the herbivore evolutionary adaptation and speciation on the relatively recently evolved species of the genus *Heracleum* is happening faster in some insect orders than in others (see Cock and Seier, Chapter 16, this volume).

Insect Species on *Heracleum mantegazzianum*

Hemiptera

The hemipteran sap suckers represent about one-third of the specialized species on giant hogweed and are usually adapted to specific plant organs.

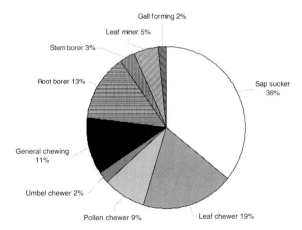

Fig. 11.2. The feeding habits (feeding guilds) of the 265 insect species found on *Heracleum* spp. Multiple entries are allowed as the larvae sometimes feed on different organs than the adults. Chewing insects feed on external plant organs and umbel chewers represent the insects chewing on the seeds or the flower stalks.

While aphids (Aphididae) such as *Paramyzus heraclei* Börner and *Cavariella theobaldi* Gillette & Brag typically feed on the leaves, *Aphis fabae* Scopoli sucks the umbels and *Anuraphis subterranea* Walker sucks from within the leaf envelope at the base of the stem. *A. subterranea* has specialized in this niche and the unwinged viviparous have a very long rostrum (both absolute and relative to the body length, 0.7 × body length; Heie, 1992) which

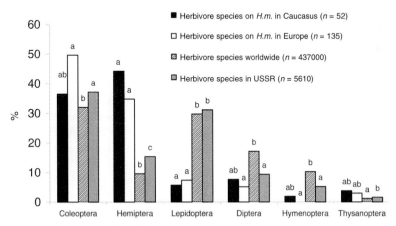

Fig. 11.3. Percentage of generalists and specialist herbivores on *H. mantegazzianum* (*H.m.*) belonging to different insect orders. Different letters above the columns for the same insect order refer to a significant difference in a 2 × 2 contingency test (*P* < 0.05). Data on the worldwide number of herbivore species according to Bernays (2003); data on herbivores on beneficial plants in the former USSR according to Kryzhanovskij (1972–1974), Narchuk and Tryapitzin (1981) and Kuznetsov (1994–1999).

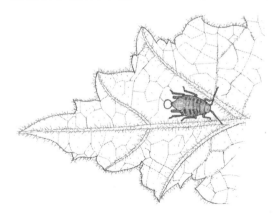

Fig. 11.4. When the green aphids *Cavariella* spp. or *Paramyzus heraclei* feed underneath the leaves of giant hogweed, they are prevented from direct feeding on the leaf veins by numerous trichomes. Drawing: S.O. Hansen.

allows penetration of the thick epidermis of the stem base of *H. mantegaz-zianum*. The leaf-feeding *C. theobaldi* and *P. heraclei* have a rostrum only half this size.

P. heraclei is yellowish green and monoecious and therefore completes its life cycle on only one host plant. *P. heraclei* is often responsible for the yellow spots on the leaves, which are probably caused by the parsnip yellow fleck virus (PYFV) (Bem and Murant, 1979) which in combination with aphid attacks killed seedlings, but only under laboratory conditions (Hansen, 2005). The plants were not killed under natural conditions, but plant growth was affected (see Seier and Evans, Chapter 12, this volume).

C. theobaldi is occasionally observed in high densities, and in late summer it often accumulates on the developing umbels. Tiny trichomes on the leaf veins prevent the insects from feeding there and consequently they keep a distance from the veins (Fig. 11.4). From April to June, *C. theobaldi* migrates from its primary host *Salix* spp. to its secondary hosts in the *Apiaceae* family (*Pastinaca* and *Heracleum* species). They migrate back in September to October when many of the above ground plant parts gradually disappear. *C. theobaldi* is also a vector for PYFV and five other viruses (Bem and Murant, 1979), none of which ever killed plants of *H. mantegazzianum* under natural conditions during our investigations.

Anuraphis subterranea is a reddish brown aphid. Its primary host is *Pyrus communis* and the secondary hosts are *Pastinaca* and *Heracleum*. On *Heracleum*, *A. subterranea* is associated with different ant species, which build a soil layer to protect the aphids inside the leaf envelope. These soil shelters prevent aphidophagous ladybirds and possibly also other predators from attacking the colonies, ameliorate the microclimate for ants and aphids, and prevent the envelopes from getting flooded (Hansen, 2005).

Aphis fabae is highly polyphagous and heteroecious, i.e. its life cycle needs host-alternation. However, it was among the arthropods that caused the greatest losses in yield of biomass, seed yield and germination capacity of a close relative species, *Heracleum sosnowskyi* Manden, which was cultivated in Poland as a fodder plant (Jurek, 1989, 1990).

Among heteropterans, *Orthops basalis* Costa (Miridae) and *Graphosoma lineatum* L. (Pentatomidae) both feed on the umbels of giant hogweed and other *Apiaceae*, where they also hibernate as imago (Wagner, 1966, 1967). Both the seeds and the umbellet stalks are consumed. Throughout most of the season the female *G. lineatum* lays its yellow eggs in clusters underneath the leaves. Experiments have demonstrated that this species can be mass-reared all the year round (Nakamura *et al.*, 1996). Severe infestations by *O. basalis* were seen on *H. sosnowskyi* in many regions of Latvia in September 2003, but this mirid also attacks other *Apiaceae* plants such as *Foeniculum vulgare* Mill. (Ulenberg *et al.*, 1986).

Lepidoptera

Among lepidopterans which are associated with the genus *Heracleum*, *Apiaceae* specialists include the stem borers in the Noctuidae, Epermeniidae seed and leaf miners, and leaf, flower and seed feeders in the Depressariidae (Berenbaum, 1990).

Among the Noctuidae, *Dasypolia templi* is distributed across Europe and in the Caucasus (Skou, 1991; Hansen *et al.*, 2006a) occurring in many different habitats, but especially in open land (Palm, 1989) and often in agricultural fields. The larvae hatch in May and start feeding on the flowers, later on the stems and roots of *Heracleum* and *Angelica* species, *Levisticum officinale* Koch and *Aegopodium podagraria* L. (Seppänen, 1970; Skou, 1991). Mating takes place in autumn and only the females overwinter. This species has a high feeding capacity but is not specific to *H. mantegazzianum*.

In the depressariid family, two genera contain several species which can potentially exert a high impact on *H. mantegazzianum*. *Depressaria* species tend to concentrate on the reproductive parts, although foliage feeding is known to occur as well (Berenbaum, 1990; Harper *et al.*, 2002). The degree of feeding specificity varies among the *Agonopterix* species. They are primarily leaf rollers, although flower and seed feeding is not uncommon (Berenbaum, 1990; Harper *et al.*, 2002).

Depressaria radiella Goeze, the parsnip webworm, feeds on the developing flowers and seeds of *Heracleum* spp., *Pastinaca* spp. and *Angelica* spp. (Palm, 1989) and occurs both in Europe and in the native area of *H. mantegazzianum* in the Western Greater Caucasus (Sampson, 1990; Bürki and Nentwig, 1997; Seier *et al.*, 2003; Hansen *et al.*, 2006a). Usually in the literature it is referred to as *Depressaria pastinacella* Duponchel. However, according to Karsholt *et al.* (2006), the valid name is *D. radiella*. Because of the ability of *D. radiella* to detoxify furanocoumarins, the caterpillar is able to tolerate high concentrations of these compounds in its diet (Berenbaum and Zangerl, 1992; Nitao *et al.*, 2003). Female moths oviposit on unopened umbels or leaf surfaces from mid May to early June (Thompson, 1978; Gorder and Mertins, 1984). The larvae hatch and start feeding on buds, developing flowers and fruits after producing a web in the occupied umbellet (Hendrix, 1979, 1984). Mature larvae exit the web and, if the stem is sufficiently large, bore into it to pupate. When the stem is unsuitable, the larvae pupate in the soil. Adults emerge 2 weeks later and overwinter until the following spring (Hendrix, 1984; Sampson, 1990).

Herbivory by the parsnip webworm may decrease the plant's reproductive success not only by the consumption of seeds, but also by reducing the rates of pollinator visitation (Lohman *et al.*, 1995, 1996). Based on field observations, large populations of the moth would be needed to control *H. mantegazzianum* since small numbers of larvae could only have a limited impact on the huge umbel production. According to Ode *et al.* (2004) the average attack rate in the Netherlands was about 30% of the plants, with an average of 1.6 individuals of *D. radiella* per plant. The mean number of webworms on *H. sphondylium* was eight times higher. In Europe, webworm populations are subject to high rates of mortality by the parasitoid *Copidosoma sosares* Walker (Hymenoptera: Encyrtidae). The parasitized individuals grow to more than double the size of normal individuals and increased feeding may reduce the seed production even further.

Remarkably similar to the life history of *D. radiella*, the larvae and pupae (Fig. 11.5) of a recently described species of the genus *Agonopterix* were found feeding on the flowering umbels of *H. mantegazzianum* at a single location in the Russian part of the Caucasus Mountains in 2003. The scientific name for this species is *Agonopterix caucasiella* (Karsholt *et al.*, 2006). The larvae of *A. caucasiella* feed on the reproductive parts of the plant and in

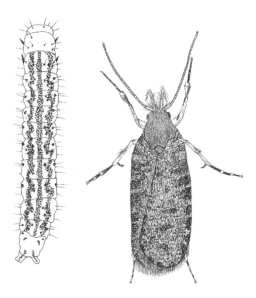

Fig. 11.5. The 6–14 mm long larva of *Agonopterix caucasiella* feeds inside a web of silk in the flower umbels of giant hogweed, protected from predators. Drawing: S.O. Hansen.

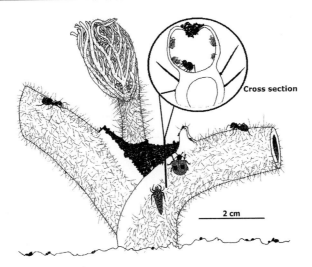

Fig. 11.6. A domatium (leaf envelope) is a hollow cavity at the stem base of 2–3-year-old *H. mantegazzianum* plants, sheltering colonies of aphids (*Anuraphis subterranea*) and ants (*Lasius niger*). The domatium is curved inwards (see cross section) and ants construct soil shelters on top of the domatium (leaf envelope, left side) when aphids are inside. Soil shelters block the entry of aphid predators like *Adalia bipunctata* (Coccinellidae). From Hansen (2005).

some cases a web was constructed around a portion of the umbellet in which the larvae fed. Larval attack was related neither to size of the umbel nor to the plant size. The habitats of *A. caucasiella* include riverbanks, abandoned fields, mountain slopes and forest clearings. Further details on morphology, systematics and nomenclature of the moth are presented in Karsholt *et al.* (2006). The biology and specialization of *A. caucasiella* need further investigations, since other *Agonopterix* species have already been proved to be suitable biocontrol agents (Julien and Griffiths, 1998).

Hymenoptera

Only a few species of herbivore hymenopterans have been found on giant hogweed (Fig. 11.3). *Tenthredo* species (Tenthredinidae) were found only in small numbers, but in contrast high numbers of different ant species were found associated with the colonies of aphids on giant hogweed. Some of these mutualistic associations appear to be nearly obligate and are, therefore, very specific. *Heracleum mantegazzianum* possesses inflated leaf envelopes at the base of the stem and these hollow cavities are perfectly suited for insects to hide inside. The aphid *Anuraphis subterranea* is continually feeding in this location, and the ant *Lasius niger* (Formicidae) seals the cavity by constructing soil shelters on top of the envelope to protect their honeydew source (Fig. 11.6). When the aphid *A. subterranea* is tended by *L. niger* on

H. mantegazzianum, this leads to increased plant growth (Hansen, 2005; Hansen *et al.*, 2006b) because other non-myrmecophilic aphid species are deterred by the ants. It is believed that such myrmecodomatia could have developed in many plant species during the co-evolution of ants and trophobionts (myrmecophilic aphids) (e.g. Offenberg, 2000). This ant–aphid relationship on giant hogweed is therefore a fascinating example of an initial stage in the evolution of the myrmecodomatia which are also frequently encountered in the tropics.

Coleoptera

Among beetles, most herbivores belong to the curculionid family. The distinctively elongated *Lixus iridis* Olivier (Curculionidae) (Fig. 11.7) feeds on the leaves and the stem of giant hogweed and other *Apiaceae* from the genera *Angelica*, *Anthriscus*, *Petroselinum*, *Daucus* and *Carum* (Hoffmann, 1954; Freude *et al.*, 1983). For oviposition, the female first drills a hole into the stem with its mouthparts, then inserts its ovipositor and lays a few eggs into the hollow stem. After hatching, the larvae feed and develop inside the stem until autumn. After the pupation, the adult beetles hatch and hibernate there (Volovik, 1988). *L. iridis* is quite common at lower altitudes of the Caucasus Mountains, but is absent at high altitudes (Hattendorf, 2005). In central Europe, this species occurs only sporadically. With moderate infestation rates, the beetles did not cause significant damage to *H. mantegazzianum* (Hattendorf, 2005). However, two studies have reported damage caused by *L. iridis* to *Levisticum officinale* (Eichler, 1951) and to *H. sosnowskyi* (Volovik, 1988), but in neither case was the degree of damage specified.

   Nastus fausti Reitter (Curculionidae) occurs in the mid mountainous forests in the alpine and subalpine zones from easternmost Europe and the Caucasus to North Ossetia and Transcaucasia. This species does not occur in western Europe. The larva feeds externally on the roots of several plant families. Since the beetle damages the roots, its impact on the plant is difficult to estimate. According to Arzanov and Davidyan (1996), little is known about

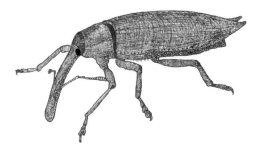

Fig. 11.7. *Lixus iridis* is 16–22 mm long and one of the largest herbivorous insects on giant hogweed. The adult feeds on the leaves and its larva develops inside the hollow stem until it pupates. Drawing: S.O. Hansen.

host plant preferences, but the species from this genus are said to be non-specific. Adult beetles are widespread in the forest belt of the Caucasus Mountains and were repeatedly found on leaves of various *Heracleum* species (Arzanov and Davidyan, 1996).

Two other curculionids are found regularly on *H. mantegazzianum*, but these also feed on other *Apiaceae* and related families. The larva of *Liophleus tessulatus* Müller feeds externally on the roots of *H. mantegazzianum*, but also on plants from other families. The adults are 7–12 mm long, feed on leaves (Sheppard, 1991) and are active during the night. *Otiorhynchus tatarchani* Reitter is polyphagous on many bush and tree species (Kryzhanovskij, 1974) and has root-feeding larvae. This species is found at altitudes of up to 1750 m a.s.l. *L. tessulatus* and *O. tartarchani* are able to breed parthenogenetically (Harde and Severa, 1981). The former species is present in western Europe, but the latter does not occur there.

The larvae of *Phytoecia boeberi* Ganglbauer (Cerambycidae) develop inside the stems or roots of giant hogweed. It has unknown host plant specificity, but other species from this genus are family-specific. We only found it in the Caucasus Mountains at the altitude of 1750 m a.s.l., where it was rather common in some localities. The females normally lay a single egg on the surface of a stem. The life cycle takes 1–2 years, with both adults and larvae overwintering inside the stems.

Diptera

Only a small proportion of dipterans are herbivorous. Most agromyzids are family-specific, and the larvae often feed on one or several plant genera belonging to the same plant family. *Melanagromyza heracleana* Zlobin is a species which was recently found on *H. mantegazzianum* in the Caucasus region and was new to science (Zlobin, 2005). The stem-feeding larvae are very common in the Caucasus but do not seem to have a negative impact on *H. mantegazzianum* (Hattendorf, 2005). It is possible that this species is not specialized in feeding only on giant hogweed, even if it has so far only been found on this species. Feeding larvae of this species generate white mines inside the hollow stem, a few millimetres wide (Fig. 11.8). This small fly develops slowly and only one generation per year is produced. The pupa overwinters inside the stem. The close relative *Melanagromyza angeliciphaga* Spencer is found on giant hogweed, but on other genera of *Apiaceae* as well (Spencer, 1972, 1976). Most impressive are the feeding traces of the agromyzid *Phytomyza sphondylii* Goureau, which makes leaf mines of about 2 cm wide and 7–10 cm long. Its larvae develop into shiny black pupae from May to June and July to September. Despite the fact that these mines are rather large, *P. sphondylii* is responsible for less than 1% damage to giant hogweed in England (Spencer, 1972, 1976; Sampson, 1990).

The 6–7 mm long larva of the carrot fly *Psila rosae* (Psilidae) is a root miner, found often on giant hogweed, but it also feeds on other *Apiaceae* (Sheppard,

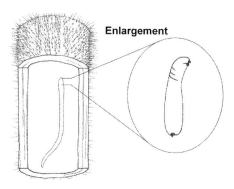

Enlargement

Fig. 11.8. The ~5 mm large larva of *Melanagromyza heracleana* mines inside the hollow stem of giant hogweed, leaving long white traces (2 mm wide), visible when the stem is sliced open. Drawing: S.O. Hansen.

1991). It hibernates as a larva and has 2–3 generations per year. Particularly the second generation is known to cause great damage to crops such as carrots. When 132 apiacean species and subspecies were tested against the carrot fly, 78 of them were proved to be new host plants, 27 were confirmed as hosts and 27 failed to support carrot flies (Hardman and Ellis, 1982).

Other herbivores

Some thrips (Thysanoptera) have been found sucking the plant cells on giant hogweed (Table 11.1). Occasionally they occur in high densities, primarily on the umbels but also the leaves. All encountered thysanopterans have been found to be polyphagous species. Among non-insect herbivores, slugs and (a few) snails (Gastropoda) were found to be regularly feeding on the ground leaves of *H. mantegazzianum*. Although they consumed a considerable amount of leaf biomass, their polyphagy classifies them as non-suitable for giant hogweed control.

Conclusions

During our own surveys we found only very few additional species over and above what has been previously made known. Considering the extensive insect collections we have made so far and the large part of the distribution range of giant hogweed we covered, it is realistic to assume that most herbivore species should have been collected. In the Caucasus, we did not find any insects feeding exclusively on *H. mantegazzianum* (Table 11.1) and this result could be called disappointing from the point of view of aiming at biological control. However, in the Caucasus we found herbivores with unknown host specificity: *Phytoecia boeberi*, *Melanagromyza heracleana* and *Agonopterix caucasiella*, two of which were so far unknown to science. If their feeding specificity turns out to be restricted to *H. mantegazzianum*, then more investigations would be worthwhile to determine their suitability for a biological control programme.

Acknowledgements

We thank Ole Karsholt, Zoological Museum, University of Copenhagen, for providing comments and advice on Lepidoptera taxonomy and nomenclature. We acknowledge valuable comments on an earlier version of this contribution by two referees and language editing by Cecily Klingler. The study was supported by the project 'Giant Hogweed (*Heracleum mantegazzianum*) a pernicious invasive weed: developing a sustainable strategy for alien invasive plant management in Europe', funded within the 'Energy, Environment and Sustainable Development Programme' (grant no. EVK2-CT-2001-00128) of the European Union 5th Framework Programme/Switzerland (BBW: EVK2-CT-2001-00128).

References

Arzanov, Yu.G. and Davidyan, G.E. (1996) Review of the genus *Nastus* Schoenherr (Coleoptera, Curculionidae) of the fauna of Crimea, European Russia and the Caucasus. *Entomological Review* 75, 134–153.

Ashwood-Smith, M.J., Ring, R.A., Liu, M., Phillips, S. and Wilson, M. (1984) Furanocoumarin resistance in the larvae of *Phytomyza spondylii* (Diptera: Agromyzidae) feeding on *Heracleum lanatum* is associated with the enzymatic breakdown of 8-methoxypsoralen. *Canadian Journal of Zoology* 62, 1971–1976.

Bem, F. and Murant, A.F. (1979) Transmission and differentiation of six viruses infecting hogweed (*Heracleum sphondylium*) in Scotland. *Annals of Applied Biology* 92, 237–243.

Berenbaum, M.R. (1990) Evolution of specialization in insect–umbellifer associations. *Annual Review of Entomology* 35, 319–343.

Berenbaum, M.R. and Zangerl, A.R. (1992) Genetics of physiological and behavioral resistance to host furanocoumarins in the parsnip webworm. *Evolution* 46, 1373–1384.

Bernays, E.A. (2003) Phytophagous insects. In: Resh, V.H. and Cardé, R.T. (eds) *Encyclopedia of Insects.* Academic Press, San Diego, California, pp. 902–205.

Bürki, C. and Nentwig, W. (1997) Comparison of herbivore insect communities of *Heracleum sphondylium* and *H. mantegazzianum* in Switzerland (*Spermatophyta: Apiaceae*). *Entomologia Generalis* 22, 147–155.

Eichler, W. (1951) Entwicklung und Lebenszyklus von *Lixus iridis. Entomologische Blätter für Biologie und Systematik der Käfer* 47, 87–95.

Emmet, A.M. (1979) *A Field Guide to the Smaller British Lepidoptera.* British Entomological and Natural History Society, London.

Emmet, A.M. (1996) *The Moths and Butterflies of Great Britain and Ireland. Volume 3. Yponomeutidae – Elachistidae.* Harley Books, Colchester, UK.

Fauna Europaea (2005) *Fauna Europaea. Version 2.1.* http://www.faunaeur.org (accessed 25 October 2005).

Freude, H., Harde, K.W. and Lohse, G.A. (1983) *Die Käfer Mitteleuropas*. Band 11. Goecke Evers, Krefeld, Germany.

Gorder, N.K.N. and Mertins, J.W. (1984) Life-history of the parsnip webworm, *Depressaria pastinacella* (Lepidoptera, Oecophoridae), in Central Iowa. *Annals of the Entomological Society of America* 77, 568–573.

Hansen, S.O. (2005) Interactions between the invasive weed *Heracleum mantegazzianum* and associated insects. PhD thesis, University of Bern, Switzerland.

Hansen, S.O., Hattendorf, J., Wittenberg, R., Reznik, S.Ya., Nielsen, C., Ravn, H.P. and Nentwig, W. (2006a) Phytophagous insect fauna of the Giant Hogweed *Heracleum mantegazzianum* in invaded areas of Europe and in its native area of the Caucasus. *European Journal of Entomology* 103, 387–395.

Hansen, S.O., Hattendorf, J. and Nentwig, W. (2006b) Mutualistic relationship beneficial for aphids and ants on giant hogweed (*Heracleum mantegazzianum*). *Community Ecology* 7, 43–52.

Harde, K.W. and Severa, F. (1981) *Der Kosmos Käferführer. Die mitteleuropäischen Käfer*. Franckh-Kosmos Verlags-GmbH & Co., Stuttgart, Germany.

Hardman, J.A. and Ellis, P.R. (1982) An investigation of the host range of the carrot fly, *Psila rosae*. *Annals of Applied Biology* 100, 1–9.

Harper, M.W., Langmaid, J.R. and Emmet, A.M. (2002) Oecophoridae. In: Emmet, A.M. and Langmaid, J.R. (eds) *The Moths and Butterflies of Great Britain and Ireland*. Harley Books, Colchester, UK, pp. 43–177.

Hattendorf, J. (2005) Impact of endophagous herbivores on the invasive weed *Heracleum mantegazzianum* and associated interactions. PhD thesis. University of Bern, Switzerland.

Heie, O.E. (1992) The Aphidoidea (Hemiptera) of Fennoscandia and Denmark. IV. Family Aphididae: Part 1 of Tribe Macrosiphini of Subfamily Aphidinae. In: *Fauna Entomologica Scandinavica*, Vol. 25. E. J. Brill/Scandinavian Science Press, Leiden, The Netherlands.

Heie, O.E. (1994) The Aphidoidea (Hemiptera) of Fennoscandia and Denmark. V. Family Aphididae: Part 2 of Tribe Macrosiphini of Subfamily Aphidinae. In: *Fauna Entomologica Scandinavica*, Vol. 28. E. J. Brill/Scandinavian Science Press, Leiden, The Netherlands.

Hendrix, S.D. (1979) Compensatory reproduction in a biennial herb *Pastinaca sativa* following insect defloration. *Oecologia* 42, 107–118.

Hendrix, S.D. (1984) Reactions of *Heracleum lanatum* to floral herbivory by *Depressaria pastinacella*. *Ecology* 65, 191–197.

Hoffmann, A. (1954) Coléoptéres Curculionides. *Faune de France* 59, 487–1208.

Imura, O. (2003) Herbivorous arthropod community of an alien weed *Solanum carolinense* L. *Applied Entomology and Zoology* 38, 293–300.

Jakob, B., Mandach, T.V. and Nentwig, W. (1998) Phytophage an *Heracleum mantegazzianum* und *Heracleum sphondylium*. *Zeitschrift für Pflanzenkrankheiten und Pflanzenschutz* 16, 223–230.

Jobin, A., Schaffner, U. and Nentwig, W. (1996) The structure of the phytophagous fauna on the introduced weed *Solidago altissima* in Switzerland. *Entomologia Experimentalis et Applicata* 79, 33–42.

Julien, M.H. and Griffiths, M.W. (1998) *Biological Control of Weeds. A World Catalogue of Agents and their Target Weeds*. CABI, Wallingford, UK.

Jurek, M. (1989) Effect of phytophages on the green matter yield of cow parsnip (*Heracleum sosnowskyi* Manden). *Biuletyn Instytutu Hodowli i Aklimatyzacji Roslin* 169, 85–94. [In Polish.]

Jurek, M. (1990) The effect of phytophages on seed weight and germination capacity of cow parsnip (*Heracleum sosnowskyi* Manden). *Biuletyn Instytutu Hodowli i Aklimatyzacji Roslin* 173–174, 117–124. [In Polish.]

Karsholt, O., Lvovsky, A. and Nielsen, C. (2006) A new species of *Agonopterix* feeding on the

Giant Hogweed in the Caucasus, with a discussion of the nomenclature of *A. heracliana* (Linnaeus) (Depressariidae). *Nota Lepidopterologica* 28, 177–192.

Kryzhanovskij, O.L. (1972–1974) *Insects and Mites: Pests of Agricultural Plants.* Vol. 1 (1972) and Vol. 2 (1974). Nauka, Leningrad. [In Russian.]

Kuznetzov, V.I. (1994–1999) *Insects and Mites: Pests of Agricultural Plants.* Vol. 3/1 (1994) and Vol. 3/2 (1999). Nauka, St Petersburg. [In Russian.]

Lohman, D.J., Zangerl, A. and Berenbaum, M. (1995) Impact of parsnip webworm on pollination ecology and fitness of *Pastinaca sativa* L. *(Apiaceae)*. *Bulletin of the Entomological Society of America* 76, 160.

Lohman, D.J., Zangerl, A.R. and Berenbaum, M.R. (1996) Impact of floral herbivory by parsnip webworm (Oecophoridae: *Depressaria pastinacella* Duponchel) on pollination and fitness of wild parsnip (*Apiaceae: Pastinaca sativa* L). *American Midland Naturalist* 136, 407–412.

Mandenova, I.P. (1950) *Caucasian Species of the Genus Heracleum.* Georgian Academy of Sciences, Tbilisi. [In Russian.]

Memmott, J., Fowler, S.V., Paynter, Q., Sheppard, A.W. and Syrett, P. (2000) The invertebrate fauna on broom, *Cytisus scoparius*, in two native and two exotic habitats. *Acta Oecologica* 21, 213–222.

Mitchell, C.E. and Power, A.G. (2003) Release of invasive plants from fungal and viral pathogens. *Nature* 421, 625–628.

Nakamura, K., Hodek, I. and Hodková, M. (1996) Recurrent photoperiod response in *Graphosoma lineatum* (L.). *European Journal of Entomology* 63, 518–523.

Narchuk, E.P. and Tryapitzin, V.A. (1981) *Insects and Mites: Pests of Agricultural Plants.* Vol. 4. Nauka, Leningrad. [In Russian.]

Nitao, J.K., Berhow, M., Duval, S.M., Weisleder, D., Vaughn, S.F., Zangerl, A. and Berenbaum, M.R. (2003) Characterization of furanocoumarin metabolites in parsnip webworm, *Depressaria pastinacella. Journal of Chemical Ecology* 29, 671–682.

Ode, P.J., Berenbaum, M.R., Zangerl, A.R. and Hardy, I.C.W. (2004) Host plant, host plant chemistry and the polyembryonic parasitoid *Copidosoma sosares*: indirect effects in a tritrophic interaction. *Oikos* 104, 388–400.

Offenberg, J. (2000) Correlated evolution of the association between aphids and ants and the association between aphids and plants with extrafloral nectaries. *Oikos* 91, 146–152.

Otte, A. and Franke, R. (1998) The ecology of the Caucasian herbaceous perennial *Heracleum mantegazzianum* Somm. et Lev. (giant hogweed) in cultural ecosystems of Central Europe. *Phytocoenologia* 28, 205–232.

Palm, E. (1989) *The Oecophorid Moths of Northern Europe (Lepidoptera: Oecophoridae) – with Special Reference to the Danish Fauna.* Fauna Books, Copenhagen. [In Danish.]

Sampson, C. (1990) Towards biological control of *Heracleum mantegazzianum* (Giant Hogweed), *Umbelliferae*. MSc thesis, Imperial College of Science, University of London.

Seier, M.K., Wittenberg, R., Ellison, C.A., Djeddour, D.H. and Evans, H.C. (2003) Surveys for natural enemies of giant hogweed (*Heracleum mantegazzianum*) in the Caucasus Region and assessment for their classical biological control potential in Europe. In: Cullen, J.M., Briese, D.T., Kriticos, D.J., Lonsdale, W.M., Morin, L. and Scott, J.K. (eds) *Proceedings of XI. International Symposium on Biological Control of Weeds*, CSIRO Entomology, Canberra. pp. 149–154.

Seppänen, E.J. (1970) *The Food-plants of the Larvae of the Macrolepidoptera of Finland.* In: Animalia Fennica 14. Werner Söderström Osakeyhtiö.

Sheppard, A.W. (1991) Biological flora of the British Isles. *Heracleum sphondylium. Journal of Ecology* 79, 235–258.

Simberloff, D. (2003) Introduced insects. In: Resh, V.H. and Cardé, R.T. (eds) *Encyclopedia of Insects.* Academic Press, San Diego, California, pp. 597–602.

Skou, P. (1991) *Handbook on the Noctuid Moths of Denmark, Norway, Sweden, Finland and Iceland*. Animal Life of Denmark, Vol. 5. Apollo Books, Stenstrup, Denmark.

Spencer, K.A. (1972) Diptera: Agromyzidae. In: *Handbooks for the Identification of British Insects*. Vol. 10, Part 5. Royal Entomological Society, London.

Spencer, K.A. (1976) The Agromyzidae (Diptera) of Fennoscandia and Denmark. In: *Fauna Entomologica Scandinavica*, Vol. 5. Scandinavian Science Press, Klampenborg, Denmark.

Thompson, J.N. (1978) Within-patch structure and dynamics in *Pastinaca sativa* and resource availability to a specialized herbivore. *Ecology* 59, 443–448.

Torchin, M.E., Lafferty, K.D. and Kuris, A.M. (2001) Release from parasites as natural enemies: increased performance of a globally introduced marine crab. *Biological Invasions* 3, 333–345.

Ulenberg, S.A., DeGoffau, L.J.W., Van Frankenhuyzen, A. and Burger, H.C. (1986) Striking infestations of insects in 1985. *Entomologische Berichten* 46, 163–171.

Volovik, S. (1988) A stem-feeding weevil as a pest of hogweed. *Zashchita Rastenii Moskva* 12, 1–31. [In Russian.]

Wagner, E. (1966) Wanzen oder Heteropteren. In: Dahl, M. and Peus, F. (eds) *Pentatomorpha. Die Tierwelt Deutschlands und der angrenzenden Meeresteile*. Vol. 54. Gustav Fischer, Jena, Germany.

Wagner, E. (1967) Wanzen oder Heteropteren. II. *Cimicomorpha*. In: Dahl, M. and Peus, F. (eds) *Die Tierwelt Deutschlands und der angrenzen Meeresteile*, Vol. 55. Gustav Fischer, Jena, Germany.

Wolfe, L.M. (2002) Why alien invaders succeed: support for the escape-from-enemy hypothesis. *American Naturalist* 160, 705–711.

Zlobin, V.V. (2005) A new species of *Melanogromyza* feeding on giant hogweed in the Caucasus (Diptera: Agromyzidae). *Zoosystematica Rossica* 14, 173–177.

12 Fungal Pathogens Associated with *Heracleum mantegazzianum* in its Native and Invaded Distribution Range

MARION K. SEIER AND HARRY C. EVANS

CABI UK Centre (Ascot), Ascot, UK

Nothing can stop them!

(Genesis, 1971)

Introduction

Pathogens and herbivores are the most important components of the natural enemy complexes associated with plant species or plant genera. Plant pathogens include viruses, bacteria, protozoa, nematodes and fungi, and it is the last which constitute by far the largest group of pathogens affecting plants in general (Agrios, 1997). Although these organisms and/or their associated disease symptoms are often less conspicuous compared to, for example, larger, more easily recognized insect feeders, pathogens play an equally important role in regulating their host populations. During history, this role and the potential devastating impact of fungal plant pathogens, once they catch up with their original (co-evolved) hosts, has frequently resulted in catastrophic outbreaks of crop diseases. This has been witnessed, for example, for potato blight, *Phytophthora infestans* (Mont.) de Bary, causing the Irish potato famine in the 19th century, as well as for coffee rust, *Hemileia vastatrix* Berk. & Broome, seriously affecting coffee production worldwide (Large, 1940). In their native ranges biotrophic fungal pathogens, such as rusts and smuts which depend entirely upon living plant tissue as their nutrient source, have often co-evolved with their plant hosts and, as a result, exhibit extreme host specificity. As well as arthropod herbivores, fungal pathogens can, therefore, also be exploited as classical biological control agents in order to regulate populations of invasive alien plant species. Traditionally classical biological control, defined as the search for host-specific natural enemies in the centre of origin of an

invasive plant and their subsequent introduction into its invaded range, has
been the domain of entomologists. However, the potential (*severe damage,
efficient dispersal*) and safety (*high specificity*) of fungal pathogens as classical
agents has recently been more and more recognized, leading to an increase in
their use (Barton, 2004). For more detailed information on the theory and
principles of classical biological control, as well as the characteristics of potential agents, see Cock and Seier (Chapter 16, this volume, and sources referred
to therein).

In order to evaluate this group of natural enemies as a management tool
for giant hogweed (*Heracleum mantegazzianum* Sommier & Levier) in its
invaded range in western Europe, the fungal pathogen complex or mycobiota
associated with this plant species, as well as its close relatives, needed to be
elucidated in both its native and introduced ranges. Subsequently, the biology
and the host range of selected fungal pathogens had to be investigated in order
to assess their potential as classical biological control agents (see also Cock and
Seier, Chapter 16, this volume). This was equally undertaken for the complex
of herbivorous arthropods associated with *H. mantegazzianum* as detailed by
Hansen *et al.* (Chapter 11, this volume).

Literature and Herbarium Surveys

A phenomenon often encountered with plant species of no immediate economic value is the paucity of knowledge concerning their associated natural
enemy complexes. This was also found to be true for giant hogweed with information about its mycobiota in the centre of origin, the Caucasus Mountains,
being particularly scarce. Literature and database searches (undertaken during
2002 and 2003), as well as the evaluation of material deposited at the herbarium of the Royal Botanic Gardens, Kew, UK (Herb K) and the fungal herbarium of CABI, Egham, Surrey, UK (Herb IMI), showed the majority of fungal
records for *H. mantegazzianum* to originate from its introduced range (Table
12.1). Furthermore, these mostly comprised generalist pathogens known to
exhibit a relatively wide host range (Sampson, 1990). In contrast, only one
fungal record was collected from the native range of giant hogweed, in
Georgia, which had initially been identified as the coelomycete pathogen
Cylindrosporium hamatum Bres. (Nakhutsrishvili, 1986). However, the taxonomic status of this pathogen was unclear, having in the past been assigned
different generic names, i.e. *Cylindrosporium*, *Phloeospora*, *Anaphysmene*,
which also have been subject to repeated revision. The study of herbarium
material from other *Heracleum* species (i.e. *H. sphondylium* L., native to
Europe) further revealed that another fungal pathogen, belonging to the genus
Septoria, had frequently been included in these records due to its morphological similarity to the aforementioned three fungal genera. Such taxonomic confusion may obviously not only concern records of fungal pathogens, but also
the identification of the respective *Heracleum* host, particularly since many
Heracleum species have also been reclassified over the years (see Jahodová *et
al.*, Chapter 1, this volume).

Table 12.1. Mycobiota associated with *H. mantegazzianum* compiled from literature, fungal databases and herbaria records.

Fungal species	Recorded from
Ascochyta heraclei Bres.	Latvia
Cylindrosporium hamatum Bres.	Georgia
Diaporthopsis angelicae (Berk.) Wehm.	Eire
Erysiphe heraclei DC.	Finland; France; Germany; Poland; Sweden; Switzerland; UK; Ukraine; USSR (former); Yugoslavia (former)
Melanochaeta aotearoae (S. Hughes) E. Müll, Harr & Sulmont	UK (England)
Periconia byssoides Pers.	No data
Peronospora heraclei Rabenh.	Germany
Phoma complanata (Tode) Desm.	Ireland (on seed)
Phomopsis asteriscus (Berk. & Broome) Grove	Ireland (on seed)
Sclerotinia sclerotiorum (Lib.) de Bary	UK (Scotland)
Sporoschisma mirabile Berk. & Broome	UK (England)
Torula herbarum (Pers.) Link	UK (Wales)

Sources: Herb IMI, CABI, Egham, UK; http://nt.ars-grin.gov/fungaldatabases/; Nakhutsrishvili (1986).

With rust pathogens *per se* being good potential biological control agents (generally specific and damaging with efficient dispersal through windborne spores), it is of interest that Nakhutsrishvili (1986) reported the rust *Puccinia heraclei* Grev. from the genus *Heracleum* in Abkahzia, Georgia. However, this author does not identify the host to species level. The rust has subsequently been recorded from *Heracleum sibiricum* L. in the Russian Republic of Mordovia (Ryzhkin and Levkina, 2004) and in Finland (Herb IMI) and is also known from *H. sphondylium* in Germany (Braun, 1982), Norway (Jørstad, 1962) and the UK (Henderson, 2004; Herb IMI).

Field Surveys

Aiming to close this apparent knowledge gap regarding the fungal pathogen complex associated with giant hogweed, as well as to identify potential candidates for biological control of this invasive plant, field surveys were conducted in its native range, the Western Caucasus region, as well as in its introduced European range.

In order to compile a complete inventory of the mycobiota associated with a specific plant host, as well as to assess the impact of individual fungal pathogens on that host, as wide an area as possible of its known geographical distribution needs to be surveyed. Furthermore, surveys should be conducted at different times of the growing season over a number of years to account for climatic factors affecting the distribution, incidence and severity of fungal infections and, thus, the fungal species composition (or mycobiota). An initial selection of fungal pathogens for further evaluation as potential biological control agents can then be based on field observations of the impact on the host, as well as on a preliminary assessment of the apparent host range, i.e. the presence or absence of specific pathogens in the field on plant species related to the host species in question (see also Cock and Seier, Chapter 16, this volume).

Five surveys were undertaken from 2002 to 2004 (jointly with some of the entomology surveys; see Hansen et al., Chapter 11, this volume) to assess the mycobiota of *H. mantegazzianum* and related species in their native western range of the North Caucasus region (Cis-Caucasus) within the Russian Federation. The geographic area covered stretched from Pyatigorsk in the east to the Krasnodar Territory and the Black Sea in the west (Fig. 12.1), and included field sites situated at altitudes between 120 and 1770 m a.s.l. On average 11 sites were visited during each survey. Population sizes of *H. mantegazzianum* varied from a few plants to larger stands and a representative number of diseased plant samples were taken at each site. Due to the short growing season when mountainous habitats are snow-free, these surveys were undertaken between late May and late August. The Georgian region of Abkhazia, though part of the native range of giant hogweed and of particular

Fig. 12.1. Areas surveyed in North Caucasus and Transcaucasia region during 2002–2004. Map courtesy of the University of Texas Libraries, The University of Texas at Austin (http://www.lib.utexas.edu/maps, accessed 1 June 2006); maps of Russia and the former Soviet Republics: Georgia maps, Georgia (shaded relief) 1999.

interest due to the reported occurrence of the rust *Puccinia heraclei* (Nakhutrishvili, 1986), could not be included in the surveys due to political and security reasons.

While *H. mantegazzianum* is not native to Armenia, an additional pathogen survey was undertaken in this southern region of the Caucasus (Transcaucasia) during July 2003 to elucidate the pathogen complex associated with other *Heracleum* species. Of particular interest was the closely related *H. sosnowskyi* Manden, which has achieved invasive status in some eastern European countries such as Latvia. This survey covered the area from Yerevan to Stepanavan in the north down to Goris in the south of Armenia (Fig. 12.1), visiting in total 11 field sites situated between 1315 and 2170 m a.s.l. Representative numbers of diseased plant samples were taken according to the variable plant population sizes found at individual sites.

Table 12.2 gives a summary inventory of fungal pathogens found to be associated with *H. mantegazzianum* and related species in their native range together with notes on their observed distribution and impact on the host. Where applicable an initial, field-based assessment of the biocontrol potential of pathogens was made based on these data. During the growing seasons of 2002–2004, several field surveys were undertaken in the introduced range of *H. mantegazzianum*, particularly in the Czech Republic, Denmark, Germany, Latvia, Switzerland and the UK. The purpose of these surveys was to assess the range and type of fungal species associated with the invasive plant and to compare these with the mycobiota in the native range. In the respective countries, giant hogweed infestations were surveyed as well as natural stands of native related species, in particular *H. sphondylium*. In Latvia, pathogen collections were made from *H. sosnowskyi*, which constitutes the dominant invasive *Heracleum* species in all the Baltic States, having been introduced as a fodder plant during former Soviet Union times (see Moravcová *et al.*, Chapter 10, and Ravn *et al.*, Chapter 17, this volume).

Areas surveyed were: the western part of the Czech Republic (mainly the Slavkovský les region; for details see Perglová *et al.*, Chapter 4, this volume); Jutland (east and west coast) and Zealand (east coast) in Denmark; county of Hessen in Germany; Cesis and Jelgava districts in Latvia; Bern area in Switzerland; in the UK, the southern English counties of Berkshire, Cornwall, Devon, Herefordshire, Hertfordshire and Surrey, as well as Wales. Pathogens were, in general, only identified to the genus level. Fungal pathogens found to be associated with *H. mantegazzianum*, *H. sosnowskyi* and *H. sphondylium* during these surveys are listed in Table 12.3.

Mycobiota associated with *Heracleum mantegazzianum* and related species in the native range

The surveys conducted in the Caucasus region revealed a rich and diverse mycobiota associated with *H. mantegazzianum* in its native range, with several fungal species recorded from this host for the first time. As expected, the presence of primary fungal pathogens attacking giant hogweed was found to be most pronounced in the early to middle parts of the growing season, while opportunistic generalist and saprophytic fungal species, utilizing moribund or senescing material, became more common towards the end of the season. Furthermore, it became apparent that the pathogen complex associated with *H. mantegazzianum* resembled that recorded from other surveyed *Heracleum* species native to the North Caucasus and Transcaucasia regions, thus confirming some literature records documented previously for the genus *Heracleum* (Nakhutsrishvili, 1986). While several pathogens were recorded from more than one *Heracleum* species, it is possible that distinct strains or pathotypes exist within each pathogen species adapted to individual host species.

The following fungal pathogens recorded from *H. mantegazzianum*, and authoritatively identified as listed in Table 12.2, were considered to be co-evolved (see Cock and Seier, Chapter 16, this volume), and thus to have potential as classical biocontrol agents of the plant in its introduced range: *Phloeospora heraclei* (Lib.) Petr.; *Septoria heracleicola* Kabát & Bubák; *Ramularia heraclei* (Oudem.) Sacc.; *Ramulariopsis* sp. nov. and *Phomopsis* sp. Specimens of these species (as well as of others) have been deposited in the dried reference collection of the internationally accredited CABI herbarium (Herb IMI) and the relevant IMI numbers are given in Table 12.2. Apart from *R. heraclei*, these pathogens have also undergone initial host-range testing on plant species closely related to *H. mantegazzianum* (see Cock and Seier, Chapter 16, this volume). A detailed account of the host-range testing method and the results of these studies will be published separately.

Phloeospora heraclei

This coelomycete fungus occurred at the majority of field sites visited and was found to be identical with the one recorded previously as *Cylindrosporium hamatum* from Georgia (see earlier). *Phloeospora heraclei* (Fig. 12.2A, B) can easily be recognized by the conspicuous, pale brown acervuli-bearing hyaline, curved conidia on the upper leaf surface of its host and the associated black crusts on the lower leaf surface (Fig. 12.2B I, II). The latter are considered to represent either feeding or survival structures; however, it cannot be ruled out that the initials of the sexual stages may later be formed in association with these crusts. This fungus is a true biotroph or obligate pathogen which is dependent on its living host, and cannot be cultured *in vitro*. In the field, *P. heraclei* was observed to cause significant damage to *H. mantegazzianum* (as well as to its other recorded *Heracleum* hosts), in the form of coalescing leaf-spots, widespread necrosis and die-back. The pathogen affected host plants of all ages; however, its impact was most severe at the seedling stage (Fig. 12.2A). Therefore, *P. heraclei* appeared to have high potential as a biological control agent: being damaging to its host at a critical phase of its life cycle, especially given the inability of giant hogweed to spread vegetatively (Ochsmann, 1996; Pyšek et al., Chapter 7, this volume). The pathogen has previously been reported in Europe from *H. sphondylium* and *H. austriacum* L. as well as from parsnip (*Pastinaca sativa* L.) (Allescher, 1903; Herb IMI). It

Fig. 12.2. Fungal pathogens on *H. mantegazzianum* and related *Heracleum* species in the Caucasus. (A) *Phloeospora heraclei* attacking *H. mantegazzianum* at the seedling stage. (B) *Phloeospora heraclei* on *H. mantegazzianum*: I. Close up of upper leaf surface showing acervuli; II. Lower leaf surface showing black crusts. (C) *Ramulariopsis* sp. nov. on *H. mantegazzianum* showing angular lesions on lower leaf surface mixed with infection of *Phloeospora heraclei* seen as black crusts (arrows). (D) *Phomopsis* sp. on *H. antasiaticum* showing moribund umbels (arrows); inset: close up of dying peduncle with lines of black pycnidia. Photo: M. Seier (B) and H. Evans (A, C, D).

has frequently been referred to as *Cylindrosporium* or *Pleospora*, as well as *Phyllachora* (Dennis, 1986); the latter due to the black crusts formed on lower leaf surfaces. Host-range testing under greenhouse conditions showed that isolates of *P. heraclei* collected from *H. mantegazzianum* attacked parsnip and coriander (*Coriandrum sativum* L.), both of which belong to genera in the same tribe as *Heracleum*, while no symptoms were observed on *H. sphondylium*. This indicates the existence of distinct pathogen strains (pathotypes) with potentially different host ranges, particularly since there are no field records of *P. heraclei* on *H. mantegazzianum* in its introduced range.

Septoria heracleicola *Kabát & Bubák*

Septoria heracleicola closely resembles *P. heraclei*, both in the disease symptoms caused and in its macromorphology. In the field, mixed infections of both pathogens are frequently encountered on *H. mantegazzianum* with *P. heraclei* being the dominant agent. Therefore, the disease is cryptic and in the field *S. heracleicola* was initially not recognized and distinguished from *P. heraclei* as a separate species. Only by closer, microscopic examination was this confusion resolved. Though potentially a suitable biocontrol agent, the distribution, abundance and impact of *S. heracleicola* on giant hogweed proved to be difficult to assess due to the similarity between these two pathogens.

In contrast to *P. heraclei*, *S. heracleicola* can readily be cultured *in vitro* and produces infective conidia. However, when maintained in culture the pathogen loses its infectivity over time, which highlights its dependency on its host for continuous reproduction. Initial host-specificity studies showed that *S. heracleicola* infected parsnip under greenhouse conditions, but the pathogen failed to attack coriander and *H. sphondylium*, although the latter is a reported host of this pathogen (Adamska, 2001).

Ramulariopsis *sp. nov.*

The cercosporoid fungus *Ramulariopsis* sp. nov. is a hitherto undescribed species and constitutes the first record of the genus *Ramulariopsis* for the family *Apiaceae*. A detailed taxonomic description of this new fungal species will be given in a forthcoming paper to be published in a mycological journal. This genus is otherwise known mainly from tropical or subtropical areas (J.C. David, Herb IMI, 2002, personal communication). After *P. heraclei*, *Ramulariopsis* sp. nov. was found to be the most abundant and widely distributed pathogen on *H. mantegazzianum*, as well as on other related *Heracleum* species in the native range. The two pathogens were often found to occur together, causing mixed infections and severely affecting individual host plants and populations (Fig. 12.2C). However, while *P. heraclei* damage on *H. mantegazzianum* was often observed relatively early in the season, i.e. at the seedling stage, *Ramulariopsis* sp. nov. generally became more prominent mid to late season, and was found to be most damaging on older, senescing leaves, thus most likely exerting a lesser overall impact on the host population.

Ramulariopsis sp. nov., which attacks all leaf stages of its host, causes highly distinctive disease symptoms in the form of angular lesions. These are most pronounced on the lower leaf surface and have a cottony appearance once sporulation occurs. While the genera *Ramulariopsis* and *Ramularia* can be distinguished by conidiophore morphology (i.e. long, hyaline, branched conidiophores of the former compared to shorter and usually, but not always, unbranched conidiophores of the latter genus) this is not feasible in a field situation, making it difficult to differentiate between this pathogen and *Ramularia heraclei* (see below).

Ramulariopsis sp. nov. grows readily *in vitro* producing conidia infective to both giant hogweed and parsnip in the host-range tests, but not to coriander or *H. sphondylium*, although this pathogen has been recorded from the latter plant species in the UK (see below). This further supports the idea of the existence of host-specific strains (pathotypes) adapted to different species within the genus *Heracleum*.

Ramularia heraclei *(Oudem.) Sacc.*

Being the less dominant pathogen and indistinguishable from *Ramulariopsis* sp. nov. in the field, the abundance and impact of *R. heraclei* on *H. mantegazzianum* in the Caucasus area could not be assessed. *R. heraclei* is a well-known and widespread pathogen on a range of genera in the *Apiaceae* throughout the Northern Hemisphere and has also been recorded on the west European native *H. sphondylium* (Dennis, 1986; Adamska, 2001; Herb IMI). The pathogen grows readily in culture; however, over time, it loses its ability to sporulate *in vitro*. Host-range testing was not undertaken for this fungal species.

Phomopsis *sp.*

Symptoms of blighted inflorescences and umbel die-back of *H. mantegazzianum*, and other *Heracleum* spp., were consistently associated with the coelomycete *Phomopsis* sp. (Fig. 12.2D). Molecular studies matched this pathogen closely to *Diaporthe angelicae* (Berk.) Wehm. (*fide* P.F. Cannon, Herb IMI, 2004). The genus *Diaporthe* is the sexual (or teleomorph) stage of *Phomopsis*. *Diaporthe angelicae*, commonly also referred to as *Diaporthopsis angelicae* (Berk.) Wehm., is a widespread and common species found on a range of different *Apiaceae*.

Initially, the distribution of *Phomopsis* sp. appeared to be localized, but later surveys showed the pathogen to be widely distributed on giant hogweed in its centre of origin. *Phomopsis* sp. attacked the main and secondary branches of umbels, as well as seeds, forming characteristic raised ridges of black pycnidia (Fig. 12.2D), causing necrosis and premature senescence. Heavy infection resulted in effete inflorescences with seed-set being completely inhibited. Thus, as with *Phloeospora heraclei*, *Phomopsis* sp. has an impact on *H. mantegazzianum* at a critical life-cycle stage, potentially preventing seed-set, thereby reducing the seed bank and the spread of this plant species, which depends entirely on sexual reproduction for its propagation (Ochsmann, 1996). The pathogen was considered, therefore, to have some potential as a biological control agent.

Phomopsis sp. is readily cultured. However, of the two characteristic conidial types, the infective α-conidia and the purportedly non-infective β-conidia, the latter was the dominant spore produced *in vitro*. Under greenhouse conditions, disease symptoms on young *H. mantegazzianum* plants were consistently produced using suspensions of α-conidia. Unfortunately, parsnip was also shown to be equally susceptible to this pathogen.

Table 12.2. Fungal pathogens recorded from *H. mantegazzianum* and its relatives during field surveys in the Caucasus region.

Fungal pathogen	Host[1]	Collected	Distribution	Part of host affected/impact	BC potential[2]
Alternaria sp.	*H. mantegazzianum*[4]	North Caucasus	Common, widespread	Leaf/insignificant	None
Alternaria-type species	*H. mantegazzianum*	North Caucasus	Very limited	Root/insignificant	None
Basidiomycete sp.	*H. mantegazzianum*	North Caucasus	Very limited	Leaf/insignificant	None
Exosporium sp.	*H. mantegazzianum*	North Caucasus	Very limited	Leaf/insignificant	None
Oidium sp. (tentatively *Erysiphe heraclei* DC.)	*H. mantegazzianum*	North Caucasus	Common, widespread	Leaf, stem/limited	None
Oidium sp. (tentatively *Erysiphe heraclei* DC.)	*H. leskovii*[4]	North Caucasus	Present at one site 710 m a.s.l.	Leaf, stem/limited	n.a.
Phloeospora heraclei (Lib.) Petr. (Herb IMI: Nos 389658, 389659)	*H. mantegazzianum*	North Caucasus	Widespread, abundant from 510 to 1770 m a.s.l.	Leaf/potentially very damaging, particularly to seedling stage early in season	High
Phloeospora sp.[3]	*H. antasiaticum*[4]	Transcaucasia	Common, abundant, from 1650 to 1700 m a.s.l.	Leaf/damaging	n.a.
Phloeospora sp.[3]	*H. pastinacifolium*[4]	Transcaucasia	Common, collected from 1350 m a.s.l.	Leaf/damaging	n.a.
Phloeospora sp.[3]	*H. ponticum*[4]	North Caucasus	Common, widespread, abundant from 1515 to 1625 m a.s.l.	Leaf/potentially very damaging	n.a.
Phloeospora sp.[3]	*H. sosnowskyi*[4]	Transcaucasia	Common, from 1315 to 1610 m a.s.l.	Leaf/common on 1st year plants, little or none on 2nd year plants	n.a.
Phloeospora sp.[3]	*H. trachyloma*[4]	Transcaucasia	Common, from 1700 to 2170 m a.s.l.	Leaf/damaging	n.a.
Phloeospora sp.[3]	*H. transcaucasicum*[4]	Transcaucasia	Present at one site 2050 m a.s.l.	Leaf/very damaging	n.a.

Phoma sp. (tentatively *P. longissima* (Pers.) Westend.) (Herb IMI: No. 389654)	*H. mantegazzianum*	North Caucasus	Restricted distribution and abundance, in heavily shaded forest habitats 920 m a.s.l.	Leaf/locally damaging	None
Phomopsis sp. (Herb IMI: Nos 392711, 392714)	*H. mantegazzianum*	North Caucasus	Common, from 120 to 1300 m a.s.l.	Inflorescence, seeds, to a lesser extent roots/ potentially damaging	High
Phomopsis sp. (Herb IMI: No. 392710)	*H. antasiaticum*	Transcaucasia	Present at one site 1450 m a.s.l.	Inflorescence, petiole stem, rachis, seeds, heavy umbel attack, significant impact on seed production	n.a.
Phomopsis sp. (Herb IMI: No. 389650)	*H. leskovii*	North Caucasus	Present at one site 620 to 650 m a.s.l.	Inflorescence/potentially very damaging	n.a.
Pseudocercosporella pastinacea (P. Karst.) U. Braun (Herb IMI: No. 391378b)	*H. antasiaticum*	Transcaucasia	Present at one site 1650 m a.s.l., this species is frequently found in association with *Ramularia heraclei*	Leaf/damaging	n.a.
Ramularia heraclei (Oudem.) Sacc. (Herb IMI: No. 389655)	*H. mantegazzianum*	North Caucasus	Distribution and abundance difficult to assess due to similarity to *Ramulariopsis* sp. (see below), positively identified from one field site 510 m a.s.l.	Leaf/limited	Limited
Ramularia heraclei (Oudem.) Sacc. (Herb IMI: No. 391378a)	*H. antasiaticum*	Transcaucasia	Distribution and abundance difficult to assess due to similarity to *Ramulariopsis*, positively identified from one location 1650 m a.s.l.	Leaf/limited	n.a.

Continued

Table 12.2. *Continued.*

Fungal pathogen	Host[1]	Collected	Distribution	Part of host affected/impact	BC potential[2]
Ramulariopsis sp. nov. (Herb IMI: Nos 389652, 389653, 389656)	*H. mantegazzianum*	North Caucasus	Widespread, abundant, recorded from 120 to 1770 m a.s.l.	Leaf/major damage later in the season	High
Ramulariopsis sp. nov. (Herb IMI: Nos. 391377, 392215)	*H. antasiaticum*	Transcaucasia	Common and abundant on all leaf stages, collected from 1450 to 1700 m a.s.l.	Leaf/damaging	n.a.
Ramulariopsis sp./ *Ramularia* sp. complex	*H. pastinacifolium*	Transcaucasia	Present at one site 1350 m a.s.l.	Leaf/limited	n.a.
Ramulariopsis sp. nov. (Herb IMI: No. 391380)	*H. ponticum*	North Caucasus	Present at one site 1300 m a.s.l.	Leaf/potentially damaging	n.a.
Ramulariopsis sp./ *Ramularia* sp. complex	*H. sosnowskyi*	Transcaucasia	Common, from 1315 to 1610 m a.s.l.	Leaf/damaging	n.a.
Ramulariopsis sp./ *Ramularia* sp. complex	*H. trachyloma*	Transcaucasia	Present at one site 1700 m a.s.l.	Leaf/very damaging	n.a.
Ramulariopsis sp./ *Ramularia* sp. complex	*H. transcaucasicum*	Transcaucasia	Present at one site 2050 m a.s.l.	Leaf/very damaging	n.a.
Septoria heracleicola Kabát & Bubák (Herb IMI: No. 389651)	*H. mantegazzianum*	North Caucasus	Irregular, cryptic, recorded from 910 to 1770 m a.s.l.	Seedlings, leaves/ possibly damaging but impact difficult to assess in the field due to overall mixed infections with *Phloeospora heraclei* as the dominant agent	Medium
Fusicladium sp. (tentatively identified)	*A. pachyptera*[4]	North Caucasus	Present at one site 1625 m a.s.l.	Leaf/damaging	n.a.
Fusicladium sp. (tentatively identified)	*A. tatianae*[4]	North Caucasus	Common, from 1340 to 1515 m a.s.l.	Leaf/damaging	n.a.

Ramularia sp. cf. *archangellicae* Lindr. (Herb IMI: No. 391379)	*A. tatianae*	North Caucasus	Present at one site 1515 m a.s.l.	Leaf/minor damage	n.a.
Asperisporium sp. (Herb IMI: No. 392214)	*E. cicutarium*[4]	Transcaucasia	Present at one field site (2050 m a.s.l.) where it was abundant and damaging, killing both leaves and whole plants, initial appearance in the field similar to black crusts formed by *P. heraclei*	Leaf/very damaging	n.a.
Ramulariopsis sp./ *Ramularia* sp. complex	*E. cicutarium*	Transcaucasia	Not identified to the genus level, on the field samples mixed infection with the pathogen above	Leaf/limited	n.a.

[1] This column does not give the comprehensive host range of the respective pathogen; refer also to chapter text.
[2] Biocontrol potential; initial field-based assessment. n.a. = not applicable.
[3] Presumably *Phloeospora heraclei*, but potentially a different strain.
[4] *Heracleum antasiaticum* Manden; *Heracleum leskovii* Grossh.; *Heracleum mantegazzianum* Sommier & Levier; *Heracleum pastinacifolium* C. Koch; *Heracleum ponticum* (Lipsky) Schischk. ex Grossh.; *Heracleum sosnowskyi* Manden; *Heracleum trachyloma* Fisch. et C.A. Mey; *Heracleum transcaucasicum* Manden; *Angelica pachyptera* Avé-Lall.; *Angelica tatianae* Bordz.; *Eleutherospermum cicutarium* (Bieb.) Boiss.

Note: Suspected viral or mycoplasmal infections, in some instances causing severe distortion of shoots/inflorescences in the form of epinasty or witches' brooms, were also observed on *H. mantegazzianum* in the centre of origin, but no data were collected concerning this pathogen group.

Mycobiota associated with *Heracleum mantegazzianum* and related species in the invaded range

The results from field surveys undertaken in the introduced range of giant hogweed confirmed that, while a range of fungal genera was found to be associated with *H. mantegazzianum* in Europe, the majority of fungal specimens identified constitute either saprophytes or opportunistic and weak pathogens having, in general, an insignificant impact on the plant. Primary pathogens such as *Phloeospora heraclei*, *Septoria heracleicola* and *Ramulariopsis* sp. nov., recorded from giant hogweed in its native range, were absent from this host in Europe. Interestingly, however, these pathogens were either recorded from *H. sphondylium* during the field surveys (i.e. *Phloeospora heraclei*, *Ramulariopsis* sp. nov.) and/or are known from the literature to attack this western European native species (i.e. *Phloeospora heraclei* (Dennis, 1986), *Puccinia heraclei* (Henderson, 2004), *Ramularia heraclei* (Dennis, 1986), *S. heracleicola* (Adamska, 2001)). This once again raises the possibility of specialized pathogen strains.

As previously mentioned for *H. mantegazzianum* in the Caucasus, *Ramulariopsis* sp. nov. constitutes a new record for *H. sphondylium* (as well as for the area of Europe in general) since to date only closely related *Ramularia* species, i.e. *R. heraclei*, have been reported to be associated with this host. In contrast to the situation in the Caucasus, the distribution of *Phloeospora heraclei* on *H. sphondylium* in the UK appears to be localized or the disease may have been overlooked due to the hirsute nature of the leaves of this host. Though frequently misidentified as a *Septoria* species, *P. heraclei* had previously been reported in the literature from *H. sphondylium* in its native range.

A slightly different situation was found in Latvia, where *Phloeospora* sp. (presumably *P. heraclei*) and *R. heraclei*, both of which are known from *H. mantegazzianum* in its native range, were found to be associated with the invasive *H. sosnowskyi*. Why these pathogens were found to attack *H. sosnowskyi* in Latvia, but have as yet not been recorded on *H. mantegazzianum* in western Europe poses the question: were they originally introduced together with the plant material; have they caught up with their hosts; or, have they 'jumped' from indigenous *Heracleum* species on to *H. sosnowskyi*?

While none of the pathogens recorded from giant hogweed in its introduced range were found to impact severely on populations of this plant species, isolates of the genera *Alternaria* and *Phomopsis* obtained from *H. sosnowskyi* and *H. mantegazzianum*, respectively, might warrant further assessment should the mycoherbicide approach be considered as one control option for these invasive species (see Cock and Seier, Chapter 16, this volume). Species of these fungal genera have previously been evaluated and used as mycoherbicides against a range of other invasive plants (e.g. *Alternaria cassiae* Jurair & Kahn against *Cassia obtusifolia* L.; *Phomopsis amaranthicola* Rosskopf & Charudattan against *Amaranthus* spp.) (Julien and Griffiths, 1998). As with the *Phomopsis* isolates characterized from *H. mantegazzianum* in the Caucasus, those collected from the invasive plant in

Table 12.3. Fungal pathogens recorded from *H. mantegazzianum* and its relatives during field surveys in its introduced range.

Fungal pathogen	Host[1]	Collected	Distribution	Impact[2]	BC potential[3]
Acremonium sp.	*H. mantegazzianum*[4]	Denmark	Widespread	No significant impact	No potential for biological control
Acremonium-type species	*H. mantegazzianum*	Denmark	Collected from one site (Vedbaek, Sealand, 8 m a.s.l.)	No impact	No potential for biological control
Alternaria sp.	*H. mantegazzianum*	Denmark	Widespread	No significant impact	No potential for biological control
Botryotrichum sp.	*H. mantegazzianum*	Denmark	Widespread	No significant impact	No potential for biological control
Cladobotryum sp.	*H. mantegazzianum*	Denmark	Widespread	No significant impact	No potential for biological control
Cladosporium sp.	*H. mantegazzianum*	Denmark	Widespread, saprophyte/secondary agent	No significant impact	No potential for biological control
Cylindrocarpon sp.	*H. mantegazzianum*	Denmark	Found at one site (Vedbaeck, Sealand, 8 m a.s.l.)	Potentially pathogenic	Unknown
Fusarium sp.	*H. mantegazzianum*	Denmark	Widespread	No significant impact	Opportunistic pathogen, potential for biological control doubtful
Fusarium sp.	*H. mantegazzianum*	UK	Found at one site at Ascot, Berkshire	No significant impact	As above
Harzia sp.	*H. mantegazzianum*	Denmark	Found at one site (Malov, Sealand, 31 m a.s.l.)	No significant impact	No potential for biological control
Microstroma-type pathogen Exobasidiales	*H. mantegazzianum*	UK	Recorded from one site at Watton at Stone, Hertfordshire	No official identification, possibly new record; impact unknown	Unknown; unusual record
Mucor sp.	*H. mantegazzianum*	Denmark	Found at one site (Arhus, Jutland, 30–32 m a.s.l.)	No significant impact	No potential for biological control
Oidium sp. (tentatively *Erysiphe heraclei* DC.)	*H. mantegazzianum*	Denmark	Widespread	Various plant parts/locally damaging	No potential for biological control
Oidium sp. (tentatively *Erysiphe heraclei* DC.)	*H. mantegazzianum*	Germany	Collected from several sites in the county of Hessen, likely to be widespread	Various plant parts/locally damaging	No potential for biological control

Table 12.3. *Continued.*

Fungal pathogen	Host[1]	Collected	Distribution	Impact[2]	BC potential[3]
Oidium sp. (tentatively *Erysiphe heraclei* DC.)	*H. mantegazzianum*	Switzerland	Collected from one site (Bern, 570 m a.s.l.), but likely to be widespread	Various plant parts/locally damaging	No potential for biological control
Penicillium sp.	*H. mantegazzianum*	Denmark	Isolated as a common mould, widespread	No significant impact	No potential for biological control
Phaeoacremonium sp.	*H. mantegazzianum*	Denmark	Widespread	No significant impact	No potential for biological control
Phoma sp.	*H. mantegazzianum*	Denmark	Widespread	No significant impact	Weak pathogen, potential for biological control doubtful
Phoma sp.	*H. mantegazzianum*	UK	Widespread	No significant impact	As above
Phomopsis sp. (Herb IMI: No. 392712)	*H. mantegazzianum*	Denmark	Widespread	No significant impact	May have potential for biological control
Sclerotinia sclerotiorum (Lib.) de Bary	*H. mantegazzianum*	Denmark	Widespread	Potentially damaging (wide host range)	Currently under investigation for its potential as a mycoherbicide
Trichoderma sp.	*H. mantegazzianum*	Denmark	Isolated from one site (Arhus, Jutland 30–32 m a.s.l.)	No significant impact	No potential for biological control
Ulocladium sp.	*H. mantegazzianum*	Denmark	Collected from one site (Esbjerg, Jutland, 6 m a.s.l.)	No significant impact	No potential for biological control
Verticillium sp.	*H. mantegazzianum*	Denmark	Widespread	No significant impact	No potential for biological control
Oidium sp. (tentatively *Erysiphe heraclei* DC)	*H. sphondylium*[4]	UK	Found at one site in North Devon, but likely to be widespread	Various plant parts/damaging	n.a.
Phloeospora heraclei (Lib.) Petr.	*H. sphondylium*	UK	Isolated from one site in Cornwall (150 m a.s.l.), appears to be either localized or cryptic, hence distribution as yet unknown	Damaging	Probably constitutes a different strain to the one present on *H. mantegazzianum*, since not recorded on this host in the UK
Ramulariopsis sp. nov. (Herb IMI: Nos. 389657, 391381)	*H. sphondylium*	UK	Appears to be common in western parts of the UK	Damaging	Relationship to the *Ramulariopsis* sp. nov. ex *H. mantegazzianum* in the Caucasus as yet unknown
Ramulariopsis sp. nov. (Herb IMI: No. 392284)	*H. sphondylium*	Czech Republic	Very localized, only found at one site in the Slaykovsky Forest	Impact unknown	n.a.
Alternaria cf. *infectoria* E.G. Simmons (Herb IMI: No. 392216)	*H. sosnowskyi*[4]	Latvia	Collected from Jelgava and Cesis districts, currently unknown how widespread in Latvia	Damaging	May have potential for biological control

				Damaging	No potential for biological control
Oidium sp. (tentatively *Erysiphe heraclei* DC)	*H. sosnowskyi*	Latvia	Collected from sites in Jelgava and Cesis districts	Damaging	
Phloeospora sp. (presumably *P. heraclei*, potentially different strain)	*H. sosnowskyi*	Latvia	Collected from Jelgava district, currently unknown how widespread in Latvia	Potentially damaging	n.a.
Ramularia heraclei (Oudem.) Sacc. (Herb IMI: No. 392217)	*H. sosnowskyi*	Latvia	Collected from Jelgava district, currently unknown how widespread in Latvia	Potentially damaging	n.a.

[1] This column does not give the comprehensive host range of the respective pathogen; refer also to chapter text.
[2] Impact is on leaves, except where stated.
[3] Biocontrol potential as a mycoherbicide; initial field-based assessment.
[4] *Heracleum mantegazzianum* Sommier & Levier; *Heracleum sosnowskyi* Manden; *Heracleum sphondylium* L.

Denmark (Table 12.3) are, molecularly, closely matched to *Diaporthe angeli-cae* (*fide* P.F. Cannon, Herb IMI, 2004). Since the species *Phomopsis aster-iscus* (Berk. & Broome) Grove has also been reported from *H. mantegazzianum* in its introduced range (Table 12.1), as well as from *Heracleum* sp. in the Caucasus (Nakhutsrishvili, 1986), a complex of *Phomopsis* species may be associated with the genus *Heracleum* which may prove difficult to elucidate.

In the Netherlands, initial studies are already underway to assess the impact of the indigenous pathogen *Sclerotinia sclerotiorum* (Lib.) de Bary on *H. mantegazzianum* with a view to developing this species as a potential mycoherbicide against this invasive alien plant (W.B. de Voogd, Bern, Switzerland, 2004, personal communication).

Conclusions

The pathogen surveys conducted in the Caucasus and in western Europe revealed that the mycobiota associated with *H. mantegazzianum* in its intro-duced range compared to its native range is impoverished. This is particularly true with respect to specialized (co-evolved), primary fungal species that are severely damaging to their hosts. Such a situation is commonly found where an alien species has been introduced into a new geographic range without its associated natural enemies (Mitchell and Power, 2003). The two fungal pathogens affecting populations of *H. mantegazzianum* most severely in its native range are *Phloeospora heraclei* and *Phomopsis* sp., which attack seedlings and umbels, respectively. Thereby both pathogens directly affect the spread of this plant species, as it is unable to reproduce vegetatively. The rust, *Puccinia heraclei*, known from *Heracleum* sp. in Georgia (Nakhutsrishvili, 1986) and, theoretically, the primary potential agent for classical biological control of giant hogweed in its invaded range, was never encountered during the current surveys on *H. mantegazzianum* or its relatives. This would suggest that the pathogen is a rare or cryptic component of the native mycobiota, pos-sibly having a restricted geographic distribution.

All the *Heracleum* species native to the Caucasus region appear to share a common mycobiota. However, this has not been evaluated fully at the more critical taxonomic and physiological levels, which would reveal if fungal species could be further subdivided into pathovars or pathotypes. Interestingly, the composition of the pathogen complex associated with *H. sphondylium* in its native western European range resembles that recorded for Caucasian *Heracleum* species. A hypothesis to explain the apparent specialization of natural enemies to the genus level rather than to the individual species level is put forward by Cock and Seier (Chapter 16, this volume). With respect to fungal pathogens, field observations, as well as greenhouse studies, strongly suggest that there are clear physiological, if not more subtle taxonomic differ-ences, between individual pathogen isolates attacking different *Heracleum* species. Such differences, possibly reflected in the existence of host-specific pathotypes, might be particularly pronounced between the collections from the Caucasus and western Europe; a hypothesis which would need to be confirmed through DNA characterization as well as detailed host-range testing. *Phloeospora heraclei*, for example, has been recorded on *H. sphondylium* throughout western Europe (Herb IMI), but does not attack *H. mantegazzianum* in its introduced range. Conversely, host-range studies showed that *P. heraclei* isolates ex *H. mantegazzianum* from its native range did not attack *H. sphondylium* under greenhouse conditions. Therefore, in the case of this pathogen at least, it can be concluded with confidence that distinct strains or pathotypes occur between *Heracleum* spp. The same may also apply to *Septoria heracleicola* and *Ramulariopsis* sp. nov., as well as to the rust *Puccinia heraclei*.

The results of the initial host-specificity testing of *Phloeospora heraclei* isolates, as well as of isolates of *Phomopsis* sp., *Ramulariopsis* sp. nov. and *S. heracleicola*, raise further questions given the apparent susceptibility of non-target species from genera closely related to *Heracleum* (see Cock and Seier, Chapter 16, this volume), i.e. *Pastinaca* and *Coriandrum*, under controlled environment conditions. Possible explanations are the well-documented phenomenon of artificial host-range extension of plant pathogens under greenhouse conditions (Evans, 1995) or the fact that loss of resistance against fungal pathogens in economically important plant species is often an unwanted side effect of commercial breeding programmes. Clearly, the scope of this project has not allowed for a more detailed characterization of the mycobiota. This would require molecular studies to better define the anomalies observed in the field and in host-range tests.

Acknowledgements

We would like to thank our CABI colleagues, Carol Ellison, Ruediger Wittenberg and Matthew Cock for their inputs and discussions throughout the project, as well as Sue Paddon and Lynn Hill for providing the plant material for the greenhouse work. We also thank all the partners of this multidisciplinary EU project for collecting diseased giant hogweed material. Scientific and logistical support during the field surveys was provided by the scientists of the Russian Academy of Sciences (St Petersburg and Pyatigorsk), particularly Dmitry Geltman, Boris Zhitar, Svyatoslav Bondarenko, Tatyana Volkovich, Vladimir Lantsov and Sergey Ya. Reznik; by George Fayvush, Kamilla Tamanyan and Mark Kalashyan from the Armenian Academy of Science, Yerevan, as well as our special colleague Myriam Poll; and by Jan Pergl and Irena Perglová from the Academy of Sciences of the Czech Republic, Průhonice. We appreciate and thank the University of Texas Libraries, Austin, Texas, USA, for making publicly available the map for Fig. 12.1. This work was supported by European Union funding under the 5th Framework programme 'EESD - Energy, Environment and Sustainable Development'; project number EVK2-2001-00128 and CABI.

References

Adamska, I. (2001) Microscopic fungus-like organisms and fungi of the Slowinski National Park. II. (NW Poland). *Acta Mycologica* 36, 31–65.

Agrios, G.N. (1997) *Plant Pathology*, 4th edn. Academic Press, San Diego, California.

Allescher, A. (1903) Fungi imperfecti: Gefarbt-sporige Sphaerioideen. In: *Rabenhorst's Kryptogamen-Flora von Deutschland, Oesterreich und der Schweiz*, Band 1 (7), 2nd edn. Kummer, Leipzig, Germany, pp. 65–128.

Barton, J. (née Fröhlich) (2004) How good are we at predicting the field host-range of fungal pathogens used for classical biological control of weeds? *Biological Control* 31, 99–122.

Braun, U. (1982) Die Rostpilze (Uredinales) der Deutschen Demokratischen Republik. *Feddes Repertorium Beihefte* 93, 213–334.

Dennis, R.W.G. (1986) *Fungi of the Hebrides*. Royal Botanic Gardens, Kew, UK.

Evans, H.C. (1995) Pathogen–weed relationships: the practice and problems of host range screening. In: Delfosse, E.S. and Scott, R.R. (eds) *Proceedings of the 8th International Symposium on Biological Control of Weeds*. DSIR/CSIRO, Melbourne, Australia, pp. 539–551.

Henderson, D.M. (2004) *Checklist of the Rust Fungi of the British Isles*. British Mycological Society, Kew, UK.

Jørstad, I. (1962) Distribution of the Uredinales within Norway. *Nytt Magasin for Botanikk* 9, 61–134.

Julien, M.H. and Griffiths, M.W. (eds) (1998) *Biological Control of Weeds. A World Catalogue of Agents and their Target Weeds*, 4th edn. CAB International, Wallingford, UK.

Large, E.C. (1940) *The Advance of the Fungi*. J. Cape, London.

Mitchell, C.E. and Power, A.G. (2003) Release of invasive plants from fungal and viral pathogens. *Nature* 421, 625–627.

Nakhutsrishvili, I.G. (1986) *Flora of Spore-producing Plants of Georgia*. N.N. Ketskhoveli Institute of Botany, Academy of Sciences of the Georgian SSR, Tbilisi. [In Russian.]

Ochsmann, J. (1996) *Heracleum mantegazzianum* Sommier & Levier (*Apiaceae*) in Deutschland. Untersuchungen zur Biologie, Verbreitung, Morphologie und Taxonomie. *Feddes Repertorium* 107, 557–595.

Ryzhkin, D.V. and Levkina, L.M. (2004) Rust fungi of the North East of the Republic Mordovia. *Mikologia i Fitopatologia* 38, 45–50.

Sampson, C. (1990) Towards biological control of *Heracleum mantegazzianum*. MSc thesis, Imperial College, University of London.

13 Defence Systems of *Heracleum mantegazzianum*

JAN HATTENDORF, STEEN O. HANSEN AND WOLFGANG NENTWIG

University of Bern, Bern, Switzerland

> *Kill them with your Hogweed hairs*
>
> (Genesis, 1971)

Introduction

In the early 19th century, giant hogweed, *Heracleum mantegazzianum* Sommier & Levier (*Apiaceae*), native to the subalpine and alpine belt of the Western Greater Caucasus, was introduced to western Europe as an ornamental plant (Ochsmann, 1996). Within 150 years it has naturalized along waterways and roads and on fallow and disturbed land all over Europe (see Pyšek *et al.*, Chapter 3, this volume; Pyšek 1994). *H. mantegazzianum* has good competitive ability and a high seed production (see Perglová *et al.*, Chapter 4, this volume) and therefore it has become an aggressive invasive weed causing problems, especially in regions where the land use is changing (Otte and Franke, 1998; Weber, 2004). The two main reasons for stopping this weed from spreading further in Europe are: (i) the reduction of biodiversity of communities and landscapes which affects the structure and function of ecosystems; and (ii) the hazard to the public because the plant sap causes a serious UV-induced photodermatitis (Lagey *et al.*, 1995).

    H. mantegazzianum has, as have plants in general, evolved defences to protect itself against attacks by herbivorous insects and vertebrates or pathogens, such as bacteria, fungi and viruses. The plant defences can either be constitutively present or they are induced in response to an attack (Karban and Baldwin, 1997). In this chapter, we will describe defence systems of *H. mantegazzianum* and estimate their relative importance for the plant. The main emphasis will be given on the defences of plants against herbivorous insects. Frequently the term 'resistance' is used to describe the ability to overcome attack, but resistance is a more general expression which also includes withstanding abiotic stresses (Agrawal, 2000).

    The defence mechanisms of plants can be manifold (Crawley, 1997). The enemy can be affected directly by mechanical barriers, or by secondary plant compounds which are toxic or reduce digestibility. Mechanical defence mechanisms against herbivores include spines, trichomes and thorns, but also enhanced tissue toughness and gluing substances on the plant surface. A hydrophobic or fast-drying cuticle can prevent plants from various pathogen infestations because fungi and bacteria often require a water film for infection. Plants have evolved a variety of chemical defence compounds such as tannins, lignins or proteinase inhibitors (quantitative defence). These compounds act dose-dependently, with higher amounts in the plant tissue decreasing the digestive value. In addition, many toxic compounds exist which are effective, even in low concentrations (qualitative defence, Nentwig *et al.*, 2004). Prominent members are the alkaloids, cyanogenic glycosides, glucosinolates, non-protein amino acids, coumarins and terpenoids (Dey and Harborne, 1997). Defence mechanisms can also occur in combination, e.g. glandular trichomes may combine mechanical defence with chemical defence by the secretion of toxic compounds.

Plant Defence by Secondary Plant Compounds

The *Apiaceae* are rich in secondary metabolites such as coumarins, essential oils, flavones, terpenes and acetylenic compounds (Bohlmann, 1971). For a long time, various species have been used as spices, vegetables or for medicinal purposes. Many of these compounds are directly or indirectly involved in plant defence (Dey and Harborne, 1997). Most characteristic for *Apiaceae* are the furanocoumarins. The fact that they are also harmful to human health (see Thiele and Otte, Chapter 9, this volume), and because they are of pharmaceutical interest, has led to intensive research on these compounds, so we will primarily focus here on these substances.

Furanocoumarins

The substances of this class are named after coumarin (Fig. 13.1), which was isolated in 1820 by Vogel from the tonka beans of the *Dipteryx odorata* Willd. tree. He named this substance after the common word, 'cumaru', for this tree in the native South American Tupi language. Coumarins are natural compounds with the 2H-benzopyran-2-one skeleton of coumarin as a characteristic feature. The vast majority have an oxygen atom at the seventh position, whereby coumarin itself is an exception. The coumarins have been divided into four classes. Coumarins with substitutions in the benzene ring are simple coumarins. An important member of this class in the biosynthetic pathway is the umbelliferone (Fig. 13.1). Simple coumarins are widespread in the plant kingdom and several hundreds of compounds have been isolated from more than 70 plant families. Furanocoumarins (also furocoumarins) are characterized by the addition of a fused furan ring. They are a typical feature of the *Apiaceae*, especially of the tribe *Peucedaneae* to which the genus *Heracleum* belongs (Molho *et al.*, 1971). Their important contribution to

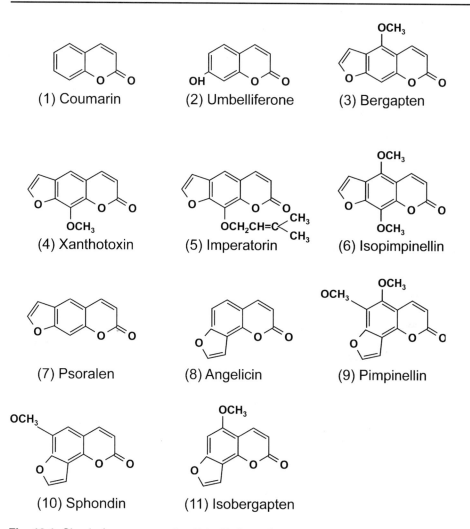

Fig. 13.1. Simple furanocoumarins (1 to 2), linear furanocoumarins (3 to 7) and angular furanocoumarins (8 to 11).

plant defence is well-documented (reviewed by Berenbaum and Zangerl, 1996). The other two classes are the pyranocoumarines, with a fused pyran ring, and the coumarins, which are substituted in the pyrone ring. Neither of these classes occur in the genus *Heracleum* (Nielsen, 1971). Here we concentrate on the furanocoumarins.

Two distinct types of furanocoumarins exist. In the linear type, the furane moiety is attached to the atoms C6 and C7 of the benzopyrane skeleton. Although linear coumarins occur sporadically in 15 families, they are ubiquitous only in the *Apiaceae* and *Rutaceae* (Murray *et al.*, 1982). The angular furanocoumarins, in which the furane moiety is attached to the atoms C7 and C8, are much less widespread. They are commonly present only in the tribes of *Apieae, Peucedaneae, Scandiceae* and *Dauceae* (Nielsen, 1971) and are an outstanding feature in the genus *Heracleum* (Murray *et al.*, 1982).

Furanocoumarins exert their toxicity in different ways. It is usually enhanced in the presence of UV-A light (320–370 nm, with a maximum at 340–360 nm) (Murray *et al.*, 1982). The phototoxic effect relies on the ability of furanocoumarins to absorb a photon, leading to a high-energy triplet state. Coumarins are capable of binding to the pyrimidine base of nuclear DNA. Photo-activated linear furanocoumarins are able to form interstrand cross-links by reacting with an additional pyrimidine base. Even though monoadducts (i.e. binding to only one pyrimidin base) are able to inhibit DNA synthesis, to produce mutations, or to cause cell death, the effects are more pronounced when cross-link formations occur. In addition, furanocoumarins, in their excited energy stage, are capable of reacting with oxygen. This may result in the generation of singlet oxygen, hydroxy radicals or superoxide anion radicals. Furthermore, furanocoumarins are able to inhibit enzymes as well as to bind to proteins and to unsaturated fatty acids (Murray *et al.*, 1982; Berenbaum and Zangerl, 1996, and references therein). An increased biosynthesis of furanocoumarins can be induced by a wide range of antagonists such as herbivorous insects, nematodes, fungi, bacteria and viruses (reviews in Murray *et al.*, 1982; Berenbaum, 1991). The details of the biosynthetic pathway of furanocoumarins are described by Stanjek and Boland (1998).

H. mantegazzianum contains a particularly high concentration of furanocoumarins. Herde (2005) investigated the coumarin contents and composition of the fruits of 36 *Apiaceae* species. *H. mantegazzianum* had, with 3.92%, the second highest concentration of furanocoumarins after *Ammi visnaga* L., which had 4.34%. Although the furanocoumarin pattern of *H. mantegazzianum* has frequently been analysed (Table 13.1), the results are difficult to compare. Besides strong differences in the solubility of the compounds in different solvents and different extraction methods, coumarins are known to differ considerably between the plant organs (Molho *et al.*, 1971; Knudson, 1983; Pira *et al.*, 1989), plant populations (Zangerl and Berenbaum, 1990; Berenbaum and Zangerl, 1998), geographical areas (reviewed by Murray *et al.*, 1982) and seasons (Knudson, 1983; Pira *et al.*, 1989). Additionally, they are also influenced by abiotic factors such as nutrient availability or UV radiation (Zangerl and Berenbaum, 1987). *Heracleum mantegazzianum* contains the linear furanocoumarins bergapten, xanthotoxin, imperatorin, isopimpinellin and psoralen (Fig. 13.1). The angular furanocoumarin, angelicin (Fig. 13.1), contributes by far the highest proportion of coumarins in the fruits. Further angular furanocoumarins are pimpinellin, and especially in the roots, sphondin and isobergapten (Fig. 13.1). Ode *et al.* (2004) and Herde (2005) found that more than half of the total furanocoumarin content of *H. mantegazzianum* belonged to angular furanocoumarins. This high proportion is surprising because all other furanocoumarin-containing plants are unexceptionally dominated by linear furanocoumarins, and angular furanocoumarins usually comprise less than 10% (Berenbaum, 1991). Among the seven other species of the tribe *Peucedaneae*, including *H. lanatum* Michx. and *H. sphondylium* L. analysed

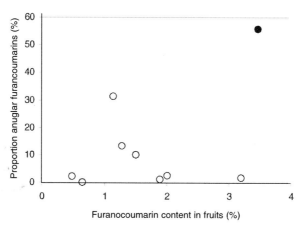

Fig. 13.2. Furanocoumarin content and proportion of angular furanocoumarins in the fruits of nine species of the tribe *Peucedaneae* (*Apiaceae*). Data gathered from Herde (2005). The black circle represents *H. mantegazzianum*.

by Herde (2005), none showed such a high proportion of angular furanocoumarins (Fig. 13.2). It is necessary to mention at this point that both analyses were conducted with plant material obtained from the invaded range of *H. mantegazzianum*. However, Komissarenko *et al.* (1965) found similar high proportions of angular furanocoumarins in the roots of plants from Russia, even if the composition was different. Molho *et al.* (1971) found different furanocoumarin patterns in the fruits of giant hogweed specimens from three botanical gardens and suggested that the invasive plants represent a group of species and do not belong only to *H. mantegazzianum* (see Jahodová *et al.*, Chapter 1, this volume).

In plants, furanocoumarins are primarily accumulated in oil canals and secretory ducts, which occur in all plant organs (Towers, 1980). These oil canals are associated with the vascular bundles. The amount of furanocoumarins, however, varies between different plant organs. In the *Apiaceae*, the highest amounts are usually located in the fruits, followed by the roots (Murray *et al.*, 1982). Pira *et al.* (1989) found the concentrations to be high in the fruit, intermediate in the leaves and low in the stems of *H. mantegazzianum*. The concentrations varied strongly during the season and the maximal concentrations were found to be asynchronous to the seasonal solar radiation. Therefore they attributed this variation to the different biological phases of the plant. Leaf and root extracts strongly inhibited the growth of the yeast *Candida albicans* Berkhout, whereas extracts of stems or stalks showed lower toxicity (Knudson, 1983). Due to the enhanced toxicity in the presence of UV radiation, furanocoumarins should be more effective on the plant's surface. In the leaves of *H. mantegazzianum*, approximately 2% of the furanocoumarins are located on the surface (Zobel *et al.*, 1990). Celery, *Apium graveolens* L., can respond to herbivory with an increased deposition of furanocoumarins on the leaf surface (Zobel and Glowniak, 1994; Stanjek *et al.*, 1997).

Table 13.1. Summary of studies screening for coumarins in *H. mantegazzianum*. Qualitative analyses: Total furanocoumarin content of each study was set as 100%. Numbers indicate the percentage of a given compound, (x) = traces; semi-quantitative analysis: ++ = high levels, + = moderate levels, (+) = traces; qualitative analyses: X = present, (X) = traces.

Reference	Country	Plant organ	Linear furanocoumarins											Angular furanocoumarins						
			Bergapten	Xanthotoxin	Imperatorin	Isopimpinellin	Psoralen	Marmesin	Columbianetin	Xanthotoxol	Oxypeucedanin	Byakangelicol	Phellopterin	Angelicin	Pimpinellin	Sphondin	Isobergapten	Isoimperatorin	Heraclesol	Unidentified
Quantitative analyses:																				
Herde (2005)	Germany	Fruits	16.7		18.6	6.0	0.5			0.3	(x)			37.5	14.2		1.1			4.9
Ode et al. (2004)	Netherlands	Fruits	21.2	7.4	4.4	6.5	4.5							55.9						
Komissarenko et al. (1965)	Caucasus	Roots	18.2	0.6	33.4										15.8	25.3	7.6			
Semi-quantitative analyses:																				
Wawrzynowicz et al. (1989)	Germany	Fruits	++	++	+	+							+							
Erdelmeier et al. (1985)	Germany	Leaves	++	++	(+)	(+)							(+)[b]	++	(+)[a]	(+)[c]	(+)[a]			
Satsyperova and Komissarenko (1978)	Caucasus	Fruits	++	+	+[a]	+	+[b]	+		+		+		++	+[a]	+[c]	+[a]		+	(+)
Qualitative analyses:																				
Berenbaum (1981)	US	Seeds	x	x	x					x				(x)						x[d]
Berenbaum (1981)	US	Leaves	(x)	(x)						(x)				(x)	x		(x)	x		x[d]
Vanhaelen and Vanhaelen-Fastré (1974)		Roots	x	x		x				x				x	x	x	x			
Molho et al. (1971)	France[e]	Fruits	x	x		x	x					x		x	x	x	x			
Molho et al. (1971)	France[e]	Leaves	x	x		x	x						x	x	x	x				
Molho et al. (1971)	France[e]	Roots	x	x		x	x					x		x		x				
Molho et al. (1971)	UK	Fruits	x	x	x	x	x	(x)	(x)					x	x	x	x			
Molho et al. (1971)	France[f]	Fruits	x	x	x	x	x	(x)	(x)					x						
Beyrich (1968)		Fruits	x	x									x	x[g]						
Karlsen et al. (1967)		Roots	x	x		x	x						x	x	x	x	x		+	
Lee et al. (1966)	USA	Roots	x	x		x									x	x	x			x

[a] = Only traces in 2 of 4 analysed plant populations.
[b] = Absent in 2 of 4 analysed plant populations.
[c] = High content in 1 of 4 analysed plant populations.
[d] = Dihydro-furanocuoumarin-glycosides.
[e] = Plant material from the botanical garden Samoëns.
[f] = Plant material from the Laboratoire d'Ecologie Générale, Brunoy.
[g] = Chief component with approximately 50%.

Furanocoumarins protect against a wide variety of organisms including vertebrates, invertebrates, fungi and bacteria, as well as DNA and RNA viruses (Murray *et al.*, 1982; Berenbaum, 1991). In arthropods, they can act as a feeding deterrent, leading to reduced growth or increased mortality (Berenbaum, 1991). Berenbaum and Zangerl (1996) reviewed studies on the biological activity of furanocoumarins and found that the degree of toxicity of the individual compounds does not follow a consistent order; the biological activity of a certain compound depends on the target organism and the feeding deterrence varies among taxa. Additionally, the importance of UV radiation is not consistent. In many studies, angelicin is less active compared to other fura-nocoumarins in the presence of UV radiation, but shows higher activity than other furanocoumarins in the absence of UV radiation. It is also difficult to state which of the various biological types of activity is most responsible for the toxic effects. Although it is generally assumed that the ability of fura-nocoumarins to bind to nuclear DNA is most responsible for the phototoxic reaction in human skin, there is some evidence that other mechanisms are also involved (Laskin *et al.*, 1985). For herbivorous insects, evidence is accumulat-ing that the inhibition of cytochrome P450 enzymes, which are involved in the detoxification of xenobiotics (see below), plays an important role (e.g. Neal and Wu, 1994). Synergistic effects of furanocoumarin mixtures have been reported from various studies. A combination of several furanocoumarins usually exhibits higher toxicity than each compound alone (Berenbaum and Zangerl, 1996; Calcagno *et al.*, 2002).

The detoxification of furanocoumarins in specialized insects is mediated via cytochrome P450 monooxygenases (Ivie *et al.*, 1983; Berenbaum, 2002). The resistance of insects is related to a faster and more efficient detoxification of this compound, whereby the induction level of P450 genes by increased enzymatic activity is of particular importance (Feyereisen, 1999; Li *et al.*, 2004). Berenbaum and Zangerl (1993) found a strong synergistic effect of the linear xanthotoxin and the angular angelicin against the black swallowtail *Papilio polyxenes* F., a lepidopteran specialist feeding on *Apiaceae*. The toxic effect was related to a reduced catabolism rate of both compounds mediated by angelicin. This resulted in enhanced levels of unmetabolized xanthotoxin. Likewise synergistic effects were found in *Depressaria radiella* (Goeze) (syn. *D. pastinacella*) (Nitao, 1989). Instead of detoxification, insects can avoid contact with toxic substances. The two aphid species, *Aphis heraclella* Davis and *Cavariella pastinacea* L., did not contain or excrete xanthotoxin after feeding on *Heracleum lanatum* (Camm *et al.*, 1976). The authors concluded that this compound is not translocated in the phloem. Berenbaum (1978) sug-gested that the spinning of a web on flower heads or the rolling of leaves might have been developed to avoid UV radiation. Furanocoumarins can also act on a multitrophic level. Larvae of *D. radiella* are less commonly parasitized by the parasitic wasp *Copidosoma sosares* (Walker) when feeding on *H. mantegazzianum* plants with a high amount of xanthotoxin (Ode *et al.*, 2004).

The high amount of furanocoumarins in the roots of many *Apiaceae* is still not understood. Certainly there are some biological activities in the absence of

UV light, especially against microorganisms. However, high contents of furanocoumarins within the *Apiaceae* are, as a rule, associated with plants occurring in light-exposed habitats and are less pronounced in woodland species
(Berenbaum, 1981), which underlines the importance of light.

Essential oils

All members of the *Apiaceae* are aromatic plants. The essential oils are
excreted in schizogenous canals in the roots, stems, leaves and inflorescences,
as well as in the oil ducts of the seeds (vittae) (Hegnauer, 1971). According to
Tkachenko (1993), the fruits of *H. mantegazzianum* contain approximately
5.5–7.0% essential oils, with octyl butyrate (32%), octyl acetate (18%) and
hexyl butyrate (9%) as major components. Further components (with contents
between 2.5% and 4%) are, in order of decreasing proportions: octyl isovaler-
ate, octanol, hexyl valerate, borneol, limonene and bornyl acetate. The levels
of octyl and hexyl butyrate are higher and the level of octyl acetate is lower in
H. mantegazzianum as compared to other *Heracleum* species (Tkachenko,
1993; Iscan *et al.*, 2004). The content of essential oils in the leaves and roots
of *H. mantegazzianum* varies highly during the vegetation period. The constituents include myrcene, α-pinene, decanol, octanol, carvacrol, cineol,
ocimene and borneol (Jain, 1969).

   Essential oils impart the characteristic odours of plants and therefore often
act as olfactory attractants or repellents to herbivorous or pollinating insects.
Limonene is used commercially in some insect repellent sprays. In contrast,
octanol is the active ingredient in several mosquito traps. Many essential oils
have antifeedant effects, e.g. octyl acetate and octyl butyrate act as deterrents
against *D. radiella* (Carroll and Berenbaum, 2002). Additionally, the essential
oils of some *Heracleum* species exhibit antibacterial, antifungal and antiviral
activities (Jain, 1969; Tkachenko *et al.*, 1995; Iscan *et al.*, 2004). Carroll *et
al.* (2000) found that the amount of the octyl esters was positively correlated
with the amount of linear furanocoumarins in the fruits of *Pastinaca sativa* L.
They suggested that the esters may also serve as carrier solvents and that they
probably enhance the penetration of furanocoumarins into the herbivore
integuments and gut walls.

Acetylenic compounds

Acetylenic compounds with a chain length of C-17 are very common in the
Apiaceae and the toxic effect of many species of this family is caused by these
compounds (Hegnauer, 1971). Two C-17 polyacetylenes (9-heptadecene-4,6-
diyn-8-ol and 2,9-heptadecadiene-4,6-diyn-8-ol) have been isolated from the
roots of giant hogweed, but with unknown bioactivity (Vanhaelen-Fastré and
Vanhaelen, 1973). Falcarindiol and panaxynol have been found in other
Heracleum species (Liu *et al.*, 1998; Nakano *et al.*, 1998). However, to what
extent acetylenic compounds contribute to plant defence in *H. mantegaz-*

zianum remains unknown. Combined with furanocoumarins, polyacetylenes have a high stimulatory effect on the oviposition of the oligophagous carrot fly, *Psila rosae* (F.), in *H. sphondylium* (Degen *et al.*, 1999).

Flavonoids

Characteristic of *Apiaceae* is a lack of true tannins, such as gallic and ellagic acids and catechins (Hegnauer, 1971). Siwon and Karlsen (1976) isolated quercitin, kaempferol and isorhamnetin from the flowers of *H. mantegazzianum*. According to Harborne (1971), the red anthocyanid pigment cyanidine and its glycoside, cyanidine 3-sambubioside, are present in the stem of giant hogweed. Although occasionally a deterrent effect of flavonoids has been reported, most herbivores are well-adapted. However, flavonoids have a strong affinity to proteins, are partially strong inhibitors of several enzyme systems and may lead to toxic effects via enzyme interaction (Harborne, 1991). In humans and rats, quercitin was found to inhibit some cytochrome P450 enzymes (Li *et al.*, 1994). Therefore the effectiveness of furanocoumarins might be enhanced in the presence of flavonoids.

Other secondary plant compounds

High levels of alkaloids are uncommon in the *Apiaceae*. For a long time, it was believed that only *Conium maculatum* L. contains alkaloids (Fairbaine, 1971). Later on, piperidine alkaloids were reported for the genus *Heracleum* (Plouviet, 1982). The caffeic acid ester, chlorogenic acid, is ubiquitous in the *Apiaceae* (Harborne, 1971). Chlorogenic acid has wide antiviral and antibacterial effects, and its contribution to anti-herbivore defences has been reported from many plant species. Triterpenes occur widely in the *Apiaceae*. However, high concentrations are rare and free triterpene acids seem to be absent in the cuticular waxes (Hegnauer, 1971).

Trichomes

'Kill them with your hogweed hairs' sang Peter Gabriel in 1971 in the Genesis song 'The Return of the Giant Hogweed'. It remains doubtful whether Genesis were trying to emphasize the meaning of trichomes in *H. mantegazzianum*. Nevertheless, the plant's hairs do make a significant contribution to its defence. Root hairs are the sole exception. They are not involved in plant protection and will not be discussed here. Ochsmann (1996) investigated the morphology of the above-ground trichomes in *H. mantegazzianum* in detail. He described five different kinds of trichomes: simple hairs, hairs on pedestals, hairs with globular-shaped tops, triangular-shaped hairs and spines. All types have in common that they are unicellular and non-glandular trichomes. The simple hairs are usually shorter than 0.5 mm. They occur predominantly along the veins on the lower surface of the leaves (see also Hansen *et al.*, Chapter 11, this volume). The

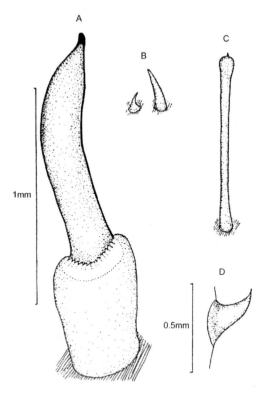

Fig. 13.3. Different plant hairs of *H. mantegazzianum*. (A) Hairs on pedestals; (B) triangular shaped hairs; (C) hairs with globular-shaped tops; (D) spines. Based on Ochsmann (1996), modified. Published with permission from the author.

second hair type (Fig. 13.3A) comprises a unicellular hair located on a reddish pedestal-like emergence. They are not trichomes in the strict sense, because the pedestal is probably derived from epidermal and subepidermal cell layers. However, to avoid confusion we will also use the term trichome in this case. This type is by far the longest, with a length of up to 5 mm. Their occurrence is restricted to the leaf petioles and flower stems. They can exhibit remarkable densities, especially on young petioles (Fig. 13.4). Their sticky surface impedes the movement of herbivores and may trap small species. The hairs with globular-shaped tops (Fig. 13.3C) emerge on the pedicles of the inflorescences and on the ovaries. They are approximately 1 mm long and possess a small conus at the end of their head. The hairs on the ovaries degenerate during fruit development. Small triangular-shaped hairs (Fig. 13.3B) occur on the pedicles. Spines (Fig. 13.3D) are restricted to the fruit margins in *H. mantegazzianum*.

   Trichomes may complement the chemical defence of a plant by secreting or storing toxic and repellent compounds. Despite extensive work on furanocoumarins, surprisingly few studies have examined them in combination with trichomes. In the only study we are aware of, Zobel and March (1993) observed furanocoumarins on the surface of trichomes on the seeds of *Daucus carota* L.

The shape of the trichomes with globular tops resembles glandular trichomes or the stinging hairs in other plant species. Therefore they might contain secondary defence compounds. On the other hand, the sizeable hairs on pedestals are situated on the plant organs with the lowest furanocoumarin contents.

Besides defence, trichomes may serve other functions as well. For instance, they can protect from UV radiation (via reflection, absorption or shading), reduce water loss (via an increased water diffusion pathway or by trapping moisture), or alter the impact of wind (Press, 1999). However, these attributes are usually related to leaf pubescence. Given that plant hairs in *H. mantegazzianum* are absent from the upper side of leaves and are not located near stomata on the lower leaf surface, it is unlikely that they contribute significantly to the plant's energy or water budget. In addition, protection against UV light would simultaneously reduce the effectiveness of the phototoxic chemical defence compounds. The spines on the fruits are an exception. They are presumably involved in the dispersal of seed via biotic vectors (see Moravcová *et al.*, Chapter 5, this volume).

Defence in Native and Invasive Populations

One prominent explanation for the success of some invasive plant species is the 'enemy release hypothesis', which predicts that plants are less controlled by their natural enemies in areas where a plant species has been introduced

Fig. 13.4. Plant hairs on petioles of *H. mantegazzianum*. Photo: J. Hattendorf.

(e.g. Keane and Crawley, 2002). Plant species which are strongly regulated by their enemies in the native range should immediately benefit from reduced attack in the invaded range. In contrast, well-defended plants gain a competitive advantage only after evolutionary adaptation (Colautti *et al.*, 2004). In the absence of enemies, selection pressure should favour re-allocation of resources from defence towards growth or reproduction, which leads consequently to an 'evolution of increased competitive ability' (EICA hypothesis – Blossey and Nötzold, 1995). Evidence is accumulating that invasive plants can undergo a rapid evolution in their new environment (Bossdorf *et al.*, 2005). Müller-Schärer *et al.* (2004) argued that plant invaders are not released from all enemies, but only from specialists. Defence traits might be associated with indirect costs in the presence of specialists. For instance, plant toxins can deter generalists, while adapted enemies use these chemicals as recognition cues or as feeding and oviposition stimulants in the native range. In the invaded range, where high levels of toxins are not associated with the attraction of specialists, plant toxins might increase in concentration, contrary to the prediction of the EICA hypothesis. In return, the plant is able to invest fewer resources in defence traits against specialists which might have higher allocation costs such as digestibility-reducing compounds or mechanical protection. This line of argumentation was supported by Joshi and Vrieling (2005). In common garden experiments, invasive populations of *Senecio jacobaea* have on average, higher alkaloid concentrations and higher resistance against generalists than native populations. However, at the same time they were less protected against specialist herbivores.

Hattendorf (2005) investigated the toxicity of plant extracts and the performance of plant hairs on leaf petioles in native and invasive populations of *H. mantegazzianum*. Trichomes are assumed to have particularly high allocation costs, because their formation depends on growth processes (Gutschick, 1999). In addition, trichomes can limit the efficiency of predators of herbivorous insects (Gassmann and Hare, 2005). In bioassays, plant extracts from invasive plant populations showed a higher toxicity against brine shrimps, *Artemia salina* (L.), as compared to plants from the native range. In contrast, plants in the invaded range had fewer and shorter trichomes. It seems that *H. mantegazzianum* exhibits the pattern predicted by Müller-Schärer *et al.* (2004). Allocation of resources at higher furanocoumarin levels might be over-compensated by lower resource investment in the more costly defence by trichomes. Ode *et al.* (2004) found similar patterns for furanocoumarins in *Pastinaca sativa*. Several furanocoumarins were found in lower concentrations in native populations in Europe as compared to populations in the USA, where the plant is invasive. Unfortunately, their study included only two European populations and therefore other explanations such as geographical distinctions might cause the differences.

Two studies investigated the variation in furanocoumarins between plant populations of *H. mantegazzianum*. Satsyperova and Komissarenko (1978) analysed fruits of four Caucasian populations. The samples covered a large area of the native range and originated from the northern and the southern slopes of the mountains (North Ossetia; Republic of Adygea; Russian

Federation: Krasnodar Krai; Abkhazia). The composition of furanocoumarins was almost identical in all cases (Table 13.1). In contrast, Molho *et al.* (1971) found significant differences in the number of compounds present in fruits from three *H. mantegazzianum* populations in Central Europe. The number of isolated furanocoumarins ranged from four to nine compounds (Table 13.1). This pattern of divergence in the invaded region as compared to native distribution range is quite surprising and could indicate different adaptation processes in the new environment. However, Molho *et al.* (1971) stated that the complex taxonomy of giant hogweeds (see Jahodová *et al.*, Chapter 1, this volume) could have led to misidentifications because similar or closely related species not belonging to *H. mantegazzianum* may have been identified as such.

Conclusion

H. mantegazzianum possesses a broad array of chemical and mechanical defence mechanisms. Seventeen furanocoumarins have been isolated from different plant organs. Of particular interest is the high proportion of angular furanocoumarins, which has never been reported from any other plant species. There is some evidence that changes in defence traits might have occurred following the introduction to new environments. Further investigations in this topic could help to understand the high invasive potential of this invasive plant.

Acknowledgements

We acknowledge valuable comments on an earlier version of this contribution by two referees and language editing by Cecily Klingler. The study was supported by the project 'Giant Hogweed (*Heracleum mantegazzianum*) a pernicious invasive weed: developing a sustainable strategy for alien invasive plant management in Europe', funded within the 'Energy, Environment and Sustainable Development Programme' (grant no. EVK2-CT-2001-00128) of the European Union 5th Framework Programme/Switzerland (BBW: EVK2-CT-2001-00128).

References

Agrawal, A.A. (2000) Specificity of induced resistance in wild radish: causes and consequences for two specialist and two generalist caterpillars. *Oikos* 89, 439–500.

Berenbaum, M.R. (1978) Toxicity of a furanocoumarin to armyworms: a case of biosynthetic escape from insect herbivores. *Science* 201, 532–534.

Berenbaum, M.R. (1981) Patterns of furanocoumarin distribution in the *Umbelliferae*: plant chemistry and community structure. *Ecology* 62, 1254–1266.

Berenbaum, M.R. (1991) Coumarins. In: Rosenthal, G.A. and Berenbaum, M.R. (eds) *Herbivores, Their Interaction with Secondary Plant Compounds*, 2nd edn. Academic Press, San Diego, California, pp. 221–250.

Berenbaum, M.R. (2002) Postgenomic chemical ecology: from genetic code to ecological interactions. *Journal of Chemical Ecology* 28, 873–896.

Berenbaum, M.R. and Zangerl, A.R. (1993) Furanocoumarin metabolism in *Papilio polyxenes*: genetic variability, biochemistry, and ecological significance. *Oecologia* 95, 370–375.

Berenbaum, M.R. and Zangerl, A.R. (1996) Phytochemical diversity: adaptation or random variation. In: Romeo, J.T., Saunders, J.A. and Barbosa, P. (eds) *Recent Advances in Phytochemistry*. Plenum Press, New York, pp. 1–24.

Berenbaum, M.R. and Zangerl, A.R. (1998) Chemical phenotype matching between a plant and its insect herbivore. *Proceedings of the National Academy of Sciences of the United States of America* 95, 13743–13748.

Beyrich, T. (1968) Vergleichende Untersuchungen über das Vorkommen von Furocumarinen bei einigen Arten von *Heracleum*. *Pharmazie* 23, 336–339.

Blossey, B. and Nötzold, R. (1995) Evolution of increased competitive ability in invasive nonindigenous plants: a hypothesis. *Journal of Ecology* 83, 887–889.

Bohlmann, F. (1971) Acetylenic compounds in the *Umbelliferae*. In: Heywood, V.H. (ed.) *The Biology and Chemistry of the Umbelliferae*. Academic Press, London, pp. 279–292.

Bossdorf, O., Auge, H., Lafuma, L., Rogers, W.E., Siemann, E. and Prati, D. (2005) Phenotypic and genetic differentiation between native and introduced plant populations. *Oecologia* 144, 1–11.

Calcagno, M.P., Coll, J., Lloria, J., Faini, F. and Alonso-Amelot, M.E. (2002) Evaluation of synergism in the feeding deterrence of some furanocoumarins on *Spodoptera littoralis*. *Journal of Chemical Ecology* 28, 175–191.

Camm, E.L., Wat, C. and Towers, G.H.N. (1976) An assessment of the roles of furanocoumarins in *Heracleum lanatum*. *Canadian Journal of Botany* 54, 2562–2566.

Carroll, M.J. and Berenbaum, M.R. (2002) Behavioral responses of the parsnip webworm to host plant volatiles. *Journal of Chemical Ecology* 28, 2191–2201.

Carroll, M.J., Zangerl, A.R. and Berenbaum, M.R. (2000) Heritability estimates for octyl acetate and octyl butyrate in the mature fruit of the wild parsnip. *Journal of Heredity* 91, 68–71.

Colautti, R.I., Ricciardi, A., Grigorovich, I.A. and MacIsaac, H.J. (2004) Is invasion success explained by the enemy release? *Ecology Letters* 7, 721–733.

Crawley, M.J. (ed.) (1997) *Plant Ecology*. Blackwell Science, Oxford, UK.

Degen, T., Buser, H.-R. and Städler, E. (1999) Patterns of oviposition stimulants for carrot fly in leaves of various host plants. *Journal of Chemical Ecology* 25, 67–87.

Dey, P.M. and Harborne, J.B. (1997) *Plant Biochemistry*. Academic Press, San Diego, California.

Erdelmeier, C.A.J., Meier, B. and Sticher, O. (1985) Reversed-phase high-performance liquid chromatographic separation of closely related furocoumarins. *Journal of Chromatography* 346, 456–460.

Fairbaine, J.W. (1971) The alkaloids of hemlock (*Conium maculatum* L.). In: Heywood, V.H. (ed.) *The Biology and Chemistry of the Umbelliferae*. Academic Press, London, pp. 361–368.

Feyereisen, R. (1999) Insect P450 enzymes. *Annual Review of Entomology* 44, 507–533.

Gassmann, A.J. and Hare, J.D. (2005) Indirect cost of a defensive trait: variation in trichome type affects the natural enemies of herbivorous insects on *Datura wrightii*. *Oecologia* 144, 62–71.

Gutschick, V.P. (1999) Biotic and abiotic consequences of differences in leaf structure. *New Phytologist* 143, 3–18.

Harborne, J.B. (1971) Flavonoid and phenylpropanoid patterns in the *Umbelliferae*. In: Heywood, V.H. (ed.) *The Biology and Chemistry of the Umbelliferae*. Academic Press, London, pp. 293–314.

Harborne, J.B. (1991) Flavonoid pigments. In: Rosenthal, G.A. and Berenbaum, M.R. (eds)

Herbivores, Their Interaction with Secondary Plant Compounds, 2nd edn. Academic Press, San Diego, California, pp. 389–429.

Hattendorf, J. (2005) Impact of endophagous herbivores on the invasive weed *Heracleum mantegazzianum* and associated interactions. PhD thesis. University of Bern, Switzerland.

Hegnauer, R. (1971) Chemical patterns and relationships of *Umbelliferae*. In: Heywood, V.H. (ed.) *The Biology and Chemistry of the Umbelliferae*. Academic Press, London, pp. 267–277.

Herde, A. (2005) Untersuchung der Cumarinmuster in Früchten ausgewählter *Apiaceae*. PhD thesis, University of Hamburg, Germany.

Iscan, G., Ozek, T., Ozek, G., Duran, A. and Baser, K. (2004) Essential oils of three species of *Heracleum*. Anticandidal activity. *Chemistry of Natural Compounds* 40, 544–547.

Ivie, G.W., Bull, D.L., Beier, R.C., Pryor, N.W. and Oertli, E.H. (1983) Metabolic detoxification: mechanism of insect resistance to plant psoralens. *Science* 221, 374–376.

Jain, S.R. (1969) Investigations on the essential oil of *Heracleum mantegazzianum*. *Planta Medica* 17, 230–235.

Joshi, J. and Vrieling, K. (2005) The enemy release and EICA hypothesis revisited: incorporating the fundamental difference between specialist and generalist herbivores. *Ecology Letters* 8, 704–714.

Karban, R. and Baldwin, I.T. (1997) *Induced Responses to Herbivory*. University of Chicago Press, Chicago, Illinois.

Karlsen, J., Van Hagen, P. and Berheim Svendsen, A. (1967) Furanocoumarins of *Heracleum mantegazzianum*. *Meddelelser fra Norsk Farmaceutisk Selskap* 29, 153–157.

Keane, R.M. and Crawley, M.J. (2002) Exotic plant invasions and the enemy release hypothesis. *Trends in Ecology and Evolution* 17, 164–170.

Knudson, E.A. (1983) Seasonal variation in the content of phototoxic compounds in giant hogweed. *Contact Dermatitis* 9, 281–284.

Komissarenko, N.F., Chenobay, I.G. and Kolesnikov, D.G. (1965) Cumarins from roots of Giant Hogweed (*Heracleum mantegazzianum* S. et L.) and Incised Hogweed (*H. dissectum* Lebed.). *Acta Institutum Botanicum Academis Scientarum URSS, Series 5* 12, 58–61. [In Russian.]

Lagey, K., Duinslaeger, L. and Vanderkelen, A. (1995) Burns induced by plants. *Burns* 21, 542–543.

Laskin, J.D., Lee, E., Yurkow, E.D., Laskin, D.L. and Gallo, M.A. (1985) A possible mechanism of psoralen phototoxicity not involving direct interaction with DNA. *Proceedings of the National Academy of Sciences of the United States of America* 82, 6158–6162.

Lee, E.C., Catalfomo, P. and Sciuchetti, L.A. (1966) Preliminary investigations on *Heracleum mantegazzianum*. *Journal of Pharmaceutical Science* 55, 521–522.

Li, W., Zangerl, A.R., Schuler, M.A. and Berenbaum, M.R. (2004) Characterization and evolution of furanocoumarins-inducible cytochrome P450s in the parsnip webworm, *Depressaria pastinacella*. *Insect Molecular Biology* 13, 603–613.

Li, X.-Y., Wang, E., Patten, C.J., Chen, L. and Yang, C.S. (1994) Effects of flavonoids on cytochrome P450-dependent acetaminophen metabolism in rats and human liver microsomes. *Drug Metabolism and Disposition* 22, 566–571.

Liu, J.H., Zschocke, S., Reininger, E. and Bauer, R. (1998) Comparison of radix *Angelicae pubescentis* and substitutes: constituents and inhibitory effect on 5-lipoxygenase and cyclooxygenase. *Pharmaceutical Biology* 36, 207–216.

Molho, D., Jössang, P., Jarreau, M.-C. and Carbonnier, J. (1971) Dérivés furannocoumariniques du genre *Heracleum*. In: Heywood, V.H. (ed.) *The Biology and Chemistry of the Umbelliferae*. Academic Press, London, pp. 337–360.

Müller-Schärer, H., Schaffner, U. and Steinger, T. (2004). Evolution in invasive plants and implications for biological control. *Trends in Ecology and Evolution* 19, 417–421.

Murray, R.D.H., Méndez, J. and Brown, S.A. (1982) *The Natural Coumarins: Occurrence, Chemistry and Biochemistry.* Wiley, Chichester, UK.

Nakano, Y., Matsunaga, H., Saita, T., Mori, M., Katano, M. and Okabe, H. (1998) Antiproliferative constituents in *Umbelliferae* plants II. Screening for polyacetylenes in some *Umbelliferae* plants, and isolation of panaxynol and falcarindiol from the root of *Heracleum moellendorffii. Biological and Pharmaceutical Bulletin* 21, 257–261.

Neal, J.J. and Wu, D. (1994) Inhibition of insect cytochromes P450 by furanocoumarins. *Pesticide Biochemistry and Physiology* 50, 43–50.

Nentwig, W., Bacher, S., Beierkuhnlein, C., Brandl, R. and Grabherr, G. (2004) *Ökologie.* Spektrum, Heidelberg, Germany.

Nielsen, B.E. (1971) Coumarin patterns in the *Umbelliferae.* In: Heywood, V.H. (ed.) *The Biology and Chemistry of the Umbelliferae.* Academic Press, London, pp. 337–360.

Nitao, J.K. (1989) Enzymatic adaptation in a specialist herbivore for feeding on furanocoumarin-containing plants. *Ecology* 70, 629–635.

Ochsmann, J. (1996) *Heracleum mantegazzianum* Sommier & Levier (*Apiaceae*) in Deutschland. Untersuchungen zur Biologie, Verbreitung, Morphologie und Taxonomie. *Feddes Repertorium* 107, 557–595.

Ode, P.J., Berenbaum, M.R., Zangerl, A.R. and Hardy, I.C.W. (2004) Host plant, host plant chemistry and the polyembryonic parasitoid *Copidosoma sosares*: indirect effects in a tritrophic interaction. *Oikos* 104, 388–400.

Otte, A. and Franke, R. (1998) The ecology of the Caucasian herbaceous perennial *Heracleum mantegazzianum* Somm. et Lev. (Giant Hogweed) in cultural ecosystems of Central Europe. *Phytocoenologia* 28, 205–232.

Pira, E., Romano, C., Sulotto, F., Pavan, I. and Monaco, E. (1989) *Heracleum mantegazzianum* growth phases and furocoumarin content. *Contact Dermatitis* 21, 300–303.

Plouviet, V. (1982) Ombellifères et familles voisines: leurs analogies et leurs distinctions biochimiques. *Monographs in Systematic Botany from the Missouri Botanical Garden* 6, 535–548.

Press, M.C. (1999) The functional significance of leaf structure: a search for generalizations. *New Phytologist* 143, 213–219.

Pyšek, P. (1994) Ecological aspects of invasion by *Heracleum mantegazzianum* in the Czech Republic. In: de Waal, L.C., Child, L., Wade, P.M. and Brock, J.H. (eds) *Ecology and Management of Invasive Riverside Plants.* Wiley, Chichester, UK, pp. 45–54.

Satsyperova, I.F. and Komissarenko, N.F. (1978) Chemical system of genus *Heracleum* L. of the USSR flora. Part 2: Pubescentia and Villosa sections. *Rastitel'nye Resursy* 14, 333–347. [In Russian.]

Siwon, J. and Karlsen, J. (1976) The isolation and identification of flavonoid aglycones from *Heracleum mantegazzianum* Somm. and Lev. *Meddelelser fra Norsk Farmaceutisk Selskap* 38, 11–12.

Stanjek, V. and Boland, W. (1998) Biosynthesis of angular furanocoumarins: mechanisms and stereochemistry of the oxidative dealkylation of columbianetin to angelicin in *Heracleum mantegazzianum* (*Apiaceae*). *Helvetica Chimica Acta* 81, 1596–1607.

Stanjek, V., Herhaus, C., Ritgen, U., Boland, W. and Städler, E. (1997) Changes in the leaf surface chemistry of *Apium graveolens* (*Apiaceae*) stimulated by jasmonic acid and perceived by a specialist insect. *Helvetica Chimica Acta* 80, 1408–1420.

Tkachenko, K.G. (1993) Constituents of essential oils from fruit of some *Heracleum* species. *Journal of Essential Oil Research* 5, 687–689.

Tkachenko, K.G., Platonov, V.G. and Satsyperova, I.F. (1995) Antiviral and antibacterial activity of essential oils from fruits of species of the genus *Heracleum* L. *Rastitel'nye Resursy* 31, 9–19. [In Russian.]

Towers, G.H.N. (1980) Photosensitizers from plants and their photodynamic action. *Progress in Phytochemistry* 6, 183–202.

Vanhaelen, M. and Vanhaelen-Fastré, R. (1974) Furanocoumarins from the root of *Heracleum mantegazzianum*. *Phytochemistry* 13, 113.

Vanhaelen-Fastré, R. and Vanhaelen, M. (1973) Polyacetyleniques en C17 des raciness d'*Heracleum mantegazzianum*. *Phytochemistry* 12, 2687–2689.

Wawrzynowicz, T., Waksmundzka-Hajnos, M. and Bieganowska, M.L. (1989) Chromatographic investigations of furanocoumarins from *Heracleum* genus fruits. *Chromatographica* 28, 161–166.

Weber, E. (2004) *Invasive Plant Species of the World: A Reference Guide to Environmental Weeds*. CABI, Wallingford, UK.

Zangerl, A.R. and Berenbaum, M.R. (1987) Furanocoumarins in wild parsnip: effects of photosynthetically active radiation, ultraviolet light, and nutrients. *Ecology* 68, 516–520.

Zangerl, A.R. and Berenbaum, M.R. (1990) Furanocoumarin induction in wild parsnip: genetics and population variation. *Ecology* 71, 1933–1940.

Zobel, A.M. and Glowniak, K. (1994) Concentrations of furanocoumarins under stress conditions and their histological localization. *Acta Horticulturae* 381, 510–516.

Zobel, A.M. and March, R.E. (1993) Autofluorescence reveals different histological localizations of furanocoumarins in fruit of some *Umbelliferae* and *Leguminosae*. *Annals of Botany* 71, 251–255.

Zobel, A.M., Brown, S.A. and Glowniak, K. (1990) Localization of furanocoumarins in leaves, fruits, and seeds of plants causing contact photodermatitis. *Planta Medica* 56, 571–572.

14 Mechanical and Chemical Control of *Heracleum mantegazzianum* and *H. sosnowskyi*

CHARLOTTE NIELSEN,[1] INETA VANAGA,[2] OLGA TREIKALE[2] AND ILZE PRIEKULE[2]

[1]*Royal Veterinary and Agricultural University, Hørsholm, Denmark;* [2]*Latvian Plant Protection Research Centre, Riga, Latvia*

They are immune to all our herbicidal battering

(Genesis, 1971)

Introduction

Much attention has been directed towards *Heracleum mantegazzianum* Sommier & Levier as an invasive plant species as it is the tallest herb in Europe, spreads rapidly especially in riparian habitats and causes severe burns to human skin (Tiley *et al.*, 1996; Nielsen *et al.*, 2005). Accordingly, much effort has been invested in controlling the plant, but with varying success. Once established, a stand is very persistent due to regeneration (see Pyšek *et al.*, Chapter 7, this volume) and large seed production (see Perglová *et al.*, Chapter 4, this volume), so that control may not be successful unless based on a long-term coordinated strategy. Experiments have shown that a small fraction of the seed production is able to survive in the soil for at least 3 years, and therefore control should continue for several years to deplete the soil seed bank (for details on its long term dynamics, see Moravcová *et al.*, Chapter 5, this volume).

Statistics for the economic costs resulting from the invasion of *H. mantegazzianum* are not available at the European level, and because the scale of control effort is highly variable between local authorities, it is difficult to estimate the costs of control even on a national or regional level. An investigation of expenditure by district councils in Scotland on control of *H. mantegazzianum* showed that costs and control varied considerably between areas, depending on the distribution and abundance of the plant. Areas of less than 1 ha cost 1–30 labour hours. For larger areas (9–19 ha), the cost varied from

1600 to 10,800 labour hours (Sampson, 1994). Similarly divergent results were obtained from an interview survey of municipalities in Denmark in 2005. Average annual expenditure on control was 10,000 € but varied between municipalities from 50 € to more than 60,000 € per year (Sørensen and Buttenschøn, 2005). For the whole area of Germany, annual costs associated with invasion by *H. mantegazzianum* were estimated to exceed 12 million € (see Thiele and Otte, Chapter 9, this volume).

These figures illustrate that on a national scale, though unevenly distributed, much time and significant amounts of money are invested, stressing the importance of an integrated control strategy to ensure that economic and labour resources are spent effectively. The degree of success achieved also depends on the methods used, frequency of treatment and the phenological stage of the plant. Recommendations of control methods should be based on scientific experiments and the experiences of practitioners, and in this chapter we aim to describe and evaluate the mechanical and chemical control options available against large invasive *Heracleum* species. Most experiences in managing invasive *Heracleum* species are from control studies of *H. mantegazzianum*. Thus, the results and recommendations for mechanical and chemical control summarized here mainly refer to *H. mantegazzianum* but may well apply to *H. sosnowskyi*, as it has a similar life strategy (see Jahodová *et al.*, Chapter 1, and Moravcová *et al.*, Chapter 10, this volume). *H. sosnowskyi* is widely distributed in the Baltic countries (see Jahodová *et al.*, Chapter 1, this volume), infesting large river catchments, and leading to changes in local biodiversity (Laivins and Gavrilova, 2003).

Mechanical Control

Mechanical control includes umbel removal, cutting of stems, mowing and root cutting. Factors such as time of treatment, frequency of treatment and cutting height affect the degree of control. Apart from root cutting, which causes immediate plant death, eventual plant mortality is a result of depletion of nutrient reserves in the roots.

Umbel removal (or cutting of the flowering stem) represents a special case of cutting that may seem very tempting in terms of the limited number of labour hours compared to other cutting techniques. However, the control method often fails to fully prevent seed production due to the high regeneration potential of the species (Nielsen *et al.*, 2005; see Pyšek *et al.*, Chapter 7, this volume).

Due to the simple and quick handling involved, cutting of leaves and flower stems is a commonly used method, particularly for control of smaller stands and scattered individuals. However, as mentioned, *H. mantegazzianum* displays a high degree of regrowth in response to removal of vegetative and reproductive parts and thus cutting is often reported to be ineffective (Lundström, 1989; Lundström and Darby, 1994; Sampson, 1994), although the unsatisfactory level of control is also likely to be related to the lack of follow-up treatments. A study in Ireland by Caffrey (1994) demonstrated that a single cutting treatment is not sufficient to prevent seed production. Plants of *H. mantegazzianum* were cut just above ground level either on 10 May or 4 July. The cutting treatment resulted in only 5–10% plant mortality, and cutting late in the growing season (4 July) induced rapid flowering and seed production (Caffrey, 1994). Similar studies have tested the plant response to cutting in early March and late May (Caffrey, 1999), at peak flowering in June (Pyšek *et al.*, 1995) or the effect of cutting height (Tiley and Philp, 2000). For all studies, a single cutting treatment resulted in reduced fecundity, but the fruits produced would still be sufficient to maintain infestation at a site (Caffrey, 1994; Pyšek *et al.*, 1995; Tiley and Philp, 2000).

Results of cutting experiments demonstrate that very frequent cutting would be required to achieve an appreciable degree of control, and it can take several years to kill non-flowering vegetative plants (Dodd *et al.*, 1994). Rubow (1990) showed that cutting with a scythe four times a year for 2 years caused no mortality among the plants. For further review of regeneration following cutting and the effects on fecundity and seed quality of *H. mantegazzianum*, see Pyšek *et al.* (Chapter 7, this volume).

In contrast to results obtained by Rubow (1990) for *H. mantegazzianum*, frequent cutting of *H. sosnowskyi* Manden in a 2-year control study in Latvia resulted in a high mortality of plants (see Ravn *et al.*, Chapter 17, this volume). Particularly in riparian habitats, where the risk of soil erosion and reinvasion is high, the management of tall invasive hogweed species should focus on both the eradication of the plant and the re-establishment of a competitive grass sward, promoted by frequent cutting. The results of the study showed that the competitive ability of the grass sward can be further enhanced by sowing of appropriate grass mixtures (see Ravn *et al.*, Chapter 17, this volume). Though less important in cultivated areas, native species should be used, ideally of local provenance. This is particularly important on sites of conservation value. We believe that these results for *H. sosnowskyi* are also likely to be applicable to control of *H. mantegazzianum*.

Another approach to obtain an acceptable level of control in a dense population is to cut only the flowering plants at mid-flowering stage. Normally plants die after seed set. However, by cutting the plants at this stage, seed production is prevented, and most plants will not die. An advantage of this method is that by leaving the larger vegetative plants untreated, first year plants will be shaded out, thereby reducing the labour effort needed the following year. When repeated carefully, this strategy should eradicate the population in a few years (Nielsen *et al.*, 2005). For larger infested sites accessible to heavy machinery, a flail mower can be used 2–3 times from April–May until flowering.

The response of *H. mantegazzianum* to root cutting differs markedly from the above-ground cutting treatments. Flowering and vegetative plants die if cut below ground level, although depth of cutting plays an important role (Tiley and Philp, 1992). The tap-roots of older plants develop a solid stem stock between the true root and the stem base bearing old leaf scars (Tiley *et al.*, 1996). When cutting the root the stem stock should be separated from the true root. The stem base and basal leaf stems can be covered by up to 5 cm of soil and the importance of cutting the stem at a sufficient depth below ground level has been demonstrated by Tiley and Philp (1997). Cutting the tap root at 15 cm below ground level prevented further regrowth and effectively killed the plant, while cutting at 5 cm below ground level (through stem tissue above the hypocotyl region) or at ground level resulted in weakened growth and flower production (Tiley and Philp, 1997). However, soil erosion may have deposited additional layers of soil on top of the plant base and such individuals should be cut at a greater depth, i.e. up to 25 cm below soil surface. The cut parts of the plants should be pulled out of the soil and left to dry. Younger plants of *H. mantegazzianum* can be dug out of the soil.

Root cutting immediately kills the plants, and if performed correctly a single treatment is sufficient (see Pyšek *et al.*, Chapter 7, this volume). However, some individuals may be missed due to high plant density and a follow-up inspection after 1–2 weeks is advisable to make sure that all plants have been killed. Although root cutting is highly effective (Hartmann *et al.*, 1995; Tiley and Philp, 1997; Meinlschmidt, 2004a), the method is also quite laborious and thus mainly suitable for single plants and small stands (< 200 individuals). Root cutting should be implemented in April or May when the soil tends to be moister and softer than in early summer.

Chemical Control

Several studies have documented that *H. mantegazzianum* is highly susceptible to herbicides. Chemical control is one of the most widely used methods against this plant, especially in districts with large infested areas, and in long-term control programmes (Sampson, 1994; Caffrey, 2001). During recent decades chemicals such as MCPA, 2,4,5-T, imazapyr and hexazinone have all been tested against *H. mantegazzianum* (Kees and Krumrey, 1983; Rubow, 1990; Dodd *et al.*, 1994), but most studies have focused on the use of glyphosate and triclopyr (e.g. Williamson and Forbes, 1982; Niesar and Geisthoff, 1999; Meinlschmidt, 2004b). Glyphosate and triclopyr are both systemic herbicides with little or no residual activity in the soil. Some formulations of glyphosate are approved for use near water, and since this herbicide is the safest and most generally used (Tiley and Philp, 1994), the following summary of chemical control studies will concentrate on the experiences with glyphosate.

The effect of glyphosate against *H. mantegazzianum* is well-documented and a single application in spring resulted in close to 100% mortality in most control experiments (Davies and Richards, 1985; Rubow, 1990; Niesar and Geisthoff, 1999). Spraying should start in April–May, but can start earlier once the plants have reached a height of 10 cm (Lundström, 1989; Dodd *et al.*, 1994). By spraying earlier, when only a minor part of the surrounding vegetation has emerged, the damage to non-target plants is less pronounced. Postponing spraying until a few weeks before flowering or later (June–July) significantly reduces the level of control; this is due to the development of a dense canopy of leaves which is difficult for the operator to penetrate and so protects smaller plants from the herbicide (Williamson and Forbes, 1982; Caffrey, 1994).

Blanket spraying might be necessary for the control of large dense stands, although in practice this treatment may damage non-target species to an unacceptable degree in amenity areas (Williamson and Forbes, 1982). Localized spot-treatments are preferred wherever possible, since spot-treatments are easier to apply, require less chemical and cause less damage to other plant species (Dodd *et al.*, 1994). For spot-treatment a knapsack sprayer fitted with a fan nozzle can be used, but various techniques such as paste, a sponge on a long handle and hand-held weed wipers are available (Williamson and Forbes, 1982; Lundström, 1989; Lundström and Darby, 1994). Weed wipers are less effective than an overall spray, but also less damaging to surrounding vegetation; thus they are suitable for small stands, isolated plants growing among sensitive vegetation, and in the vicinity of watercourses (Lundström, 1989; Rubow, 1990; Barratt, 2003).

In addition to assessing the susceptibility of *H. mantegazzianum* to herbicide treatment, chemical control experiments have evaluated important factors such as application time, dose and non-target effects on surrounding vegetation. Experience has shown that an early application kills mainly adult plants (Lundström, 1989), but may leave the seedlings unaffected as they germinate later in spring. In order to control newly established seedlings, as well as suppressing persistent adult plants, repeated applications may be necessary. Two to four repeated treatments are frequently needed in management and control programmes (Lundström, 1989; Caffrey, 2001), but the additional applications prevent the re-establishment of replacement ground cover. Glyphosate used at the dose recommended by the manufacturers is sufficient to obtain satisfactory control of *H. mantegazzianum* (Caffrey, 1994, 2001) and increasing the dose prolongs recovery time for other herbaceous plants and grasses. Recolonization following glyphosate treatment is slow, allowing *H. mantegazzianum* seedlings to germinate, and increasing the risk of re-infestation and soil erosion. Glyphosate is hazardous for all surrounding vegetation and repeated herbicide treatments could be replaced with cutting methods to promote the growth of grazing-tolerant species.

Although chemical control is considered effective and cheap, increasing public opposition towards the use of herbicides has led national and local authorities to ban this use within their authority. In Denmark, a voluntary agreement between national, regional and local authorities came into effect in 2003, which banned the use of pesticides on all publicly owned property. It soon became evident, however, that an exception from the general ban, restricted to the use of glyphosate against *H. mantegazzianum*, was necessary in order to provide the local authorities with adequate tools for control of the plant. The use of herbicides is controversial due to possible effects on ground water quality, whereas the EU Water Framework Policy (European Commission, 2000) sets out to achieve good standards of ground water and surface water. When *H. mantegazzianum* stands are treated with glyphosate it should not be regarded as the sole management option, but rather as the first step in an overall strategy of killing plants persisting over several years (for information on the age structure of *H. mantegazzianum* populations see

Perglová *et al.*, Chapter 4, this volume), thus preventing the production of further seeds (Williamson and Forbes, 1982). In a long-term strategy an initial herbicide treatment could be followed by the sowing of grass-mixtures (see Ravn *et al.*, Chapter 17, this volume) and the use of mechanical methods or combined mechanical and chemical methods (see below) to re-establish a dense vegetation cover.

Integrated Control Options

Successful control can be achieved by any one or a combination of the above-mentioned methods, but also integrating different treatments can lead to effective control. Integrating different treatments could be quicker or more economically feasible since certain methods are more appropriate for larger than small plants (and vice versa). For example, plants which have grown too large to be sprayed may be cut down with a long-handled knife and the shorter regrowth subsequently treated with herbicide (Lundström, 1989; Tiley and Philp, 1992, 1994).

The combination of soil cultivation and herbicides has proven efficient, e.g. for the control of *H. sosnowskyi* spreading into agricultural land in Latvia (see Ravn *et al.*, Chapter 17, this volume). Several different soil cultivation methods are available and can be easily integrated with other control methods such as herbicide application or cutting. The appropriate method of soil cultivation depends on the situation in a given field and the intended subsequent use of the cultivated area. Effective control of *H. sosnowskyi* by soil cultivation is not possible without substantial disturbance of existing vegetation, and due to the use of heavy machinery the method is restricted to large areas accessible by road.

In a Latvian study, soil cultivation was combined with mechanical control by cutting and the application of a selective herbicide targeted for broad-leaved species (Vanaga and Gurkina, 2006). The experiment was laid out as a randomized block design in an abandoned field. Experimental plots were ploughed in spring and after soil cultivation a grass mixture consisting of native grass species and cultivars was sown. The developing grass sward was either treated once with the herbicide against broadleaved species or cut several times during the vegetative period. The cutting treatment was repeated in the following year. An overview of the two combined treatments is provided in Table 14.1. Prior to ploughing and chemical or mechanical control all plots were disc-harrowed. The herbicide used was a formulated mixture of MCPA (54 g a.i./l), fluroxypyr (27 g a.i./l) and clopyralid (267 g a.i./l) registered for use against broadleaved species in Latvia. The experimental plots were laid out in a field (N57° 19′ 28″, E25° 47′ 40″) that had not been cultivated for 5 years and was heavily infested with *H. sosnowskyi* (25–30 plants/m^2). The effect of the combined methods was measured three times each year in the growing season as establishment success of new seedlings, and changes in the number of *H. sosnowskyi* plants regenerating. Results of the last assessment after 2 years are shown in Table 14.1.

Table 14.1. Effect of soil cultivation and weed control on the density of *H. sosnowskyi* plants (means of adult plants and seedlings). Plot size was 30 m² from which three plots of 0.25 m² were sampled. Changes in surrounding vegetation following control measures are shown in the last two columns (mean grass cover and density of broadleaved individuals – *H. sosnowskyi* plants are not included in the assessment of broadleaved plant species). The grass cover was assessed on a 0–9 scale, where 0 is no grass and 9 equals 80–90% grass cover. All assessments were carried out on 11 August (Year 1) and 1 September (Year 2). The letter subscripts express the 5% significance results of pair-wise comparisons of the three different treatments within each year.

| Treatment in year 1 | | | Treatment in year 2 | | | *H. sosnowskyi* seedlings/m² | | Adult *H. sosnowskyi* plants/m² | | Grass cover (0–9 scale) | | Density of broadleaved species (individuals/m²) | |
Soil cultivation	Grass sown	Additional weed control	Weed control	Early weed control	n	Year 1	Year 2	Year 1	Year 2	Year 1	Year 2	Year 1	Year 2
None	–	None	None	Mech. × 1	6	26.4a	22.2a	4.9a	6.4a	–	–	73.3b	38.0b
Ploughed	√	Mech. × 5	Mech. × 3	Mech. × 1	6	6.4b	16.4a	1.3b	0.0b	4.2a	7.2b	113.1a	104.7a
Ploughed	√	Chemical	Chemical	Mech. × 1	6	4.0b	0.2b	1.8b	2.2b	4.4a	8.9a	56.7b	0.2c

For both years the combined treatment of ploughing and chemical or mechanical control significantly reduced the density of surviving *H. sosnowskyi* plants. Complete control was not achieved – maybe due to the vigorous regrowth of *H. sosnowskyi* after soil cultivation in early spring. Both herbicide and cutting damage may not kill the rootstock, and damaged plants prevented from flowering can survive until the following year (as reported by Tiley and Philp (1997) for *H. mantegazzianum*). In the treatment of ploughing combined with frequent cutting, there was 100% mortality of established plants in the second year, whereas ploughing in combination with chemical control reduced the average density of *H. sosnowskyi* to 2.2 plants/m^2 (Table 14.1). In both *H. mantegazzianum* and *H. sosnowskyi*, the vast majority of seeds are found in the upper soil layer of 0–5 cm (Krinke *et al.*, 2005; Moravcová *et al.*, Chapter 10, this volume), and soil cultivation by ploughing is likely to affect the germination success of the seeds. Such an effect seemed evident in the assessment of seedling densities of *H. sosnowskyi* in the Latvian study. For both control treatments the number of seedlings per m^2 was significantly reduced in the first year. For the plots treated with herbicide the reduction in seedlings was further enhanced by the second season application of the herbicide against broadleaved species, resulting in only 0.2 seedlings/m^2 (Table 14.1). In contrast, seedling density increased unexpectedly in the second year for the plots controlled by cutting. This could be due to the slight reduction in the density of broadleaved species that may have provided sufficient light and space for an increased density of seedlings. Sowing of grass mixtures resulted in a well-established grass sward with up to 90% cover after 2 years of growth (sward density score: 7.2–8.9; Table 14.1). On mechanically treated plots the broadleaved species increased in density compared to untreated plots. Ploughing may have induced the germination of broadleaved species from the soil seed bank. However, this effect was suppressed in the chemically treated plots where the herbicide reduced the number of broadleaved species to 0.2 plants/m^2 (Table 14.1). Results are available for only 2 years, but the experience so far emphasizes that this kind of integrated control is especially suitable for former agricultural land, and sowing grass mixtures minimizes the risk of soil erosion and re-invasion.

Management of *Heracleum mantegazzianum*

Before a management programme is started, it is worthwhile to consider how resources should be optimally invested. Such a management programme should contain clearly defined objectives of the management (e.g. eradication or containment), identification of all stands and single plants in the area as well as habitats vulnerable to invasion, and the availability of sufficient resources in terms of money, labour and equipment (Nielsen *et al.*, 2005). It is necessary to consider which means of control is most suitable for a given site. Selection of the appropriate control method depends on a number of factors, including the number of plants and the size of the targeted area, distance to water courses, access to the area and ground carrying capacity (for heavy machinery), land use, recreational value, etc.

Table 14.2. Summary of control options depending on population size.

| Control option | Population size | | |
	< 200 individuals	200–1000 individuals	> 1000 individuals
Cutting	Manual cutting, for example, by scythe	Manual cutting by scythe or flail mowing	Flail mowing
Umbel removal	Emergency treatment of plants succeeding in flowering	Emergency treatment of plants succeeding in flowering	Emergency treatment of plants succeeding in flowering
Root cutting	Suitable for small populations	Laborious but efficient	Very laborious
Chemical control	Spot treatment	Spot treatment wherever possible; overall spray of dense populations	Overall spray of dense populations
Soil cultivation and chemical control	Not recommended	Not recommended	Suitable for larger stands that must be accessible for heavy machinery

In some cases integrated methods are a better option than using a single method repeatedly during the growing season. Table 14.2 summarizes mechanical and chemical control methods described in this chapter and their recommended use depending on the size of the targeted population of *H. mantegazzianum*. Root cutting is mainly restricted to small areas, whereas the combination of soil cultivation and chemical control requires large areas. Cutting and chemical control can be used in most infested areas provided that the size of the area matches the equipment available for control. As already mentioned, it is not advisable to rely on umbel removal only in a control programme. Timing is crucial when using this method (see Pyšek *et al.*, Chapter 7, this volume) and umbel removal should rather be considered as a follow-up treatment after cutting or chemical application. For other methods the time period for control is more limited, although timing also influences overall control efficiency.

Table 14.3 depicts the most suitable period for starting the control for selected methods based on the fact that perennating plants start sprouting from the root in March, and flower in late June or early July in the temperate zone of Europe (see Perglová *et al.*, Chapter 4, this volume). Root cutting and chemical control may start in April when plants are in their early development phase (Sørensen, 2002). Spraying with herbicides should be applied 1–3 times and completed before flowering, while root cutting can be carried out throughout the summer if soil moisture and increasing root size allow digging in the soil (Table 14.3). Cutting above ground level may start when plants have grown 50–100 cm tall and should be repeated at least once or twice in the following months, depending on the density of the stand. Cutting frequencies up to six times per growing season are reported, but even in a dense stand three times should be sufficient for efficient control. However, additional cuttings are by no means wasted effort, as increased cutting intensity promotes the growth

Table 14.3. Seasonal schedule for different control options. The shaded cells depict when control should take place and 'x' indicates suggestions for the intervals between control and follow-up treatments. The schedule is valid for regions where perennating plants sprout from the root in March and flower in late June or early July. Abbreviations: disc, disc harrowing; plo, ploughing; so, sowing of grass species; ch, chemical control; cu, control by cutting.

	April	May	June	July	August
Cutting			x x	x x	x
Umbel removal				x x	x
Root cutting		x x	x		x
Chemical control	x	x	x		
Soil cultivation and chemical control		disc plo so	ch		
Soil cultivation and mechanical control		disc plo so	cu	cu	cu

of grazing-tolerant vegetation. Umbel removal should be done at peak flowering in late June or early July. Some plants may subsequently regenerate, and these flowers should be removed approximately 4 weeks later (Table 14.3). To make sure that the applied treatments have resulted in an appreciable level of control, all treated sites should be inspected for re-sprouting plants and repeated visits will enable the effectiveness of control methods to be measured (Child and Wade, 2000). They will also enable maps to be kept up to date and to assess further spread.

Management experiments have shown that *H. mantegazzianum* can be eradicated by an efficient and determined effort (Wade *et al.*, 1997). Over several years, costs of eradication are significantly lower than simply preventing further spread of existing stands. Small infestation sites can be eradicated with a few hours of work, while preventing established stands increasing in area every year would cost much more time. Examples of local authorities controlling the plant for 10–15 years without obtaining a satisfactory level of control (Sørensen and Buttenschøn, 2005) emphasize that control measures should be coordinated, correctly performed and repeated for several years. If control is conducted 2–4 times in spring and early summer (Table 14.3) practical experiences have shown that plant numbers are reduced by 75% within 1–4 years (Fig. 14.1). Root cutting and chemical control kill the majority of plants in the first year, while an intense cutting regime deals with most of the population in 4 years. In the following years only very limited resources are needed to eradicate the last plants of the stand. Once an infestation has been dealt with, regular surveys for at least 5 years must be undertaken and action taken to eradicate any newly germinating individuals.

Finally, the value of committed fieldworkers understanding the importance of preventing seed set of all plants should not be ignored. Therefore, the manager of the control programme must be highly motivated to achieve successful control. The project officer must be responsible for the mapping and monitoring of plants, and planning of a long-term control strategy. Practical experience has shown that members of the public are willing to play an active

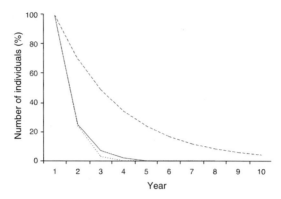

Fig. 14.1. Comparison of reduction in numbers of *H. mantegazzianum* plants for three different control methods (- - - cutting; ···· root cutting; —— chemical control) over a period of 10 years in Denmark. Percentage of plants present at the beginning of the control scheme is shown on the *y*-axis. If the control methods are applied repeatedly during the growing season, root cutting and chemical control eradicate the plants in a growing site within 4–5 years, whereas cutting at above-ground level may take 8–10 years. Estimates are either by counting the number of controlled plants each year until eradication has been achieved (root cutting) or by evaluating the reduction in invaded area each year after the control (cutting and chemical control) has been applied.

role reporting observations of the plant. An awareness-raising campaign will not only help the authorities to identify the detailed distribution of the plant in the area, but also contribute to a greater understanding of the problems following invasion by *H. mantegazzianum* and the need for coordinated long-term control strategies.

The recommendations listed above are based on results from studies on *H. mantegazzianum*, but given their similar life strategy, these management outlines may be valid for *H. sosnowskyi* as well.

Conclusions

Across Europe, much effort is invested in controlling *H. mantegazzianum* and *H. sosnowskyi*. When starting a control programme the objective of the campaign should be clearly defined. Preventing the spread of seeds may seem less laborious than eradication, but over a period of years the costs of eradication are in many cases significantly lower than trying to prevent the spread of seeds. Mapping all stands and isolated plants before control is started will provide an overview to the extent of the infestation and help with planning an appropriate control strategy. Examples of local authorities controlling the plant for more than a decade without obtaining a satisfactory level of control demonstrate that years of repeated control are wasted if no systematic approach is adopted. Different control options are easily combined, thus exploiting the advantages of various options depending on the density of the stand and size of infested area. Stands of large invasive *Heracleum* species can be eradicated by an efficient and determined effort, but all treated sites should be inspected, and if necessary treated, the following year for re-sprouting plants and for a further 5 years for newly germinated plants to ensure complete success.

Acknowledgements

The authors would like to thank Matthew Cock and Johannes Kollmann for critical reading of the manuscript. The study was supported by the project 'Giant Hogweed (*Heracleum mantegazzianum*) a pernicious invasive weed: developing a sustainable strategy for alien invasive plant management in Europe', funded within the 'Energy, Environment and Sustainable Development Programme' (grant no. EVK2-CT-2001-00128) of the European Union 5th Framework Programme.

References

Barratt, T. (2003) *Controlling Invasive Plants in the Tweed Catchment*. The Tweed Forum, Roxburghshire, UK.

Caffrey, J.M. (1994) Spread and management of *Heracleum mantegazzianum* (Giant Hogweed) along Irish River corridors. In: de Waal, L.C., Wade, P.M., Child, L.E. and Brock, J.H. (eds) *Ecology and Management of Invasive Riverside Plants*. Wiley, Chichester, UK, pp. 67–76.

Caffrey, J.M. (1999) Phenology and long-term control of *Heracleum mantegazzianum*. *Hydrobiologia* 415, 223–228.

Caffrey, J.M. (2001) The management of Giant Hogweed in an Irish river catchment. *Journal of Aquatic Plant Management* 39, 28–33.

Child, L.E. and Wade, P.M. (2000) *The Japanese Knotweed Manual. The Management and Control of an Invasive Alien Weed*. Packard, Chichester, UK.

Davies, D.H.K. and Richards, M.C. (1985) Evaluation of herbicides for control of Giant Hogweed *Heracleum mantegazzianum* and vegetation re-growth in treated areas. *Annals of Applied Biology* 6, 100–101.

Dodd, F.S., de Waal, L.C., Wade, P.M. and Tiley, G.E.D. (1994) Control and management of *Heracleum mantegazzianum* (Giant Hogweed). In: de Waal, L.C., Child, L.E., Wade, P.M. and Brock, J.H. (eds) *Ecology and Management of Invasive Riverside Plants*. Wiley, Chichester, UK, pp. 111–126.

European Commission (2000) Directive 2000/60/EC of the European Parliament and of the Council of 23 October 2000 establishing a framework for Community action in the field of water policy. *Official Journal* (OJ L 327), 22 December.

Hartmann, E., Schuldes, H., Kübler, R. and Konold, W. (1995) *Neophyten. Biologie, Verbreitung und Kontrolle ausgewählter Arten*. Ecomed, Landsberg, Germany.

Kees, H.V. and Krumrey, G. (1983) *Heracleum mantegazzianum* – Zierstaude, Unkraut und 'Giftpflanze'. *Gesunde Pflanzen* 35, 108–110.

Krinke, L., Moravcová, L., Pyšek, P., Jarošík, V., Pergl, J. and Perglová, I. (2005) Seed bank of an invasive alien, *Heracleum mantegazzianum*, and its seasonal dynamics. *Seed Science Research* 15, 239–248.

Laivins, M. and Gavrilova, G. (2003) *Heracleum sosnowskyi* in Latvia: sociology, ecology and distribution. *Latvijas Vegetacija* 7, 45–65. [In Latvian.]

Lundström, H. (1989) New experiences of the fight against the Giant Hogweed, *Heracleum mantegazzianum. Swedish Crop Protection Conference* 1, 51–58.

Lundström, H. and Darby, E.J. (1994) The *Heracleum mantegazzianum* (giant hogweed) problem in Sweden: suggestions for its management and control. In: de Waal, L.C., Child, L.E., Wade, P.M. and Brock, J.H. (eds) *Ecology and Management of Invasive Riverside Plants.* Wiley, Chichester, UK, pp. 93–100.

Meinlschmidt, E. (2004a) *Der Riesen-Bärenklau.* Faltblattreihe Integrierter Pflanzenschutz, Heft 5: Unkrautbekämpfung. Sächsische Landesanstalt für Landwirtschaft, Dresden, Germany.

Meinlschmidt, E. (2004b) *Praxisbericht zur Bekämpfung eines Riesen-Bärenklaubestandes im Kreis Döbeln.* Report available through Lakuwa Rural Management and Water Management, GmbH Dahlen, Grossböhla.

Nielsen, C., Ravn, H.P., Nentwig, W. and Wade, M. (eds) (2005) *The Giant Hogweed Best Practice Manual. Guidelines for the Management and Control of an Invasive Weed in Europe.* Forest and Landscape Denmark, Hørsholm, Denmark.

Niesar, C.M. and Geisthoff, N. (1999) Bekämpfung des Riesenbärenklaus mittels Glyphosaten. *AFZ/Der Wald* 22, 1173–1175.

Pyšek, P., Kučera, T., Puntieri, J. and Mandák, B. (1995) Regeneration in *Heracleum mantegazzianum*: response to removal of vegetative and generative parts. *Preslia* 67, 161–171.

Rubow, T. (1990) Giant Hogweed: importance and control. 7. *Danske Planteværnskonference*, pp. 201–209. [In Danish.]

Sampson, C. (1994) Cost and impact of current control methods used against *Heracleum mantegazzianum* (giant hogweed) and the case for investigating a biological control programme. In: de Waal, L.C., Wade, M., Child, L.E. and Brock, J.H. (eds) *Ecology and Management of Invasive Riverside Plants.* Wiley, Chichester, UK, pp. 55–65.

Sørensen, M.A. (2002) Management Strategy for the Non-chemical Control of Giant Hogweed along the Stream Seest Mølleå. Forest and Landscape Engineer thesis, The Danish Forestry College, Nødebo, Denmark. [In Danish.]

Sørensen, M.A. and Buttenschøn, R.M. (2005) *The Extent of Prevention of Giant Hogweed by the Danish Municipalities.* Videnblad, 6.0-20. Forest and Landscape, Hørsholm. [In Danish.]

Tiley, G.E.D. and Philp, B. (1992) Strategy for the control of giant hogweed (*Heracleum mantegazzianum*) on the River Ayr in Scotland. *Aspects of Applied Biology* 29, 463–466.

Tiley, G.E.D. and Philp, B. (1994) *Heracleum mantegazzianum* (giant hogweed) and its control in Scotland. In: de Waal, L.C., Child, L.E., Wade, M. and Brock, J.H. (eds) *Ecology and Management of Invasive Riverside Plants.* Wiley, Chichester, UK, pp. 101–109.

Tiley, G.E.D. and Philp, B. (1997) Observations on flowering and seed production in *Heracleum mantegazzianum* in relation to control. In: Brock, J.H., Wade, P.M., Pyšek, P. and Green, D. (eds) *Plant Invasions: Studies from North America and Europe.* Backhuys, Leiden, The Netherlands, pp. 123–137.

Tiley, G.E.D. and Philp, B. (2000) Effects of cutting flowering stems of Giant Hogweed *Heracleum mantegazzianum* on reproductive performance. *Aspects of Applied Biology* 58, 77–80.

Tiley, G.E.D., Dodd, F.S. and Wade, P.M. (1996) Biological flora of the British Isles. *Heracleum mantegazzianum* Sommier & Levier. *Journal of Ecology* 84, 297–319.

Vanaga, I. and Gurkina, J. (2006) Methods of control for *Heracleum sosnowskyi* in Latvia. *Proceedings of the International Scientific Conference 'Strategy and Tactics of Plant Protection'.* The Institute of Plant Protection NAS of Belarus, Minsk, Belarus, pp. 81–84.

Wade, P.M., Darby, E.J., Courtney, A.D. and Caffrey, J.M. (1997) *Heracleum mantegazzianum*: a problem for river managers in the Republic of Ireland and the United Kingdom. In: Brock, J.H., Wade, P.M., Pyšek, P. and Green, D. (eds) *Plant Invasions: Studies from North America and Europe*. Backhuys, Leiden, The Netherlands, pp. 139–151.

Williamson, J.A. and Forbes, J.C. (1982) Giant Hogweed (*Heracleum mantegazzianum*): its spread and control with glyphosate in amenity areas. In: *Proceedings 1982 British Crop Protection Conference – Weeds*. British Crop Protection Council, Farnham, UK, pp. 967–972.

15 Control of *Heracleum mantegazzianum* by Grazing

RITA MERETE BUTTENSCHØN[1] AND CHARLOTTE NIELSEN[2]

[1]*Danish Centre for Forest, Landscape and Planning, Vejle, Denmark;*
[2]*Danish Centre for Forest, Landscape and Planning, Hørsholm, Denmark*

> *They are invincible*
> (Genesis, 1971)

Introduction

Grazing is used in many countries to control tall herbs, including *Heracleum mantegazzianum* Sommier & Levier and pastureland weeds. In 2004, 20% of Danish municipalities used cattle and sheep grazing to control *H. mantegazzianum* (Sørensen and Buttenschøn, 2005). Very few studies document the effect of livestock grazing on *H. mantegazzianum* (Andersen, 1994; Andersen and Calov, 1996; Sørensen, 2002), but it can be observed that *H. mantegazzianum* spreads in moist, tall herb communities on abandoned land, whereas it is rare or absent on pastureland. Different species of ruminants, hindgut fermentators and omnivores are reported to eat *H. mantegazzianum* and other *Heracleum* species, but acceptability is variable between species, and regional differences are reported within species. In this chapter we summarize and update available knowledge on animal grazing of *H. mantegazzianum*.

Biomass Production and Nutritional Value of *Heracleum mantegazzianum*

H. mantegazzianum has a high biomass production, which led to its widespread use as a forage plant for livestock in eastern Europe. Data from invasive stands on open land show dry matter production of 5.7–7.1 t/ha above ground and 2.4 t/ha below ground (Tiley *et al.*, 1996; Otte and Franke, 1998). In comparison, the studies of Ruskova (1973) showed yields of up to 94 t fresh matter/ha (approximately 15 t dry matter/ha) under forage cropping conditions. Undesirable tainting of milk odour and taste (Satsyperova, 1984) and the availability of alternative forage types such as grass and maize

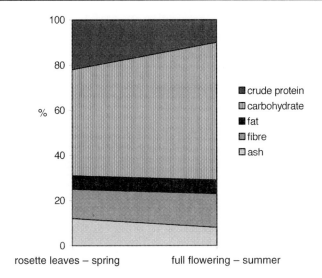

Fig. 15.1. Trend of change in the content of main groups of nutrients in *H. mantegazzianum* from spring to summer. The figure is based on data from The Latvian Agricultural University and the Latvian Agricultural Advisory Centre (I. Vanaga, Riga, 2003, personal communication).

silage has led to the disuse of *Heracleum* spp. as forage plants during the last 10–15 years.

Only limited data are available concerning the nutritional value of *H. mantegazzianum*. Since *H. mantegazzianum* is a large herb with winter senescence of foliage, the chemical composition of plant tissues changes with phenological stage. Data from Latvia (I. Vanaga, Riga, 2003, personal communication) make it possible to draw up a partial picture of the nutrient content of *H. mantegazzianum* (Fig. 15.1).

Foliage of *H. mantegazzianum* is readily digestible. Between 70% and 75% of the dry matter (DM) content is available as water-soluble carbohydrate, fat and crude protein – equal to or greater than the content of comparable livestock forage plants such as hay, grass and maize silage, sugarbeets and turnips (Sørensen, 1965). This implies a high organic matter digestibility (OMD), in the 80–90% range. As a sufficient energy supply is > 60% OMD for herbivore livestock, and > 70% for pigs, clearly foliage of *H. mantegazzianum* is a feasible fodder for herbivores and pigs. The crude protein of the foliage during early vegetative growth is very high, 20% or more, declining to 12–14% by mid-summer (Fig. 15.1, Satsyperova, 1984; Otte and Franke, 1998). This is similar to the content in forage-legumes and greater than the content in other forage plants for livestock (Sørensen, 1965). The sufficiency level for livestock production except intensive dairy production is at or below 12% crude protein in DM. The fibre content is very low and at a constant level. The fat content is sufficient, as > 2% is recommended to supply the necessary essential fatty acids (Sørensen, 1965).

Some information on mineral content of full flowering H. mantegazzianum in mid-summer is available (Otte and Franke 1998), and this suggests
that H. mantegazzianum at this stage can supply sufficient P, Ca and K for
grazing animals. H. mantegazzianum normally grows in relatively nutrient-
rich environments on near neutral soils (Tiley et al., 1996). In near neutral
soils (pH 5.5–7.0) the amounts of soluble plant mineral nutrients reach an
optimal combination for herbivores. Phosphorus, which often is a limiting
factor in animal nutrition on acidic or alkaline soils, has its peak solubility in
this pH range (Swift et al., 1979) with up to 20% phosphorus in the labile
pool. In acidic soils the labile pool of phosphorus gradually decreases to less
than 5% of the total as soil pH decreases from 5.5 to 3.5, whilst < 10% is
available for plant nutrition in alkaline soils. This suggests that H. mantegazzianum, on typical sites would contain sufficient phosphorus for grazing
animals. Magnesium and a number of other essential cations also reach a solubility optimum within the pH range 5.5–7.0 (Fitter and Hay, 1987). Against
this background and the evidence on mineral contents of plants growing under
different soil pH conditions (Swift et al., 1979; Buttenschøn and Buttenschøn,
1982; Buttenschøn et al., 2001), we suggest that the mineral content of H.
mantegazzianum in general is adequate for livestock nutrition.

However, the high OMD and crude protein content of H. mantegazzianum is not without problems for grazing livestock, when grazing takes place
in dense stands where H. mantegazzianum contributes a large part of the
forage uptake. The main problem is associated with hyperactivity in the micro-
biological breakdown, with subsequent gas production in the fermentation
chambers of the large herbivores (J. Buttenschøn, Vejle, 2006, personal communication). In particular, rumen fermentation is liable to this condition, and
bloat may arise as a consequence. Another digestion problem is associated
with the high crude protein content, which may lead to irritation and inflammation of the gut and diarrhoea. Offering straw to the livestock may prevent
both conditions, since fodder with high fibre content prevents the conditions
arising.

Choosing Livestock for Control by Grazing

Large mammal grazers have evolved three basic principles for dealing with cell-
wall breakdown:

1. Ruminants (e.g. sheep, cattle) have pre-stomach fermentation and an emol-
liating tank, where micro-organisms break down cell walls. The food bulk is
emolliated in fluid, regurgitated, re-chewed and finally sent to the animal's
proper digestive system, resulting in a long gastro-intestinal passage time.
2. Hindgut fermentators (e.g. horses) pass food through the normal digestive system first, i.e. acid-enzyme breakdown and absorption of metabolites, then microorganisms ferment the food in the wide hindgut where
further absorption takes place, leading to a medium gastro-intestinal passage
time.

3. Omnivores (e.g. pigs) largely depend on their own digestive effort and select for low cell-wall content in fodder, resulting in a short gastro-intestinal passage time.

The different digestion strategies have implications for diet selection of the livestock (Table 15.1), as has the size of the animals.

Large herbivores have been described over a gradient from grazers to browsers (Hofmann, 1989) according to the degree that they rely on field-layer vegetation (grazers) and foliage and twigs of woody species (browsers) as their nutrition source. This approach, however, is rather rigid and does not allow for actual habitat differences, adaptation thereto by the animals, or offspring learning from adults, which appears to be vital for good performance in a given habitat (Illius and Gordon, 1993; Provenza and Cincotta, 1993). This means that the grazing characteristics of livestock should be seen in the context of evolved genotypes and phenotypic adaptation to the actual grazing scenario. We have summarized some main characters of livestock in Table 15.1. Other traits, which determine the utilization of different plant species by livestock, are smell and taste. *H. mantegazzianum* has a bitter taste (Terney, 1993). Horses in particular, and to a lesser extent cattle, avoid many bitter-tasting plants. To what extent this influences the uptake of *H. mantegazzianum* needs to be further assessed.

In choosing livestock for control of *H. mantegazzianum*, we need to assess the suitability of the livestock in relation to the typical *H. mantegazzianum* site – moist, but not wet, with near neutral pH, nutrient- and humus-rich soils and a high mineralization rate (see Thiele *et al.*, Chapter 8, this volume). In practice, this excludes none of the livestock in question. The critical question is do all the livestock species eat the plant in sufficient amounts to be efficient in its control?

Sheep and goats seek out *H. mantegazzianum* and prefer it to rough grasses and sedges (Terney, 1993; Andersen, 1994; Caffrey, 1994; Dodd *et al.*, 1994; Lundström and Darby, 1994; Andersen and Calov, 1996; Tiley *et al.*, 1996; Sørensen, 2002). Several authors mention cattle and pig grazing as effective control measures (Terney, 1993; Dodd *et al.*, 1994; Tiley and Philp, 1994; Tiley *et al.*, 1996). Our practical experiences in Denmark with a strike-force cattle herd or sheep and cattle herd substantiate this. Terney (1993) mentions horse grazing as a possible control measure of *H. mantegazzianum*, but reports from practice suggest that horses do not eat substantial amounts of the plant.

Sheep prefer young lush foliage of *H. mantegazzianum* to older leaves with higher fibre content and less energy and protein, but older leaves, stem and flowers are also eaten, and their stems broken to reach them (Terney, 1993). Seedlings, however, are reported to be generally avoided by sheep in Danish grazing control practice (K. Møller, Municipality of Vejle, 2003, personal communication). The sheep's preference ranking, young foliage > old foliage > stems, may apply to other livestock species that eat *H. mantegazzianum*, because it optimizes the energy input and nutritive value of the diet.

Table 15.1. Grazing characteristics of the main husbandry animal grazers (after Buttenschøn and Buttenschøn, 1982; Tolhurst and Oates, 2001).

Food preferences	Grazing pattern	Comments	*Heracleum mantegazzianum*
Cattle, grazers: Graze a broad spectrum of plant species and communities. Prefer fresh vegetation in vegetative growth, but include variable amounts of tall, coarser vegetation with stems, inflorescences; browse, senescent leaves and litter. Preference ranking sweet > neutral > acidic > acrid.	Graze preferred vegetation to 3–6 cm high lawns in a mosaic pattern. Eat substantial amounts of coarse vegetation. Graze most woody species as an integrated part of grazing. Have coprophobic behaviour – dung pats are avoided for months.	Perform well on wet ground, but may cause poaching of soil. Broad face limits selectivity. Slashing and ripping off vegetation, rather than biting it off. Limited effect on woody species encroachment – fragmented woodland develops over time. Low-cost fencing.	*H. mantegazzianum* lies within the grazing choice of cattle. The relatively high grazing height may prolong the vegetative survival of the plant. The acrid taste/smell has not been reported to reduce uptake. Coprophobic behaviour provides grazing-free areas.
Horses, grazers: Graze a broad spectrum of plant species and communities. Prefer fresh vegetation in vegetative growth, but include variable amounts of tall, coarser vegetation with stems, inflorescences; browse, senescent leaves and litter. Preference ranking sweet > neutral > acidic >> acrid. Selectively avoid several poisonous species.	Graze down to 2–3 cm height. Develop large lawns with short vegetation and leave other areas practically ungrazed. Eat large amounts of coarse vegetation. Limited browsing of woody species, many of which are avoided. Have coprophobic behaviour – dung pats are avoided for months and much dung is excreted in latrine areas.	Perform well on wet ground but poaching may be a problem – poaching is more likely with shod horses. Vegetation is bitten off. May accelerate development of woodland fragments. Are susceptible to a variety of poisonous plants. Low-cost fencing.	Reports on *H. mantegazzianum* herbivory by horses vary. The smell and taste appear to repel substantial uptake. Toxic substances may be ingested in substantial amounts before the hindgut fermentation breaks them down. Coprophobic behaviour provides larger grazing-free areas.
Sheep, grazers–browsers: Graze selectively on the plant species or plant parts. Avoid plants or plant parts with high fibre content. Eat inflorescences preferentially. Leave high tussocks and coarse vegetation ungrazed. Preference ranking: neutral = sweet > acrid > acidic.	Graze down to 1–2 cm height. Graze in small-scale areas and over time develop a small-scale pattern lawn mosaic frequently in a net of neglected vegetation. Eat large amounts of browse and young herbs selectively. Do not avoid own dung.	Do not perform well on wet ground. The narrow face allows sheep to graze deep into scrub. Have a large effect on woody species and may conditionally control woodland fragment development. Eat inflorescences of many herbs and limit seeding. High-cost fencing.	Sheep eat *H. mantegazzianum* by preference. Eat inflorescences. The very close grazing will enhance root-resource exhaustion.

Continued

Table 15.1. *Continued.*

Food preferences	Grazing pattern	Comments	*Heracleum mantegazzianum*
Goats, browsers–grazers: Graze selectively on the plant species or plant parts. Graze high vegetation – tall lush grass, high herbs; browse, but not senescent leaves and litter. Preference ranking: neutral = sweet > acrid > acidic.	Can graze close to ground, but prefer to graze in all vegetation horizons that are present. Eat large amounts of browse selectively. Do not avoid own dung.	Do not perform well on wet ground and are susceptible to damp and cold weather. The narrow face allows goats to graze deep into scrub. Effective browsers that can control woodland development. High browse consumption and resting under scrub may enhance local development of nitrophilous flora. High-cost fencing.	Goats eat *H. mantegazzianum* by preference. Eat inflorescences. The differentiated grazing height is a draw-back at low stocking rates.
Pigs, grazers–omnivores: Graze selectively on plants or plant parts with high-energy content – fresh leaves, fruits, acorn, seeds and roots. May de-root high and coarse vegetation structures with attractive roots. Preference ranking not known.	Graze and root at and below soil level. Do not avoid own dung.	Perform well on wet ground. Pigs are well equipped to root deep into the ground and eat roots. Extensive rooting may disturb the ecosystem undesirably. Also eat elements of soil fauna.	Young *H. mantegazzianum* leaves are eaten by pigs, which also eat the roots.

Effect of Grazing on *Heracleum mantegazzianum* and its Management

In a grazing study on a mesotrophic meadow, *H. mantegazzianum* was significantly reduced within the first 2–3 years of sheep grazing (Andersen, 1994) and the species disappeared almost completely from the grazed area by the seventh year (Andersen and Calov, 1996). By this time, the cover of *H. mantegazzianum* in the grazed area was below 1% and the species was only present in one of two initially dense stands, while in two ungrazed control areas it had 40–50% cover. During the first 3 years of the experiment, the number of other plant species increased from below 30 to about 70, but by the seventh year had dropped back to just over 30. There was some initial increase in the number of grass species, followed by a non-significant decrease, but the trend in species number was mostly accounted for by herbs and, more locally, woody species (Andersen and Calov, 1996).

Andersen and Calov (1996) also sampled the seed bank in the grazed and ungrazed areas in the sixth and seventh year of the experiment. Seedlings of

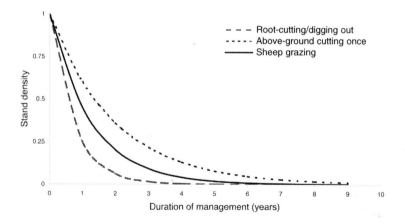

Fig. 15.2. Extinction-rate trend for *H. mantegazzianum* under different control measures. The trends are based on an assumption of constant extinction rate from year to year. The extinction rate by root cutting and above-ground cutting is taken from the results of Nielsen (2005), while the extinction rate for sheep grazing is from the results of Andersen (1994), Andersen and Calov (1996) and Sørensen (2002). Unfortunately Andersen made no count of *H. mantegazzianum* individuals during the first 3 years of grazing, but states that no mature individuals were found after the third grazing season. In the same study, no vegetation observations were made during the fourth to sixth grazing season, and then no *H. mantegazzianum* individuals were found in observations after the sixth grazing season.

large herbs including *H. mantegazzianum* emerged in large quantities in trials using the soil from ungrazed areas, but no germination of *H. mantegazzianum* was obtained from the soil taken from grazed plots (for details on seed bank in *H. mantegazzianum*, see Chapter 5).

In his short-term comparative study of control methods, Sørensen (2002) found a marginally significant reduction in the number of *H. mantegazzianum* seedlings ($P = 0.1$), an increase in small vegetative plants ($P < 0.05$) and a reduction in large flowering individuals ($P < 0.001$) in grazed compared to ungrazed plots at a waste site. This comparison did not consider cover or biomass changes, just numbers of individuals, and only refers to changes in the first season after the first 4–5 months of management. Yet it provides solid evidence that grazing prevented plants of *H. mantegazzianum* from flowering (for details on the effect of grazing on delayed flowering in *H. mantegazzianum*, see Perglová *et al.*, Chapter 4, this volume).

The above studies show that sheep grazing is an efficient control measure even in the first season. However, to achieve an appreciable level of control, grazing must be continued until *H. mantegazzianum* disappears from the pasture sward. The seed bank will normally be depleted within 2–4 years (Chapter 5; Krinke *et al.*, 2005; Nielsen, 2005), whereas storage roots may persist at least twice that period (R.M. Buttenschøn and C. Nielsen, unpublished observations). In initially dense stands the management strategy may be changed from intensive to extensive measures after 2–3 years of grazing

Table 15.2. Suitability-criteria of livestock species in relation to *H. mantegazzianum* control. The scale used is: ++ = very suitable, + = suitable, (+) = suitable, but not good, − = not suitable. The fencing costs are mentioned in more detail in connection with the subheading 'economics'. The minimum area demand is based on estimates on minimal number of individuals to sustain a herd with in-herd fodder utilization learning: two or three adults with offspring, and the carrying capacity of a 'typical' *H. mantegazzianum* site.

	Sheep	Goat	Cattle	Horse	Pig
Consumption of *H. mantegazzianum*	++	++	++	(+)	++
Suitability for strike-force grazing	++	++	(+)	−	(+)
Dry pastureland	++	++	++	++	+
Moist pastureland	+	+	++	+	+
Wet pastureland	−	−	++	+	++
Rich grass–sedge flora	+	+	++	++	
Rich herb flora	−	+	++	++	
Rich moss–lichen flora	+	+	++	+	
Minimum strike force area unit (ha)	0.05	0.05			
Minimum area (ha) needed to sustain a herd	0.5	0.5	2	2	0.5
Fencing costs	High	High	Low	Intermediate	High
Amenity-user conflicts – comfort of public with animal species, potential injury by animals	Low	Low	Intermediate	High	Intermediate
Dog conflicts	High	Intermediate	Intermediate	Low	Intermediate

(Andersen and Calov, 1996). In general, eradication of *H. mantegazzianum* from a site takes 5–10 years of grazing management (Fig. 15.2).

The impacts of grazing on other vegetation, and the likely vegetation mix after grazing, are relevant when considering management options. Several years of sheep grazing may result in rather species-poor, nitrogen-demanding vegetation of disturbed sites with a few dominant species such as *Urtica dioica* L. or *Juncus effusus* L. This is because a typical site invaded by *H. mantegazzianum* is on humus- and nutrient-rich, moist soils where species less preferred from a conservation management viewpoint tend to increase. Therefore, the choice of animal species needs to be related to site-specific management goals (Table 15.2).

Management Strategies

Grazing management of *H. mantegazzianum* should be timed to catch the early foliage, i.e. during April or at the latest early May, depending on regional differences in seasons. The grazing should exhaust the root resources by either continuous or recurrent periods of biomass removal, and can be discontinued in early autumn when re-sprouting has stopped.

The management strategy for grazing control of *H. mantegazzianum* depends on the stand area and density. Two main management strategies can be applied: strike-force grazing and summer grazing. Strike-force grazing consists of grazing with a very high stocking rate for several short periods distributed over the growing season. Summer grazing in this context is continuous, normal pastureland grazing from late spring to mid autumn with a stocking rate that is balanced in accordance with the site productivity and other possible aims of the management, e.g. patchy, uneven structured vegetation to promote specific flora or fauna species. Tether grazing and strip-grazing systems have the same effect as strike-force grazing.

In dense stands of *H. mantegazzianum*, it is necessary to start with a series of heavy grazing interventions over short periods, three or four times per year. Short, intensive grazing periods have an effect similar to cutting, as productive foliage is completely eaten and regrowth draws substantial resources from the root. This type of grazing can be implemented using a strike-force herd of animals, which are moved from site to site according to a planned timetable, with regular return to all target sites during the growing season. Heavy grazing will leave the area open to light and there is a side effect of heavy poaching. Poached soil with sufficient supply of light and increased nutrient mineralization (Buttenschøn *et al.*, 2001) will provide good conditions for germination of *H. mantegazzianum* seeds. The effect of the strike-force management should be monitored and discontinued once the stands of *H. mantegazzianum* have been reduced to a minor contribution to the vegetation of the site, because the secondary aim of the control management normally is to re-establish a dense sward of semi-natural grassland vegetation, and this cannot be accomplished on a site heavily disturbed by grazers. Development of a dense sward also prevents or slows down the reinvasion of the site by *H. mantegazzianum* (see Chapters 14 and 17). As an alternative, the strike-force concept may be continued, but with reduced stocking rate – this could be used in combination with other control measures on numerous, but small sites, e.g. in urban environments where small patches of land at several sites are prone to *H. mantegazzianum* invasion.

Where the stand of *H. mantegazzianum* is open and less dense, summer grazing is preferred to strike-force grazing, because from the beginning it supports the development of a dense, wear-resistant pastureland sward. Summer grazing can only be performed on areas a few hectares or more in size and reduces the man-hour input as the livestock need not be handled and moved around. If there is semi-natural pastureland adjacent to the controlled site, its inclusion into the grazing regime should be considered as it can promote re-vegetation of the controlled site and development towards a stable pastureland.

Grazing control should include monitoring for inflorescence development and, if necessary, cutting of umbels to prevent seed production on the site.

Herds used to graze *H. mantegazzianum* should include some individuals familiar with the plant. The importance of the social contact and learning process that takes place in an animal herd with mother–offspring contact is crucial. Animals not acquainted with the vegetation in question tend to overeat plant species that they acquire a taste for, and thus can be subject to poisoning or digestive disorders. In the case of *H. mantegazzianum*, bloating and diarrhoea is a latent problem, as noted above.

In Denmark several different models are used to organize control of *H. mantegazzianum* based on grazing (Sørensen, 2002):

1. The local authority provides its own livestock. In Denmark control using a herd owned by the local authority is often carried out on public land, as part of job training activities or projects involving handicapped people. This approach is expensive, but funds may be obtained from other public bodies engaged in addressing social problems.

2. A private nature management society owns the herd of livestock and provides it on a member partnership basis. Such partnerships often emerge in larger townships, or their vicinity, and relate to common land for larger development complexes or public amenity lands, such as parks, green connection corridors or riverside lawns. The society finances the herd and the members manage it. There are examples of such societies operating with success over more than a decade (Hansen, 1997). This solution normally involves no public expenses, but fencing subsidies may be available.

3. Control on public land based on a grazing contract between a public authority and a farmer. In Denmark the farmer typically gets the grazing use of the land free of charge, while the authority provides fencing.

4. Grazing by contractors, where a local livestock breeding society or a group of farmers offer solutions for nature managers or control of alien species using mobile livestock herds. The contractor is usually paid for this service. Herds are moved from site to site during the season. This solution is frequently used for 'strike-force' grazing. In Denmark contractor grazing is partly funded through European Community (EC) area subsidies, but additional public funds are usually necessary.

Within the EC the basis for granting area subsidies varies from one member state to another, so that national applications of contract grazing must be adapted accordingly. In practice, mainly sheep have been used for the control of *H. mantegazzianum* by contract grazing, but goats, cattle or pigs may be used instead.

Seed Translocation by Animals

Fischer *et al.* (1996) have shown that seeds of many species are dispersed in the fleece of sheep. They concluded that seed surface structure and height of the inflorescences were the main determinants for successful translocation of seeds on fleeces. They ranked seed surface traits hooked > bristly > coarse > weakly coarse > smooth and noted a large increase in seed carriage when the seed exposure height in the vegetation was above 40 cm, i.e. when the exposure height reached the ventral part of the body or higher. Seeds can be carried in the fleece for long periods. Fischer *et al.* (1996) found medium to low numbers of seeds of a number of *Apiaceae* in the fleece, e.g. *Torilis japonica* (Houtt.) DC., *Daucus carota* L., *Anthriscus sylvestris* (L.) Hoffm., *Angelica sylvestris* L., *Pimpinella saxifraga* L. and *Heracleum sphondylium* L. Of these only *P. saxifraga* and *D. carota* occur naturally in the calcareous grasslands studied; the other species were picked up as the sheep were moved between pastures. Other livestock will also carry seeds in their pelt, but this varies with the pelt's character. Fruits of *H. mantegazzianum* are borne well above the sheep's back, which enhances transfer of seeds to the fleece, and the coarse rough surface of the seeds results in efficient retention in the fleece, and hence dispersal. This can be a problem if grazed stands of *H. mantegazzianum* develop seeds and the grazing livestock are then moved to another site.

Besides the translocation of seeds in pelts, ingestion of seeds may result in endozoochorous spread within a pasture or from one pasture to the next. It is not known whether seeds of *H. mantegazzianum* will germinate after passage through livestock intestines. Two other species of *Umbelliferae*, *P. saxifaga* and *Anthriscus sylvestris*, have been found to germinate in cattle dung pats (R.M. Buttenschøn and J. Buttenschøn, unpublished data), so this possibility merits evaluation.

If grazing occurs in fruiting stands of target species such as *H. mantegazzianum*, holding the animals in pens for 2 days before transfer to another pasture minimizes endozoochorous dispersal, but not translocation of seeds in pelts between pastures (Frost and Launchbaugh, 2003). Accordingly, plants of *H. mantegazzianum* in grazed areas should be prevented from developing fruits, which can be achieved by starting grazing in spring and cutting off and removing inflorescences if the grazing does not prevent flowering (see Chapter 7).

Livestock Poisoning

The toxic compounds in *H. mantegazzianum* are furanocoumarins (synonyms: furocoumarins, psoralens; see Hattendorf *et al.*, Chapter 13, this volume), which may cause skin to become hypersensitive on contact and exposure to sunlight (see Thiele and Otte, Chapter 9, this volume; Cooper and Johnson, 1984). This group of chemicals is widespread in species of *Apiaceae* and numerous other plants. Satsyperova (1984) has summarized information on the presence of different furanocoumarins in the foliage and seeds of *H. mantegazzianum* and found notably higher contents in seeds than in foliage. She also reports that furanocoumarins affect the balance of the steroid hormones promoting production rates in livestock and interfere with the fertility of the female livestock, at least when fed appreciable amounts as silage fodder. Ultraviolet light induces chemical changes in the active compounds, resulting in compounds that cause hypersensitive inflammation (see Chapter 9). Thus poisoning by furanocoumarins is not a poisoning in the lay sense of the word, i.e. poisoning by interference with vital bio-functions or poisoning through overdose. There is no evidence to suggest that light hypersensitivity may be transferred by ingestion of affected tissue and thereby pose a threat in the food chain – the concentration in animal tissue is low in comparison to the carrier plants, and biodegradation and excretion of the active compounds is an ongoing process. Furanocoumarins can be found in milk of livestock that have fed on *H. mantegazzianum* and are reported to give the milk an unpleasant anise smell (Satsyperova, 1984). Hence, cows, sheep or goats producing milk for human consumption should not be used for control of *H. mantegazzianum* in dense stands.

In humans, particularly humans with a low degree of skin pigmentation, furanocoumarins in combination with sunlight exposure of the skin cause inflammation of the skin (dermatitis) and blistering of the skin, affecting the epidermis and hypodermis (Nielsen *et al.*, 2005; Chapter 9). The blisters subside within a few days leaving brownish pigmentations in the skin. Hypersensitivity to UV light may persist for months or years. There is limited documented evidence that furanocoumarins also cause light-hypersensitive dermatitis in livestock. Andrews *et al.* (1985) were able to reproduce dermatitis on sheared skin and the oral mucosa of white sheep. Practical experiences of a strike-force shepherd suggest that labial dermatitis does occur on unpigmented sheep following consumption of substantial amounts of *H. mantegazzianum* (Møller, 2002). Due to hairlessness, thin epidermis and thin mucosa, the lips and the labio-oral interface is the most exposed part of the susceptible animals. Normally the dense hair cover and thicker epidermis and oral-cavity mucosa in livestock (compared to humans) would prevent the development of dermatitis. Nevertheless, as a safeguard measure, densely haired and dark skinned breeds of livestock should be used for grazing in dense stands of *H. mantegazzianum*, to reduce the contact of exposed skin with the plants' furanocoumarins and decrease the possibility of light-hypersensitive dermatitis. Using individuals experienced to *H. mantegazzianum* also reduces this risk, as the experienced to inexperienced individuals learning (often adult to offspring learning process; 'Choosing livestock' section above) reduces the hazard of over-eating unfamiliar plants. If dermatitis does develop in livestock grazing *H. mantegazzianum* the afflicted animals should be removed from pasture and kept in a stable (i.e. in shade) until the dermatitis is cured.

Economic Aspects of Control by Grazing

It is very difficult to make uniform assessments of control costs for *H. mantegazzianum* between different sites, let alone between different countries. Differences in stand density, stand size, stand shape, topography and character of sites makes it impossible to reach a standardized cost–resource estimate. Sørensen and Buttenschøn (2003) compared different management measures in 36 sites ranging from 0.09 to 2.3 ha (Table 15.3). While the costs of cutting remain more or less the same per area unit with increasing area of managed plot, the fencing costs are reduced significantly with increasing size. Animal husbandry costs are also markedly reduced with increasing pasture size.

Table 15.3. Comparative costs of different methods of control of *H. mantegazzianum*. Whereas root-cutting was applied only to small stands with low density, manual cutting and sheep grazing was applied to small and large stands with high and low density (after Sørensen and Buttenschøn, 2003). Cost-ratio = 1 per m² equals 0.2 € and per individual equals 0.17 €. The differences between the two ratio denominations are due to relations of stand size and density and the practical choice of management measure.

Method	Individual plants treated/person hour	Cost-ratio/m²	Cost-ratio/individual plant
Manual cutting	1000	2	1.6
Root-cutting	100	6	1.0
Sheep grazing	–	1	1.3

Conclusions

Managed animal grazing is an effective control method for larger areas infested with *H. mantegazzianum* and on areas that are inaccessible for machines and difficult to manage by manual methods. Grazing at appropriate stocking rates prevents flowering and seed setting of *H. mantegazzianum*, if started by mid spring, during the growth of vegetative leaf rosettes. Grazing should normally continue for 10 years, until the seed bank is depleted and root stocks die from exhaustion of resources.

Herds of pastured animals should contain individuals that are familiar with *H. mantegazzianum*, as this provides a safeguard against over-eating and poisoning. Dark skinned and densely pelted livestock should be used to minimize the risk of dermatitis.

More controlled, comparative studies are needed to assess the efficiency of different livestock species for the control of *H. mantegazzianum*.

Acknowledgements

The authors would like to thank Matthew Cock for improving the English. The study was supported by the project 'Giant Hogweed (*Heracleum mantegazzianum*) a pernicious invasive weed: developing a sustainable strategy for alien invasive plant management in Europe', funded within the 'Energy, Environment and Sustainable Development Programme' (grant no. EVK2-CT-2001-00128) of the European Union 5th Framework Programme.

References

Andersen, U.V. (1994) Sheep grazing as a method of controlling *Heracleum mantegazzianum*. In: de Waal, L.C., Child, L.E., Wade, P.M. and Brock, J.H. (eds) *Ecology and Management of Invasive Riverside Plants*. Wiley, Chichester, UK, pp. 77–91.

Andersen, U.V. and Calov, B. (1996) Long-term effects of sheep grazing on Giant Hogweed (*Heracleum mantegazzianum*). *Hydrobiologia* 340, 277–284.

Andrews, A., Giles, C.J. and Thomsett, L.R. (1985) Suspected poisoning of a goat by giant hogweed. *The Veterinary Record* 116, 205–207.

Buttenschøn, J. and Buttenschøn, R.M. (1982) Grazing experiments with cattle and sheep on nutrient poor, acidic grasslands and heath. *Natura Jutlandica* 21, 1–48.

Buttenschøn, R.M., Buttenschøn, J., Petersen, H. and Ejlersen, F. (2001) Husbandry and grazing. In: Pedersen, L.B., Buttenschøn, R.M. and Nielsen, T.S. (eds) *Grazing and Management of Nature Areas – Effects on Nutrients Cycling and Biodiversity.* Park og Landskabsserien 30, Forest & Landscape Denmark, Hørsholm, Denmark, pp. 25–45. [In Danish.]

Caffrey, J.M. (1994) Spread and management of *Heracleum mantegazzianum* (Giant Hogweed) along Irish river corridors. In: de Waal, L.C., Child, L.E., Wade, P.M. and Brock, J.H. (eds) *Ecology and Management of Invasive Riverside Plants.* Wiley, Chichester, UK, pp. 67–76.

Cooper, M.R. and Johnson, A.W. (1984) *Poisonous Plants in Britain and Their Effects on Animals and Man.* Ministry of Agriculture, Forestry and Fisheries, Reference Book 161. Her Majesty's Stationery Office, London.

Dodd, F.S., de Waal, L.C., Wade, P.M. and Tiley, G.E.D. (1994) Control and management of *Heracleum mantegazzianum* (Giant Hogweed). In: de Waal, L.C., Child, L.E., Wade, P.M. and Brock, J.H. (eds) *Ecology and Management of Invasive Riverside Plants.* Wiley, Chichester, UK, pp. 11–126.

Fischer, S.F., Poschlod, P. and Beinlich, B. (1996) Experimental studies on the dispersal of plants and animals in calcareous grasslands. *Journal of Applied Ecology* 33, 1206–1222.

Fitter, A.H. and Hay, R.M.K. (1987) *Environmental Physiology of Plants*, 2nd edn. Academic Press, London.

Frost, R.A. and Launchbaugh, K.L. (2003) Prescription grazing for rangeland weed management: a new look at an old tool. *Rangeland* 25, 43–47.

Hansen, B. (1997) *Nature-Management Societies in Conservation Management.* Park-og Landskabsserien nr. 15. Forest and Landscape Denmark, Hørsholm. [In Danish.]

Hofmann, R.R. (1989) Evolutionary steps of ecophysiological adaptation and diversification of ruminants: a comparative view of their digestive system. *Oecologia* 78, 443–457.

Illius, A.W. and Gordon, I.J. (1993) Diet selection in mammalian herbivores: constrains and tactics. In: Hughes, R.N. (ed.) *Diet Selection: An Interdisciplinary Approach to Foraging Behaviour.* Blackwell, Oxford, UK, pp. 157–181.

Krinke, L., Moravcová, L., Pyšek, P., Jarošík, V., Pergl, J. and Perglová, I. (2005) Seed bank in an invasive alien *Heracleum mantegazzianum* and its seasonal dynamics. *Seed Science Research* 15, 239–248.

Lundström, H. and Darby, E. (1994) The *Heracleum mantegazzianum* (Giant Hogweed) problem in Sweden: suggestions for its management and control. In: de Waal, L.C., Child, L.E., Wade, P.M. and Brock, J.H. (eds) *Ecology and Management of Invasive Riverside Plants.* Wiley, Chichester, UK, pp. 93–100.

Møller, K. (2002) Giant Hogweed. *Tidskrift for Dansk Fåreavl* 5, 4–6. [In Danish.]

Nielsen, C., Ravn, H.P., Nentwig, W. and Wade, M. (2005) *The Giant Hogweed Best Practice Manual. Guidelines for the Management and Control of an Invasive Weed in Europe.* Forest and Landscape Denmark, Hørsholm, Denmark.

Nielsen, S.T. (2005) The return of the giant hogweed: population ecology and management of *Heracleum mantegazzianum* Sommier & Levier. MSc thesis, Department of Biological Sciences, University of Aarhus, Denmark.

Otte, A. and Franke, A. (1998) The ecology of the Caucasian herbaceous perennial *Heracleum mantegazzianum* Somm. et Lev. (Giant Hogweed) in cultural ecosystems of Central Europe. *Phytocoenologia* 28, 205–232.

Provenza, F.D. and Cincotta, R.P. (1993) Foraging as a self-organizational learning process:

accepting adaptability at the expense of predictability. In: Hughes, R.N. (ed.) *Diet Selection: an Interdisciplinary Approach to Foraging Behaviour*. Blackwell, Oxford, UK, pp. 78–101.

Ruskova, V.M. (1973) The rhythms of growth and development of species of *Heracleum* introduced to Moscow. *Byulleten Glavnogo Botanicheskogo Sada* 87, 46–49. [In Russian.]

Satsyperova, I.F. (1984) *Hogweeds in the Flora of the USSR: New Forage Plants*. Nauka, Leningrad. [In Russian.]

Sørensen, M.A. (2002) Management strategy for the non-chemical control of giant hogweed along the stream Seest Mølleå. Forest and Landscape Engineer thesis, The Danish Forestry College. [In Danish.]

Sørensen, M.A. and Buttenschøn, R.M. (2003) *Control of Giant Hogweed. Control Methods and Economy*. Videnblad nr. 6.0-1, Forest and Landscape, Hørsholm, Denmark. [In Danish.]

Sørensen, M.A. and Buttenschøn, R.M. (2005) *The Extent of Prevention of Giant Hogweed by the Danish Municipalities*. Videnblad nr. 6.0-20, Forest and Landscape, Hørsholm, Denmark. [In Danish.]

Sørensen, P.H. (1965) *Animal Feeding*. DSR Forlag, København, Denmark. [In Danish.]

Swift, M.J., Heal, O.W. and Anderson, M. (1979) Decomposition in Terrestrial Ecosystems. In: *Studies in Ecology*, Vol. 5. Blackwell, Oxford, UK.

Terney, O. (1993) Giant hogweed – a threat to Danish nature. *Bio-Nyt* 83, 1–18. [In Danish.]

Tiley, G.E.D. and Philp, B. (1994) *Heracleum mantegazzianum* (Giant Hogweed) and its control in Scotland. In: de Waal, L.C., Child, L.E., Wade, P.M. and Brock, J.H. (eds) *Ecology and Management of Invasive Riverside Plants*. Wiley, Chichester, UK, pp. 101–109.

Tiley, G.E.D., Dodd, F.S. and Wade, P.M. (1996) Biological flora of the British Isles: *Heracleum mantegazzianum* Sommier & Levier. *Journal of Ecology* 84, 297–319.

Tolhurst, S. and Oates, M. (2001) *The Breed Profiles Handbook*. Grazing Animal Project, English Nature, UK.

16 The Scope for Biological Control of Giant Hogweed, *Heracleum mantegazzianum*

MATTHEW J.W. COCK[1] AND MARION K. SEIER[2]

[1]*CABI Switzerland Centre, Delémont, Switzerland;* [2]*CABI UK Centre (Ascot), Ascot, UK*

> *Still they are invincible . . .*
>
> (Genesis, 1971)

Introduction

Biological weed control is the use of natural enemies of plants for their control. Natural enemies included within this scope are normally characterized as developing completely on individual plants (although some groups may use alternate hosts at different parts of their life cycle or in different generations). Thus, the groups considered for use in biological control are normally diseases, particularly fungi (see Seier and Evans, Chapter 12, this volume), and arthropods, especially insects and mites (see Hansen *et al.*, Chapter 11, this volume), rather than grazing animals such as sheep and cattle (see Buttenschøn and Nielsen Chapter 15, this volume). There are two main weed biological control strategies in use: the so-called classical approach involving the introduction of exotic natural enemies, and the regular augmentative release of natural enemies.

Augmentative biological control is normally based on the use of indigenous fungi, formulated and applied as a mycoherbicide, although alien species could also be considered. They may or may not be narrowly host-specific – indigenous fungi with a wider host range can be used since the formulation and ecosystem targeted application process can be managed so as to minimize adverse effects on non-target plants. Such an approach is effectively an environmentally friendly parallel to the use of a chemical herbicide. It needs to be regularly repeated, wherever the target weed requires control, and there are significant financial implications to its use (Charudattan, 1991; Evans *et al.*, 2001).

Classical biological control is used to control alien plants by the introduction of narrowly host-specific natural enemies from the native range of the target weed, and this is the focus of the research carried out under this project and reported here. It offers a long-term, sustainable solution to alien weed problems because the introduced biological control agents are able to reproduce and disperse, finding and controlling all populations of the target weed wherever it occurs without further intervention. Because it is inherently safe and cost-efficient, it is the preferred method of control in countries with a history of tackling invasive alien plants, such as Australia (McFadyen, 2000).

Classical weed biological control by the introduction of co-evolved, host-specific arthropods and pathogen natural enemies has been practised for more than 100 years, with more than 350 species released against over 100 targets in more than 70 countries (Julien and Griffiths, 1998). Major successes have been achieved against invasive alien plants in prairies, pasture, forestry, arable land, wetlands and open water (McFadyen, 2000). Australia, New Zealand, USA (including Hawai'i), Canada and South Africa have been particularly active in recent years (Julien and Griffiths, 1998; McFadyen, 1998, 2000).

Although European Union countries have been the source of nearly 400 weed biological control agents used elsewhere (Julien and Griffiths, 1998; Shaw, 2003), weed biological control using exotic biological control agents has yet to be implemented in Europe, except that releases have been made in the former USSR (Julien and Griffiths, 1998). In the UK, there was an experimental release of *Altica carduorum* (Guérin-Méneville), a chrysomelid defoliator of *Cirsium arvense* (L.) Scop. (*Asteraceae*) from France and Switzerland (Baker *et al.*, 1972); this has been referred to as an attempt at biological control, but there was no extensive release programme or serious attempts to identify suitable source areas. A research programme on the possibilities for biological control of bracken (*Pteridium aquilinum* L.), a globally distributed weed of upland pasture, demonstrated that two insects from South Africa were suitably host-specific and permission was given for field-cage releases in the UK (Fowler, 1993). However, the programme was not completed because of lack of funding, and concern about potential changes in the ecology of bracken infestations, open stands of which had recently been identified as an important habitat for the nationally endangered high brown fritillary butterfly, *Argynnis adippe* (L.) (Nymphalidae) (Warren and Oates, 1995; Joy, 1998). As yet, no other European programmes have approached this stage, although projects against Japanese knotweed, *Fallopia japonica* (Houtt.) Ronse Decraene (*Polygonaceae*), in the UK (R.H. Shaw, Ascot, 2005, personal communication) and *Acacia* spp. (*Mimosaceae*) in Portugal (H. Marchante, Coimbra, 2005, personal communication) are well advanced.

Several alien plant species have been identified as suitable targets for biological control in Europe, including giant hogweed, *Heracleum mantegazzianum* Sommier & Levier (*Apiaceae*) (Shaw, 2003; Sheppard *et al.*, 2006). When the GIANT ALIEN project was formulated, giant hogweed appeared to be a particularly appropriate target for biological control because: (i) it is an alien species in unmanaged land subject to little regular disturbance (see Thiele *et al.*, Chapter 8, this volume; Pyšek and Pyšek, 1995; Müllerová *et al.*, 2005); (ii) it is a weed of riparian habitats where conventional herbicide control options are very limited (see Pyšek *et al.*, Chapter 3 and Nielsen *et al.*, Chapter 14, this volume; Pyšek and Prach, 1994); (iii) the plant was considered biennial (see Perglová *et al.*, Chapter 4, this volume; Tiley *et al.*, 1996) and its seed bank was suspected to be short-lived (see Moravcová *et al.*, Chapter 5, this volume) so that results might be expected quickly; (iv) it was considered part of the natural tall-herb flora of subalpine meadows in the Western Greater Caucasus where it did not dominate other plants to a great extent (see Otte *et al.*, Chapter 2, this volume); (v) other *Apiaceae* were known to have associated specialist herbivores and pathogens (Berenbaum, 1981, 1983; Bürki and Nentwig, 1998; Sheppard, 1991); and (vi) the presence of characteristic varied furanocumarins, biochemically active secondary compounds (Molho *et al.*, 1971; Nielsen, 1971; Hattendorf *et al.*, Chapter 13, this volume), suggested the likelihood of specialized herbivores and pathogens adapted to this biochemistry (Berenbaum, 1981, 1983), and perhaps using the secondary chemicals as cues for oviposition and germination. As the research programme described in this monograph was formulated, it was anticipated that the main factors determining the frequency of giant hogweed in its native range would be natural enemy attack combined with plant competition. Furthermore, amongst these damaging natural enemies there were expected to be several highly specialized insects and fungi that would be suitably host-specific and which could be introduced into Europe without significant risk to any non-target economic or indigenous plant species.

The core of a weed biological control programme is detailed laboratory and field studies to assess the risk of any potential biological control agent causing significant damage to non-target economic and indigenous plants (Harley and Forno, 1992). The approach that has been demonstrated to work is the centrifugal phylogenetic host-specificity approach first formally outlined by Wapshere (1974). In this approach, it is assumed that the plants most at risk from narrowly host-specific fungi and insects are those phylogenetically most closely related to the target. Many years of experimental work and field observation have now confirmed this (McFadyen, 2000; Pemberton, 2000). Traditionally, the principal concern has been that potential weed biological control agents should present no threat to economically important plants, especially crops. One result of this is that weed biological control agents have been introduced that were capable of attacking indigenous plants closely related to the target. Relatively recent reports of the possibility of population level impact to threatened and endangered species in the USA (Louda *et al.*, 1997; Pemberton, 2000) have raised public and scientific concern about this risk, and now all weed biological control initiatives should include an assessment of the risks of attack to non-target indigenous plants. In the simplest scenario, there is no risk, i.e. the potential biological control agents are monospecific (specific to the target species only). More commonly biological control agents are narrowly specific – physiologically capable of attacking some non-target species, but ecologically very unlikely to do so. However, even if closely related non-target species are at risk of attack, this does not mean that population-level effects will be negative (e.g. Baker *et al.*, 2004), but clearly they will be much more difficult to predict.

Our approach to host-range testing potential biological control agents for *H. mantegazzianum* was based on a proposed test plant list (M.K. Seier, unpublished), which tried to take into account phylogenetic criteria, but also included plant species from outside the order *Apiales* which, like *H. mantegazzianum*, are known to produce furanocoumarins (Nielsen, 1971). The phylogeny of the *Apiaceae* is not firmly established, and recent DNA-based studies have shown significant differences from the traditional morphological approach (Downie *et al.*, 1998, 2000a, b, 2001). However, a group of genera including *Heracleum* (apioid superclade Group B or '*Heracleum*' clade of Downie *et al.* (2001)) continue to match closely, although perhaps not exactly, to the subtribe *Tordyliinae* (= *Tordylieae*). Moreover, *Heracleum* is not a monophyletic genus – as is the case for more than half the current genera of *Apiaceae* examined (Downie *et al.*, 1998, 2000a, b). Unfortunately, *H. mantegazzianum* has not as yet been included in these studies. *Pastinaca*, which includes parsnip (*P. sativa* L.), is very close to the core species of *Heracleum*, including *H. sphondylium* L. (Theobald, 1971; Downie *et al.*, 1998). *Coriandrum*, which includes coriander (*C. sativum* L.), is also in the apioid superclade of Downie *et al.* (2001) and placed close to the '*Heracleum*' clade in many of the DNA-based phylogenies (Downie *et al.*, 1998, 2000a, b, 2001).

Thus, in this study, potential biological control agents were first screened on species within the traditional genus *Heracleum*, i.e. the western European native *H. sphondylium*, and within the *Tordyliinae*. Natural enemies which attacked these key test plants were eliminated, unless there was reason to think that the results were due to laboratory artefacts, in which case more realistic ecological tests were set up to try and assess the risk of damage to non-target species, principally at the individual plant level, rather than the population level.

In Chapters 11 (Hansen *et al.*, this volume) and 12 (Seier and Evans, this volume) the results of the surveys of insects and fungal pathogens associated with giant hogweeds in the Caucasus Mountains are reported. In this chapter we consider the scope for their use as biological control agents for *H. mantegazzianum* in Europe.

Survey Results and Host-specificity Observations

In spite of a range of insects (see Hansen *et al.*, Chapter 11, this volume) and fungal pathogens (Seier and Evans, Chapter 12, this volume) attacking roots, stems, petioles, leaves and inflorescences, in almost all cases, natural enemy impact on *H. mantegazzianum* in the north-western Greater Caucasus was minimal. No outbreaks of insect or fungal natural enemies causing impact were observed, apart from apparently significant impact of *Phloeospora heraclei* (Lib.) Petr. observed on seedlings and locally severe damage to umbels due to *Phomopsis* sp. (see Seier and Evans, Chapter 12, this volume). Dominant stands of *H. mantegazzianum* do occur in disturbed areas of the north-western Greater Caucasus, and human interference particularly through grazing cattle is probably the main factor controlling giant hogweed in such stands.

Observations of insects on giant hogweed and other *Heracleum* spp. in the Caucasus Mountains (see Hansen *et al.*, Chapter 11, this volume) showed that nearly all associated insects were either polyphagous or oligophagous, associated with several *Heracleum* spp. and perhaps other *Apiaceae* which could not be surveyed. Further laboratory studies on selected oligophagous specialists showed a wider range of food plant acceptance within the *Apiaceae*. None of the associated insect species appeared to be monophagous, i.e. feeding only on *H. mantegazzianum*.

However, of the insect natural enemies found during the project surveys (see Hansen *et al.*, Chapter 11, this volume), several deserve further comment. *Nastus fausti* Reitter is a weevil associated with *H. mantegazzianum*, whose larvae feed on roots and adults feed on leaves; it is not considered to be narrowly host-specific. A cerambycid beetle, *Phytoecia boeberi* Ganglbauer, the larvae of which attack stems or roots, is considered likely to be family-specific. The larvae of a small moth, *Agonopterix caucasiella* Karsholt (Depressariidae), feed on umbels, in a similar way to the closely related parsnip pest, *Depressaria radiella* (Goeze), which is already present in Europe. Finally, the larvae of an agromyzid fly, *Melanagromyza heracleana* Zlobin, mine inside the stems, but do not seem to have any impact on the plants. There is no additional information on host range from field observations or laboratory testing (see Hansen *et al.*, Chapter 11, this volume) to suggest that any of these agents would be suitably host-specific for consideration in a biological control programme. Indeed, the pattern of specificity to a few species of *Apiaceae* found amongst other natural enemies does not give any particular cause for optimism regarding host-specificity of these insects, and the indications from the literature are that the two beetles are not narrowly specific. The umbel-feeding moth already has a closely related, heavily parasitized ecological homologue in Europe, and although it would be of interest to make further observations on its host-specificity, this moth is unlikely to be an effective biological control agent. The lack of impact of the agromyzid fly on its host argues that, even if it were narrowly host-specific, it would not be effective as a biological control agent.

Similarly, in Chapter 12 (Seier and Evans, this volume), surveys for pathogens associated with *H. mantegazzianum* and other *Heracleum* spp. in the Caucasus Mountains showed that the species of *Heracleum* surveyed share a common mycobiota. A common characteristic in many fungal species is that within-species diversity may include more specific fungal strains, but this was not investigated in depth here because our laboratory studies showed that the selected pathogens isolated from *H. mantegazzianum* in the Greater Caucasus could produce symptoms on parsnip and coriander (see Seier and Evans, Chapter 12, this volume).

Thus, the studies carried out under the GIANT ALIEN project and summarized here revealed that none of the natural enemies assessed, either insects or fungal pathogens, exhibited sufficient specificity to be considered safe for introduction into Europe. Non-target plant species shown to come under attack by the selected insects and fungal pathogens were other *Heracleum* spp. and/or parsnip and, to a lesser extent, coriander (see Seier and Evans, Chapter 12, this volume).

Countries with a long-standing experience in classical biological control such as Australia have previously introduced classical agents including pathogens known to cause symptoms on non-target species in tests under greenhouse conditions. Such decisions were taken on a pest-risk assessment basis, weighing the benefits of controlling an invasive weed against the potential risk of non-target damage, e.g. introduction of the rusts *Maravalia cryptostegiae* (Cummins) Ono and *Puccinia melampodii* Diet. & Holw. into Australia to control *Cryptostegia grandiflora* Roxb. ex R.Br. (*Asclepiadaceae*) and *Parthenium hysterophorus* L. (*Asteraceae*), respectively (Seier, 2005a). However, with classical biological control of weeds never having been implemented before in Europe, and the lack of relevant experience, any introduction of such potentially controversial agents must be considered inappropriate at this point in time.

Classical Biological Control of Weeds in Europe

For Europe, the European and Mediterranean Plant Protection Organization (EPPO) has developed standards for the 'safe use of biological control' (EPPO, 2001) based on the IPPC Code of Conduct for the import and release of exotic biological control agents (IPPC, 1996). However, neither the code nor the EPPO standards are legally binding and the latter also explicitly exclude microorganisms. Within the EU the legislation that currently covers potential introductions of fungal pathogens for weed biological control is EU directive 91/414 (1991). This directive deals with the 'placing of plant protection products on the market' and classifies microorganisms as 'active substances of plant protection products'.

Initially developed to protect humans and the environment from harmful chemicals, these EU regulations assign microbial agents a pesticide status with immense registration costs, and demonstration of 'product' efficacy paramount for registration. They are, therefore, totally inappropriate for dealing with fungal agents as classical biological control agents for weeds, which generally carry no human health risk and, in contrast to a marketable product, are not for sale – being intended to establish permanently in the environment after initial releases for 'the public good' (Harley and Forno, 1992). As a direct result of the EU directive, in the UK the approval of any potential application for the importation and release of exotic fungal pathogens falls within the responsibilities of the Pesticide Safety Directorate (PSD). Despite a recent PSD pilot scheme offering a free pre-consultation and reduced registration fees for applications dealing with alternative control measures at around 33,000 € (R.H. Shaw, Ascot, 2005, personal communication), the cost to register a fungal pathogen as a classical biological control agent for giant hogweed would still be out of reach for most public-funded project budgets.

Switzerland is not a member of the EU, but would expect to follow closely the EU directive when addressing the issue of pathogens as classical agents during the current revision of its legislation for plant protection products (H. Dreyer, Bern, 2005, personal communication; FOE, 2005).

In sharp contrast, applications for the release of arthropods as classical biological control agents are not currently covered by EU legislation. These 'beneficial pests' are dealt with in the UK free of charge by the Department of

Environment, Food and Rural Affairs (Defra) Biotechnology Unit supported by the Advisory Committee on Releases into the Environment (ACRE), an institution much better placed to take such decisions. In Portugal, the national bodies to give permission for introduction of exotic species (and to enforce the law on 'prohibition of intentional introduction of species') are the Institute for Nature Conservation (ICN) where conservation interests are concerned and the Forestry Authority (DGF) where forestry, aquaculture etc. interests are affected. Releases in the environment have to be authorized at the ministerial level based on a proposal prepared by ICN in consultation with DGF (H. Marchante, Coimbra, 2006, personal communication). In Germany, the introduction of exotic arthropods is not regulated, nor is their release into populated areas. The legal document covering the release of exotic arthropods into 'the wild' is the 'Federal Nature Conservation Act' from 2002, which due to Germany's federal system is enforced by individual federal states and regional authorities. This document specifically spells out that the impact of any such release on other EU-member countries needs to be considered beforehand (F. Klingenstein, Bonn, 2005, personal communication; FOEN, 2005).

Much can be learnt from countries such as Australia, New Zealand and South Africa if classical biological control of invasive alien weeds using fungal pathogens is to become a viable concept for Europe. Most of all, the problems in the current legislation dealing with microbial classical agents need to be addressed and, ideally, these organisms would be dealt with under a different non-pesticide legislation. Appropriate national bodies, such as Defra and ACRE in the UK, should be tasked with the evaluation of applications for the import and release of classical fungal agents and with granting approval for the implementation.

Beyond this, consultation processes between European states need to be in place for the introduction of classical biological control agents, whether arthropods or pathogens, as their spread will not be halted by national borders (Cock *et al.*, 2006). The introduction process could be significantly facilitated by the adoption of harmonized protocols and legislation for the introduction of classical biological control agents in Europe. Current initiatives are developing information standards for introduction and release of biological control agents in Europe (Bigler *et al.*, 2005), but this needs to be built upon.

Fortunately, due to recent initiatives a dialogue with the European authorities concerning the issues highlighted above has led to the establishment of an EU Concerted Action Project REBECA. This project will review components of the current directive 91/414, and should lead to guidance on how better to accommodate and facilitate the use of biological control in Europe; it is important that this initiative fully addresses the use of pathogens and insects as weed biological control agents.

Evolutionary History as an Explanation for the Absence of Specialized Natural Enemies

The studies carried out in this project have confirmed that *H. mantegazzianum* is an invasive alien species in unmanaged land that is subject to little regular disturbance (see Thiele *et al.*, Chapter 8, this volume), and is a weed of riparian habitats where conventional herbicide control options are very limited (see Pyšek *et al.*, Chapter 3, and Nielsen *et al.*, Chapter 14, this volume). However, it is not, as previously assumed, a rigid biennial but a monocarpic perennial; immature plants may continue to accumulate reserves for several years until it is able to flower (see Perglová *et al.*, Chapter 4, this volume); nevertheless it is still relatively short-lived (see Pergl *et al.*, Chapter 6, this volume). Results are still accumulating in the seed-bank duration study, but it is clear already that the great majority of seeds in the soil do not last more than 3 years (see Moravcová *et al.*, Chapter 5, this volume; Krinke *et al.*, 2005).

However, ecological observations and studies carried out by the project in the north-western Greater Caucasus have shown that in its native range, although stream-side infestations in forests at lower altitudes appear natural, everywhere else giant hogweed is always associated with human land-use change, including sites where it occurs above the current tree line (see Otte *et al.*, Chapter 2, this volume). Furthermore, its frequency in these habitats depends partly on continuing human intervention, particularly grazing by cattle and mixed herds and soil enrichment by animal dung (see Otte *et al.*, Chapter 2, this volume), and, contrary to our expectations, is not due to a combination of insect damage (see Hansen *et al.*, Chapter 11, this volume), pathogen damage (see Seier and Evans, Chapter 12, this volume) and plant competition.

The surveys and laboratory studies carried out under this project have shown that for both insects and pathogens, the natural enemies in the Caucasus Mountains are a mixture of generalist species (attacking plants of more than one family) and narrowly specific species (attacking several species within the genus *Heracleum* or closely related genera). No natural enemy species were shown to be monospecific to *H. mantegazzianum* (see Hansen *et al.*, Chapter 11, and Seier and Evans, Chapter 12, this volume), however, with respect to fungal pathogens field observations and greenhouse inoculations strongly indicate the existence of host-specific pathogen strains which could have a narrower host range than initially perceived (see Seier and Evans, Chapter 12, this volume). Nevertheless, this aspect requires a detailed assessment and as long as indigenous European *H. sphondylium* and economic plants of the *Apiaceae* such as parsnip and coriander can potentially be at risk from all these narrowly specific natural enemies, we conclude that none should be considered for introduction into Europe or North America.

However, while there appears to be little scope for classical biological control of *H. mantegazzianum* at the moment, the mycoherbicide approach may still be an option in an integrated management strategy for its control in its introduced range. As already mentioned by Seier and Evans (Chapter 12, this volume), the indigenous pathogen *Sclerotinia sclerotiorum* (Lib.) de Bary is currently under evaluation as a mycoherbicide agent and other fungi, such as a *Phomopsis* sp. recorded from *H. mantegazzianum* in its introduced range, may warrant further studies.

The native range of *H. mantegazzianum* extends to both the north and the south sides of the Western Greater Caucasus. For security reasons, the southern side in Abkhazia, Georgia, could not be surveyed during this project. Hence, it is possible that there could be additional natural enemies restricted to this part of the natural range, such as the rust *Puccinia heraclei* Grev. reported to occur on *Heracleum* sp. in this area (Nakhutsrishvili, 1986), one or more of which are species-specific. This deserves to be evaluated when the political situation makes this possible, although it could be argued that a potential biological control agent that could not spread naturally to both sides of the Greater Caucasus is less likely to be effective when introduced.

Based on other weed biological control programmes, the absence of suitable host-specific natural enemies is an unusual and unexpected conclusion. There are very few other weed biological control survey and testing programmes that failed to find suitably host-specific biological control agents. *Bryophyllum delagoense* (Eckl. & Zeyh.) Schinz, mother-of-millions (*Crassulaceae*) in Australia may be another case, since natural enemies attack horticultural varieties of closely related *Kalanchoë* spp. (M. Julien, Montpellier, 2005, personal communication). Other examples demonstrate that there are suitable biological control agents for countries with no plants closely related to the target species, but not for countries with species congeneric with the target. Thus, our generalization may be partly an artefact, inasmuch as it is only in the last 10 years that significant concerns have arisen over potential non-target effects and the great majority of weed biological control agents were assessed and released on less stringent requirements of host-specificity. Nevertheless, our conclusion is still unexpected, in view, for example, of the diverse secondary compounds, including simple, linear and angular furanocoumarins found in different combinations in different *Heracleum* spp. (Molho *et al.*, 1971; Nielsen, 1971; see Hattendorf *et al.*, Chapter 13, this volume), which might have stimulated and maintained host-specificity (Berenbaum, 1983).

There is no obvious explanation for the lack of monospecific natural enemies, but here we hypothesize that it may relate to the speciation history of the genus *Heracleum* and its natural enemies in the Caucasus Mountains during successive glaciation events during the Pleistocene.

The Caucasus is a recognized hotspot of biodiversity (Myers *et al.*, 2000) and one of only five ecoregions outside the tropics with more than 5000 plant species (Kier *et al.*, 2005). *Apiaceae* is the fourth most speciose family in the Colchis (south-western Greater Caucasus) after *Asteraceae*, *Poaceae* and *Fabaceae*, with 148 species, many of which are endemic (Kikvidze and Ohsawa, 2001). Furthermore, tall herbaceous vegetation (altherbosa), which is generally uncommon, but typical of the Colchis timberline and subalpine meadows, has been considered the typical habitat of *H. mantegazzianum* (but see Otte *et al.*, Chapter 2, this volume), and is high in endemics (Kikvidze and Ohsawa, 2001). Although *Heracleum* is a circumboreal genus, traditionally reported to comprise 60–70 species (Heywood, 1971), the Caucasus Mountains are the centre of diversity and endemism for the genus *Heracleum* (Mandenova, 1950).

The key reason for the high biodiversity in the Caucasus Mountains is that they border two glacial refugia: the Colchis refugium just south of the Greater Caucasus and bordering the Black Sea (which was shallower during glaciation

(Velichko and Kurenkova, 1990)) and the Hyrcanian or Lycia Refugium on the south border of the Caspian Sea (which was somewhat deeper and more extensive than today (Velichko and Kurenkova, 1990)) to the extreme southeast of the Greater Caucasus (Tarasov et al., 2000; Kikvidze and Ohsawa, 2001; Zazanashvili et al., 2004; Conservation International, 2005).

Climate throughout the region is variable, with annual rainfall ranging from as little as 150 mm in the eastern part of the hotspot on the Caspian coast to more than 4000 mm in the coastal mountains along the Black Sea (Zazanashvili et al., 2004; Conservation International, 2005). Hence there is considerable scope for evolutionary pressure towards locally adapted populations, and the high endemic diversity of Apiaceae suggests this has happened repeatedly in the family.

North of the Greater Caucasus is the broad North Caucasus Plain, the eastern part of which is below sea level. Here, steppes in the west grade to semi-desert ecosystems, and then to desert in the east (Zazanashvili et al., 2004; Conservation International, 2005). To the south the barriers to the highlands of Turkey and Iran are much less significant (Zazanashvili et al., 2004), indeed the Transcaucasian Depression between the Greater Caucasus and the Lesser Caucasus would seem to be a greater barrier.

There is evidence that the normal response to climate changes such as glaciation is for many species to move, sometimes over vast distances, so as to stay in areas with compatible climate and ecology (Coope, 1979, 1994, 1995; Jansson and Dynesius, 2002). However, where there are substantial eco-geographical barriers to movement, such as around the Greater Caucasus, the scope for long-distance movement may be limited. Thus, species may be limited to short-distance movements up or down the mountains to stay within an acceptable climate, which at the glacial maximum means that species would be confined to one of the two very restricted refugia at each end of the Greater Caucasus. Species would move back up into the mountains during the interglacial periods, and disperse to the east from the Colchis Refugium or west from the Hyrcanian Refugium. Kingdom (1990) has argued and documented how a similar process has led to the accumulation of endemic species on the isolated geologically old mountains of Africa, such as the Usambaras in Tanzania.

During glaciation, in these refugia of limited size, small populations biased by Allee effects would hybridize and adapt under strong natural selection, giving rise to 'new' partially differentiated populations. During interglacial periods these new populations would spread back into the mountains, and the new populations from the two refugia would meet, mingle, hybridize or develop isolating mechanisms. The different evolving populations might retreat into one or both refugia. Thus, one can imagine that over several glaciation events, there was a dynamic situation with populations evolving, merging, hybridizing and segregating, so that today Heracleum (amongst other genera) is a large, diverse genus in the Caucasus Mountains (Mandenova, 1950), with a mixture of distributions, with significant variation and possible hybrids whose status is seldom clear. Furthermore, Molho et al. (1971) found that different populations, purportedly of H. mantezzianum, showed significant differences in the make up of their furanocoumarin content (see Hattendorf *et al.*, Chapter 13, this volume).

Against this background, consider the situation for natural enemies and how natural selection might act on them. One hypothesis is that natural enemies would co-evolve with their host plants, giving rise to an equal (or greater) diversity of natural enemies for each original herbivore or pathogen. There are indications of such patterns from apparent evolution in temperate glacial refugia (e.g. Mayr, 1970). A similar explanation was proposed for the rich tropical fauna of Amazonia (Haffer, 1969; Brown *et al.*, 1974; Brown, 1982), with widely spaced refugia, and several generations per year, facilitating rapid evolution during the isolation of glacial maxima. More recently, evidence based on molecular phylogeography is accumulating to suggest that most speciation occurred earlier in the glaciation cycles or even before the Pleistocene (Klicka and Zink, 1997; Hewitt, 2000; Paulo *et al.*, 2001). Repeated population isolation during the Pleistocene reinforced speciation processes that were initiated already during the Pliocene, or produced intraspecific population genetic subdivision (Ribera and Vogler, 2004).

Compared with the Amazon, speciation mechanisms would not work as quickly or efficiently in the Greater Caucasus with two refuges in close proximity at each end of a relatively small and isolated mountain range, in a temperate region with often only one generation a year. Co-evolution is unlikely to describe what happened at the species level. Monospecific and narrowly specific natural enemies would be confined during glaciation to a limited population of hosts in one or both refugia. Monospecific natural enemies would be more prone to extinction due to chance events than narrowly specific natural enemies able to breed on several closely related species. During interglacial periods these selected populations would spread back into the mountains and encounter 'new' but similar populations of the same or closely related species from the other refugium. Incipient or actual host-specificity mechanisms would be likely to break down under such dynamic pressures. Natural enemies specializing on single populations of a *Heracleum* sp. would be selected against, compared with those that were able to feed on several similar *Heracleum* spp. even though the biochemistry of these plant populations varied (and was probably also evolving).

This hypothesis explains what might have happened during the Pleistocene in this system in the Caucasus Mountains. Molecular taxonomic studies might unravel some of the phylogeny of *Heracleum* in the region, but if the phylogeny involved anastomatoses as different 'species' and 'forms' met and hybridized leading to introgression or merging of populations, one could not expect such studies to reveal a straightforward strict phylogeny – this would break down in such a dynamic and interacting situation and the approach might only be able to deliver a relatedness classification.

What testable predictions can be derived from this hypothesis? One is that strict monospecificity will not evolve amongst natural enemies of an actively evolving and hybridizing group of populations and species. One possible example of this is the natural enemies of *Hieracium* spp. (*Asteraceae*). This enormous genus still needs a lot of study and DNA-based analysis to under-stand the species and phylogeny, but it is believed that there is a great deal of hybridization in its indigenous (and exotic) range, and that it appears to be actively evolving in the mountainous regions of western Eurasia where the highest concentration of species occurs (Gottschlich, 1996). Surveys of insects as possible biological control agents for introduced species of *Hieracium* in New Zealand and North America have not been designed to answer this par-ticular question, but while there is evidence for specificity to subgenera and species groups, there is little if any evidence of any insect herbivores being strictly monophagous (G. Grosskopf, Delémont, 2006, personal communica-tion).

Conclusions

There are still some areas that need clarification before concluding that there is no prospect of finding host-specific natural enemies for biological control of giant hogweed. Observations are incomplete on several insects, which merit further evaluation before they can be definitively eliminated as suitable for use in biological control. Also, surveys should be carried out in Abkhazia, Georgia, when this can be done safely. Even so, if the hypothesis put forward in the previous section about evolution of host-specificity in this system were correct, then there is little cause for optimism that suitably host-specific natural enemies would be found. However, to make these studies would be a test of one pre-diction from our hypothesis, and so should be done if only for the further insight into the evolution of host-specificity.

Could these results have been predicted a priori? Given the experience with all other weed biological control projects, and that our hypothesis to explain the results is neither tested nor established, we see no way that these results could have been predicted in advance. The research had to be done to get to this point. This does not negate the potential for weed biological control in Europe – there are many exotic plant species that should be considered in future (Sheppard *et al.*, 2006), and there is no reason to think that suitably host-specific natural enemies could not be found for these.

In interpreting our results, the additional insight from having worked as part of a multi-disciplinary 'model' team has been extremely beneficial. Questions about ecology and population genetics could be answered simulta-neously. However, it would have been helpful if the overall project had included more work on the other giant hogweeds, particularly DNA-based studies to understand better their evolutionary history and phylogeny.

It has also become clear that the current European and national legislation does not facilitate the use of weed biological control agents in Europe, and in some cases is clearly a barrier (Seier, 2005b). Insect agents are not covered by European legislation and so would need to be evaluated for introduction under national legislation. Some European countries have relevant legislation, although not always intended for use with weed biological control, but others have none. It is important that the new EU Concerted Action Project,

REBECA, starting in 2006, will address this issue, to facilitate weed biological control in Europe in the future.

Acknowledgements

We thank our CABI colleagues, Harry Evans, Ruediger Wittenberg, Dick Shaw and Carol Ellison for inputs and discussions throughout the project, as well as Mic Julien and Andy Sheppard (CSIRO European Laboratory) for advice and suggestions. Working as part of a multidisciplinary EU project has been stimulating, and we thank all the partners for their inputs. This work was supported by European Union funding under the 5th Framework programme 'EESD – Energy, Environment and Sustainable Development'; project number EVK2-2001-00128 and CAB International.

References

Baker, C.R.B., Blackman, R.L. and Claridge, M.F. (1972) Studies on *Haltica carduorum* Guerin (Coleoptera: Chrysomelidae) an alien beetle released in Britain as a contribution to the biological control of the creeping thistle, *Cirsium arvense* (L.) Scop. *Journal of Applied Entomology* 9, 819–830.

Baker, J.L., Webber, N.A.P. and Johnson, K.K. (2004) Non-target impacts of *Aphthona nigriscutis*, a biological control agent for *Euphorbia esula* (leafy spurge), on a native plant *Euphorbia robusta*. In: Cullen, J.M., Briese, D.T., Kriticos, D.J., Lonsdale, W.M., Morin, L. and Scott, J.K. (eds) *Proceedings of the XI International Symposium on Biological Control of Weeds*. CSIRO Entomology, Canberra, Australia, pp. 247–251.

Berenbaum, M. (1981) Patterns of furanocoumarin distribution and insect herbivory in the Umbelliferae: plant chemistry and community structure. *Ecology* 62, 1254–1266.

Berenbaum, M. (1983) Coumarins and caterpillars: a case for coevolution. *Evolution* 37, 163–179.

Bigler, F., Bale, J.S., Cock, M.J.W., Dreyer, H., Greatrex, R., Kuhlmann, U., Loomans, A.J.M. and van Lenteren, J.C. (2005) Guidelines on information requirements for import and release of invertebrate biological control agents in European countries. *CAB Reviews: Perspectives in Agriculture, Veterinary Science, Nutrition and Natural Resources* 1(1), 10 pp. *Biocontrol News and Information* 26, 115N–123N.

Brown, Jr, K.S. (1982) Historical and ecological factors in the biogeography of aposematic neotropical butterflies. *American Zoologist* 22, 453–471.

Brown, Jr, K.S., Sheppard, P.M. and Turner, J.R.G. (1974) Quaternary refugia in tropical America: evidence from race formation in *Heliconius* butterflies. *Proceedings of the Royal Society of London, Series B* 187, 369–378.

Bürki, C. and Nentwig, W. (1998) Comparison of herbivore insect communities of *Heracleum sphondylium* and *H. mantegazzianum* in Switzerland (*Spermatophyta*: Apiaceae). *Entomologia Generalis* 22, 147–155.

Charudattan, R. (1991) The mycoherbicide approach with plant pathogens. In: TeBeest, D.O. (ed.) *Microbial Control of Weeds*. Chapman and Hall, New York, pp. 24–57.

Cock, M.J.W., Kuhlmann, U., Schaffner, U., Bigler, F. and Babendreier, D. (2006) The usefulness of the ecoregion concept for safer import of invertebrate biological control agents. In: Bigler, F., Babendreier, D. and Kuhlmann, U. (eds) *Environmental Impact of Invertebrates* for *Biological Control of Arthropods: Methods and Risk Assessment*. CABI, Wallingford, UK, pp. 202–221.

Conservation International (2005) Biodiversity hotspots. Caucasus. http://www.biodiversityhotspots.org/xp/Hotspots/caucasus (accessed 30 December 2005).

Coope, G.R. (1979) Late Cenozoic fossil Coleoptera: evolution, biogeography, and ecology. *Annual Review of Ecology and Systematics* 10, 247–267.

Coope, G.R. (1994) The response of insect faunas to glacial–interglacial climatic fluctuations. *Philosophical Transactions of the Royal Society of London, Series B* 344, 19–26.

Coope, G.R. (1995) Insect faunas in ice age environments: why so little extinction? In: Lawton, J.H. and May, R.M. (eds) *Extinction Rates*. Oxford University Press, Oxford, UK, pp. 55–74.

Downie, S.R., Ramanath, S., Katz-Downie, D.S. and Llanas, E. (1998) Molecular systematics of *Apiaceae* subfamily *Apioideae*: phylogenetic analyses of nuclear ribosomal DNA internal transcribed spacer and plastid rpoC1 intron sequences. *American Journal of Botany* 85, 563–591.

Downie, S.R., Katz-Downie, D.S. and Watson, M.F. (2000a) A phylogeny of the flowering plant family *Apiaceae* based on chloroplast DNA rpl16 and rpoC1 intron sequences: towards a suprageneric classification of subfamily *Apioideae*. *American Journal of Botany* 87, 273–292.

Downie, S.R., Watson, M.F., Spalik, K. and Katz-Downie, D.S. (2000b) Molecular systematics of Old World *Apioideae* (*Apiaceae*): relationships among some members of tribe *Peucedaneae* sensu lato, the placement of several island-endemic species, and resolution within the apioid superclade. *Canadian Journal of Botany* 78, 506–528.

Downie, S.R., Plunkett, G.M., Watson, M.F., Spalik, K., Katz-Downie, D.S., Valiejo-Roman, C.M., Terentieva, E.I., Troitsky, A.V., Lee, B.-Y., Lahham, J. and El-Oqlah, A. (2001) Tribes and clades within *Apiaceae* subfamily *Apioideae*: the contribution of molecular data. *Edinburgh Journal of Botany* 58, 301–330.

EPPO (European and Mediterranean Plant Protection Organisation) (2001) EPPO Standards: Safe use of biological control. Import and release of exotic biological control agents. *EPPO Bulletin* 31, 29–35.

Evans, H.C., Froehlich, J. and Shamoun, S.F. (2001) Biological control of weeds. In: Pointing, S.B. and Hyde, K.D. (eds) *Bio-exploitation of Filamentous Fungi. Fungal Diversity Research Series* 6, 349–401.

FOEN (Federal Office of the Environment, Switzerland) (2005) Révision de l'ordonnance sur la dissémination dans l'environnement. http://www.umwelt-schweiz.ch/buwal/fr/medien/presse/artikel/20051222/01214/index (accessed 26 January 2006).

Fowler, S.V. (1993) The potential for control of bracken in the UK using introduced herbivorous insects. *Pesticide Science* 37, 393–397.

Gottschlich, G. (1996) *Hieracium*. In: Sebald, O., Philippi, G. and Wörz, A. (eds) *Die Farn- und Blütenpflanzen Baden-Württembergs, Band 6: Spezieller Teil (Spermatophyta, Unterklasse Asteridae) Valerianaceae bis Asteraceae*. Verlag Eugen Ulmer, Stuttgart, Germany, pp. 393–529.

Haffer, J. (1969) Speciation in Amazonian forest birds. *Science* 165, 131–137.

Harley, K.L.S. and Forno, I.W. (1992) *Biological Control of Weeds: a Handbook for Practitioners and Students*. Inkata Press, Melbourne, Australia.

Hewitt, G. (2000) The genetic legacy of the Quaternary ice ages. *Nature* 405, 907–913.

Heywood, V.H. (1971) Systematic survey of Old World *Umbelliferae*. In: Heywood, V.H. (ed.) *The Biology and Chemistry of the Umbelliferae. Botanical Journal of the Linnean Society* 64, Supplement 1, 31–41.

IPPC (International Plant Protection Convention) (1996) *Code of Conduct for the Import and Release of Exotic Biological Control Agents. International Standards for Phytosanitary Measures Publication No. 3.* FAO, Rome.

Jansson, R. and Dynesius, M. (2002) The fate of clades in a world of recurrent climatic change: Milankovitch oscillations and evolution. *Annual Review of Ecology and Systematics* 33, 741–777.

Joy, J. (1998) *Bracken for Butterflies.* Butterfly Conservation, Dedham, UK. [A5 leaflet.]

Julien, M.H. and Griffiths, M.W. (eds) (1998) *Biological Control of Weeds. A World Catalogue of Agents and their Target Weeds*, 4th edn. CAB International, Wallingford, UK.

Kier, G., Mutke, J., Dinerstein, E., Ricketts, T.H., Küper, W., Kreft, H. and Barthlott, W. (2005) Global patterns of plant diversity and floristic knowledge. *Journal of Biogeography* 32, 1107–1116.

Kikvidze, Z. and Ohsawa, M. (2001) Richness of Colchic vegetation: comparison between refugia of south-western and East Asia. *BMC Ecology* 1 (6). http://www.biomedcentral.com/1472-6785/1/6 (accessed 30 December 2005).

Kingdom, J. (1990) *Island Africa: The Evolution of Africa's Rare Animals and Plants.* Collins, London.

Klicka, J. and Zink, R.M. (1997) The importance of recent ice ages in speciation: a failed paradigm. *Science* 277, 1666–1669.

Krinke, L., Moravcová, L.,Pyšek, P., Jarošík, V., Pergl, J. and Perglová, I. (2005) Seed bank in an invasive alien *Heracleum mantegazzianum* and its seasonal dynamics. *Seed Science Research* 15, 239–248.

Louda, S.M., Kendall, D., Connor, J. and Simberloff, D. (1997) Ecological effects of an insect introduced for the biological control of weeds. *Science* 277, 1088–1090.

Mandenova, I.P. (1950) *Caucasian Species of the Genus Heracleum.* Georgian Academy of Sciences, Tbilisi. [In Russian.]

Mayr, E. (1970) *Populations, Species and Evolution.* Belknap Press, Cambridge, Massachusetts.

McFadyen, R.E.C. (1998) Biological control of weeds. *Annual Review of Entomology* 43, 369–393.

McFadyen, R.E.C. (2000) Successes in biological control of weeds. In: Spencer, N.R. (ed.) *Proceedings of the X International Symposium on Biological Control of Weeds 4–14 July 1999.* Montana State University, Bozeman, Montana, pp. 3–14.

Molho, D., Jössang, P., Jarreau, M.-C. and Carbonnier, J. (1971) Dérivés furanno-coumariniques du genre *Heracleum* et plus spécialement de *Heracleum spengelianum* Wight & Arn. et *Heracleum ceylanicum* Gardn. ex C.B. Clarke étude phylogénique. In: Heywood, V.H. (ed.) *The Biology and Chemistry of the Umbelliferae. Botanical Journal of the Linnean Society* 64, Supplement 1, 337–360.

Müllerová, J., Pyšek, P., Jarošík, V. and Pergl, J. (2005) Aerial photographs as a tool for assessing the regional dynamics of the invasive plant species *Heracleum mantegazzianum. Journal of Applied Ecology* 42, 1042–1053.

Myers, N., Mittermeier, R.A., Mittermeier, C.G., Da Fonseca, G.A.B. and Kent, J. (2000) Biodiversity hotspots for conservation priorities. *Nature* 403, 853–858.

Nakhutsrishvili, I.G. (1986) *Flora of Spore-producing Plants of Georgia* (Summary). N.N. Ketskhoveli Institute of Botany, Academy of Sciences of the Georgian SSR, Tbilisi, Georgia. [In Russian.]

Nielsen, B.E. (1971) Coumarin patterns in the *Umbelliferae.* In: Heywood, V.H. (ed.) *The Biology and Chemistry of the Umbelliferae. Botanical Journal of the Linnean Society* 64, Supplement 1, 325–336.

Paulo, O.S., Dias, C., Bruford, M.W., Jordan, W.C. and Nichols, R.A. (2001) The persistence of Pliocene populations through the Pleistocene climatic cycles: evidence from the phylogeography of an Iberian lizard. *Proceedings of the Royal Society of London, Series B* 268, 1625–1630.

Pemberton, R.W. (2000) Predictable risk to native plants in weed biological control. *Oecologia* 125, 489–494.

Pyšek, P. and Prach, K. (1994) How important are rivers for supporting plant invasions? In: de Waal, L.C., Child, L.E., Wade, P.M. and Brock, J.H. (eds) *Ecology and Management of Invasive Riverside Plants*. Wiley, Chichester, UK, pp. 19–26.

Pyšek, P. and Pyšek, A. (1995) Invasion by *Heracleum mantegazzianum* in different habitats in the Czech Republic. *Journal of Vegetation Science* 6, 711–718.

Ribera, I. and Vogler, A.P. (2004) Speciation of Iberian diving beetles in Pleistocene refugia (Coleoptera, Dytiscidae). *Molecular Ecology* 13, 179–193.

Seier, M.K. (2005a) Exotic beneficials in classical biological control of invasive alien weeds: friends or foes? In: Alford, D.V. and Brackhaus, G.F. (eds) *Plant Protection and Plant Health in Europe: Introduction and Spread of Invasive Species. Berlin, Germany, 9–11 June 2005*. Symposium Proceedings 81. British Crop Protection Council, Alton, UK, pp. 191–196.

Seier, M.K. (2005b) Fungal pathogens as classical biological control agents for invasive alien weeds – are they a viable concept for Europe? In: Nentwig, W., Bacher, S., Cock, M.J.W., Hansörg, D., Gigon, A. and Wittenberg, R. (eds) *Biological Invasions: from Ecology to Control*. Neobiota 6. Institute of Ecology of the TU Berlin, pp. 165–176.

Shaw, R.H. (2003) Biological control of invasive weeds in the UK: opportunities and challenges. In: Child, L.E., Brock, J.H., Brundu, G., Prach, K., Pyšek, P., Wade, P.M. and Williamson, M. (eds) *Plant Invasions: Ecological Threats and Management Solutions*. Backhuys, Leiden, The Netherlands, pp. 337–354.

Sheppard, A.W. (1991) Biological flora of the British Isles. *Heracleum spondylium* L. *Journal of Ecology* 79, 235–258.

Sheppard, A.W., Shaw, R.H. and Sforza, R. (2006) Top 20 environmental weeds for classical biological control in Europe: a review of opportunities, regulations and other barriers to adoption. *Weed Research* 46, 93–117.

Tarasov, P.E., Volkova, V.S., Webb III, T., Guiot, J., Andreev, A.A., Bezusko, L.G., Bezusko, T.V., Bykova, G.V., Dorofeyuk, N.I., Kvavadze, E.V., Osipova, I.M., Panova, N.K. and Sevastyanov, D.V. (2000) Last glacial maximum biomes reconstructed from pollen and plant macrofossil data from northern Eurasia. *Journal of Biogeography* 27, 609–620.

Theobald, W.L. (1971) Comparative anatomical and developmental studies in the *Umbelliferae*. In: Heywood, V.H. (ed.) *The Biology and Chemistry of the Umbelliferae. Botanical Journal of the Linnean Society* 64, Supplement 1, 179–197.

Tiley, G.E.D., Dodd, F.S. and Wade, P.M. (1996) Biological flora of the British Isles. *Heracleum mantegazzianum* Sommier and Levier. *Journal of Ecology* 84, 297–319.

Velichko, A.A. and Kurenkova, A.A. (1990) Landscapes of the Northern Hemisphere during the Late Glacial Maximum. In: Soffer O. and Gamble G. (eds) *The World at 18,000 BP*. Unwin Hyman, London, pp. 255–265.

Wapshere, A.J. (1974) A strategy for evaluating the safety of organisms for biological weed control. *Annals of Applied Biology* 77, 201–211.

Warren, M.S. and Oates, M.R. (1995) The importance of bracken habitats to fritillary butterflies and their management for conservation. In: Smith, R.T. and Taylor, J.A. (eds) *Bracken: An Environmental Issue*. International Bracken Group Special Publication No. 2. Aberystwyth University, Aberystwyth, UK, pp. 178–181.

Zazanashvili, N., Sanadiradze, G., Bukhnikashvili, A., Kandaurov, A. and Tarkhnishvili, D. (2004) Caucasus. In: Mittermeier, R.A., Gil, P.R., Hoffmann, M., Pilgrim, J., Brooks, T., Mittermeier, C.G., Lamoreux, J. and Da Fonseca, G.A.B. (eds) *Hotspots Revisited: Earth's Biologically Richest and Most Endangered Terrestrial Ecosystems*. CEMEX, Mexico City, Mexico, pp. 148–152.

17 Revegetation as a Part of an Integrated Management Strategy for Large *Heracleum* Species

HANS PETER RAVN,[1] OLGA TREIKALE,[2] INETA VANAGA[2] AND ILZE PRIEKULE[2]

[1]*Royal Veterinary and Agricultural University, Hørsholm, Denmark;* [2]*Latvian Plant Protection Research Centre, Riga, Latvia*

Stamp them out! We must destroy them!

(Genesis, 1971)

Integrated Management Strategies in the Control of Invasive Species

Several large members of the genus *Heracleum* (*Apiaceae*) have been culti-vated as ornamentals outside their native range, and some have also been used as melliferous plants and fodder crops. A group of closely related, stout species of the genus *Heracleum* introduced into Europe are called 'large', 'tall' or 'giant' hogweeds (Nielsen *et al.*, 2005). Three of these species have become invasive in Europe: *H. mantegazzianum* Sommier & Levier, *H. sosnowskyi* Manden. and *H. persicum* Desf. ex Fischer (see Jahodová *et al.*, Chapter 1, this volume). Control methods to eradicate these invasive species (see Pyšek *et al.*, Chapter 7; Nielsen *et al.*, Chapter 14; and Cock and Seier, Chapter 16, this volume) need to be followed up with appropriate post-control man-agement (Buttenschøn and Nielsen, Chapter 15, this volume; Zavaleta *et al.*, 2001). When soil is left denuded after successful control of the weed, it is extremely vulnerable to reinvasion. Preventive precautions should be taken to support the establishment of vegetation cover of competitive native species. To achieve this goal, an integrated weed management strategy (IWMS) is a convenient and necessary approach to managing invasions by large *Heracleum* species in Europe. Integrated weed management strategy is an optimal combination use of all appropriate management methods (Julien and White, 1997). When applying an integrated weed management strategy, flex-ibility in the choice of methods is needed to match the management objectives for each location. In previous chapters some well known components of integrated weed management that can be used against large invasive *Heracleum* species have been presented (Pyšek *et al.*, Chapter 7; Nielsen *et al.*, Chapter 14; and Cock and Seier, Chapter 16, this volume). In practice, chemical, mechanical and grazing control methods should be used individually or in combination, depending on the economical, ecological and social situation of the target site, region and country.

In an integrated management strategy against plant invasions, Hobbs and Humphries (1995) advise conservation managers to focus on the invaded ecosystem rather than the invading species alone. Ecosystems vary in their susceptibility to invasion and this susceptibility can be altered by management activities. It is generally accepted that disturbance is a major factor affecting the invasibility of natural ecosystems (Rejmánek, 1989; Hobbs, 1991; Hobbs and Huenneke, 1992). Natural disturbances such as floods and storms may be important in some cases, but human disturbances such as changed grazing or fire regimes, fragmentation, nutrient enrichment and road construction are generally more common. Establishment of gas pipelines and electricity lines through forests are a major cause of soil disturbance and creating wind tunnels. However, ecosystem structure and function can also change significantly due to an invading species. If this happens, control of the invader will not in all cases restore the ecosystem (Hobbs and Humphries, 1995). Sheley and Krueger-Mangold (2003) draw attention to the fact that without careful selection and skilled implementation of chemical, mechanical and grazing control methods, severe negative effects on non-target species and ecosystem health can occur, and the process of invasion can accelerate. Simmons (2005) advocates the use of intrinsic biological attributes of selected native plant species within the threatened community to suppress or exclude the invasive species. According to Turnbull *et al.* (2000), the success of an invasive plant species may simply depend on the limitation by seed of an equivalent or superior native competitor.

An encouraging transplant study shows that addition of specific native plants may lead to competitive displacement of the invader. In Curlew Valley, northern Utah, USA, the dynamics of the community with an invasive, annual grass *Bromus tectorum* L. and two native plant species, *Elymus elymoides* Raf. and *Artemisia tridentata* Nutt., was examined. The results indicate that *E. elymoides* suppresses *B. tectorum*, thereby indirectly facilitating establishment of *A. tridentata*, which is only competitive in areas that are free of the invasive *B. tectorum* (Booth *et al.*, 2003). Success of an invader may not imply that all native plant species are poor competitors compared to exotic species. Turnbull *et al.* (2000) suggest that in seed sowing experiments, many native species would successfully establish in unoccupied patches. Successful establishment of exotic invasive species could therefore be reduced or eliminated by sowing seeds of native competitors that are locally absent (Simmons, 2005). In an experiment on a rural roadside near Austin, Texas, USA, oversowing native *Gaillardia pulchella* Foug. at 10 g of seeds/m^2 in October resulted in a 72% reduction of aboveground dry mass of the invasive *Rapistum rugosum* (L.) All. the following May without significant suppression of co-occurring native species (Simmons, 2005).

Therefore, we carried out studies to assess whether the use of indigenous plants could be used to manage seedling establishment of large hogweeds.

Revegetation Options Following the Eradication of Large *Heracleum* Species

Successful eradication of stands of invasive weeds leads to increased vulnerability to reinvasion by the species that was eradicated or other invasive species. Many invasions change ecosystem functioning (Richardson *et al.*, 2000) and resulting site characteristics may be favourable to invasive species in general. Areas from which an invasive species has been eradicated may be subject to a changed disturbance regime and usually nutrition levels will have increased. If the soil is left denuded, the areas will be easily accessible for invasive weeds from the surrounding areas (Zavaleta *et al.*, 2001).

During the search for natural regulating factors of *H. mantegazzianum* in the north-western Greater Caucasus (see Cock and Seier, Chapter 16, this volume), it was found that *H. mantegazzianum* becomes established quickly on newly denuded river banks (see Otte *et al.*, Chapter 2, this volume). Where the melting snow in spring causes flooding of river banks and leaves the soil surface without vegetation, *H. mantegazzianum* is one of the most successful early colonizers. In the regions where *Heracleum* is introduced it often creates single-species stands along rivers and streams (see Pyšek *et al.*, Chapter 3, this volume). This is, however, not the only habitat type where large *Heracleum* species may out-compete native vegetation. In general, areas with a high level of disturbance and intensive human activities are more likely to be invaded by large *Heracleum* species (see Thiele *et al.*, Chapter 8, this volume).

It has been suggested that successful establishment of an invasive species does not always imply its competitive superiority over all native plant species in the area. On a local scale successful establishment of native plants in unoccupied patches following sowing of seeds indicates that there is a potential for increasing diversity even where invasive plant species have established (Turnbull *et al.*, 2000). Therefore, the outcome of invasion may depend on propagule pressure, i.e. the amount of available seed of species native to the invaded site. Use of cover crops following the control of invasive species can therefore be a practical tool for post-control management of affected sites.

Cover crops may influence environmental conditions and alter the outcome of competition in different ways. An agricultural study in no-till maize (*Zea mays* L.) from southern New England, USA, shows that a cover crop of *Vicia grandiflora* Scop. with rye (*Secale cereale* L.) suppresses weeds better than rye alone. Although *V. grandiflora* produced less biomass and nitrogen than *V. villosa* Roth., it survived better under suboptimal seedbed conditions and produced crops for several seasons from a single sowing (DeGregorio *et al.*, 1995). Cover crops could act by decreasing the level of resources available to re-invading species and alter the outcome of the competition between native and invasive plants. This may be especially efficient if resources are reduced, for which native species are better competitors. During restoration of sedge meadow wetlands in mid-west and northwest USA, *Phalaris arundinacea* L., a perennial circumboreal grass, is recognized as an aggressive invasive plant species (Perry and Galatowitsch, 2003). *Echinochloa crus-galli* (L.) P. Beauv. is able to suppress *P. arundinacea* in these areas. However, *E. crus-galli* also suppresses desired, native wetland species such as *Carex hystericina* Muhl. Perry and Galatowitsch (2003) conclude that although some cover crops prevent native species becoming established, others may affect the outcome of competition in a more desireable way. Ecologists seeking cover crops for invasive species control should first determine the environmental conditions under which native species are more successful (Perry and Galatowitsch, 2003).

Soil cultivation negatively affects the survival and performance of large *Heracleum* species. Seedlings and young plants are killed by ploughing and if seeds are covered by more than 25 cm of soil, germination is hindered. Not all infested areas give opportunities for post-control management by tillage; at riversides and in open forests other methods must be applied. In central European beech (*Fagus sylvatica* L.) forests in Bern, Switzerland, and Denmark it was observed that *H. mantegazzianum* occasionally invades temporary gaps in the understorey following the fall of mature trees. When branches of adjacent trees narrow the gap, *H. mantegazzianum* disappears as it cannot compete for light with young beech trees. Eventually, they disappear and the gap closes again (H.P. Ravn, unpublished observations; Fig. 17.1). The capability to out-shade *H. mantegazzianum* will depend on the tree species. Thus, it has been observed in Denmark that where a stand of oak, *Quercus robur* L., meets a stand of beech and hornbeam, *Carpinus betulus*

Fig. 17.1. Forest gap in *Fagus sylvestris* forest closing on a population of *H. mantegazzianum*, Bern, Switzerland. Photo: H.P. Ravn.

Fig. 17.2. Forest trees may out-shade the large *Heracleum* species. In this picture from Denmark the borderline between the open *Quercus robur* forest (centre and right) and the more shading *Fagus sylvatica* and *Carpinus betulus* forest (to the far left) is also the borderline for out-shading of *H. mantegazzianum*. Photo: H. Lerdorf.

L., *H. mantegazzianum* will survive in the oak forest only (Fig. 17.2) (H. Lerdorf, Denmark, 2005, personal communication). In a similar vein, Pieret *et al.* (2005) demonstrated that under an open *Alnus/Salix* forest the seed rain from *H. mantegazzianum* and the density of its seedlings were significantly higher under open canopy than under a closing canopy due to succession towards forest. Shade tolerance may vary between the *Heracleum* species; it has been observed that *H. sosnowskyi* is more shade tolerant than *H. mantegazzianum* (Z. Gudzinkas, Vilnius, 2005, personal communication).

Experimental Sowing of Grass Mixtures for Revegetation of Sites Invaded by *Heracleum sosnowskyi*

Study site and methods

Heracleum sosnowskyi was grown as fodder crop in several countries of the former Soviet Union (see Jahodová *et al.*, Chapter 1, this volume). After this practice stopped between 1947 and 1950, *H. sosnowskyi* became a major weed in adjacent arable fields, along roads and in national parks. A recent investigation showed that an area of 12,000 ha is currently covered with

H. sosnowskyi in Latvia (Olukalns *et al.*, 2005). Serious problems are faced along riverside areas, where the application of chemical treatment is prohibited and access by machinery limited.

Cover crops were experimentally used to manage sites where other methods of control of large *Heracleum* species had resulted in denuded land with bare soil, vulnerable to reinvasion. This chapter summarizes these experiments, which were conducted in the Gauja National Park, Latvia (N 57° 20′ 45″, E 25° 39′ 06″), a region extensively invaded by *H. sosnowskyi* (Fig. 17.3). The national park is located in the catchment of Gauja River, district of Vidzeme, in north-east Latvia. The flora of the Gauja catchment includes 650 species, 341 genera and 198 plant families – more than half the total number of species in Latvia (Tabaka and Baronina, 1979). Study sites were located alongside the River Vaive, a tributary of the River Gauja. The extremely organic fertile soil with neutral pH on the riverside is favourable for rapid growth of *H. sosnowskyi*, which dominates local plant communities due to its large size and high reproductive output. The cover of *H. sosnowskyi* in nitrophilous plant communities alongside the Vaive reaches up to 85% and the invasion reduced the number of other species to six (*Aegopodium podagraria* L., *Urtica dioica* L., *Artemisia vulgaris* L., *Filipendula ulmaria* (L.) Maxim., *Antrhiscus sylvestris* (L.) Hoffm., *Phalaris arundinacea*) (Rasins and Fatare, 1986; Laivins and Gavrilova, 2003).

An experiment was set up to evaluate whether frequent cutting of *H. sosnowskyi* stands is an effective control method, and whether the establishment of dense grass cover can be promoted by sowing grass mixtures following the removal of *H. sosnowskyi* from experimental plots. The study site was a riverside slope with a tree layer of *Alnus* spp., *Fraxinus*

Fig. 17.3. Study sites were located beside the River Vaive, Latvia. The organic fertile soil is favourable for rapid growth of *H. sosnowskyi*, which dominates in local plant communities due to its large size and high reproductive output. Photo: O. Treikale.

Table 17.1. Plant density in experimental plots with *H. sosnowskyi* Manden suppressed by cutting and grass mixtures after three cuts by lawn mower in the first growing season. Heavy infestation corresponds to 8–17 plants/m^2 of adult *H. sosnowskyi* before first cut, light to 1–7 plants/m^2. See text for detail on the two grass mixture treatments; no grasses were sown in control plots. Means ± SE are given ($n = 5$). The data were analysed using ANOVA. Note that the statistical analysis needs to be interpreted with caution because of the design involving pseudoreplicates. Cover was not tested for LSD because statistical analysis for the numbers of plants shows the same differences between variants.

Infestation level/ treatment	Total number of all plants/m^2	*H. sosnowskyi* seedlings		Grass cover	
		Number/m^2	Cover (%)	Number of shoots/m^2	Cover (%)
Heavy/Control	390 ± 146.7	158 ± 78.9	40.5	138 ± 121.0	35.4
Heavy/Grass mixture 1	729 ± 27.2	140 ± 64.9	19.2	517 ± 72.2	70.9
Heavy/Grass mixture 2	655 ± 131.6	218 ± 90.9	33.3	387 ± 126.2	59.1
Light/Control	342 ± 115.4	138 ± 63.4	40.4	122 ± 104.8	35.7
Light/Grass mixture 1	963 ± 218.1	240 ± 100.7	27.9	653 ± 186.0	67.8
Light/Grass mixture 2	581 ± 34.4	172 ± 35.3	29.6	341 ± 46.1	58.7
LSD$_{95}$ (*P* value)	377.4 (0.02)	218.6 (0.89)		342.2 (0.02)	

excelsior L.., *Salix* spp., *Cerasus padus* (L.) Delarbre, shrub *Rubus idaeus* L. (= *strigosus* Michx.) and the following herbaceous species typically present before the invasion of *H. sosnowskyi*: *Humulus* spp., *Aegopodium podagraria*, *Artemisia vulgaris*, *Achillea millefolium* L., *Arctium lappa* L., *Cirsium arvense* L., *Chelidonium majus* L., *Galium verum* L., *Glechoma hederacea* L., *Linaria vulgaris* (L.) Miller, *Taraxacum officinale* L., *Trifolium repens* L., *Viola tricolor* L., *Urtica dioica*, *Impatiens parviflora* L., *Lamium amplexicaule* L., *Sisymbrium officinale* L., *Stellaria media* L., *Capsella bursa-pastoris* L., *Polygonum convolvulus* L., *Myosotis arvensis* L., *Veronica* spp., *Agropyron repens* (L). P. Beauv., *Poa* spp. and *Carex* spp.

Plots of 2 × 1 m were established with three treatments in ten replicates (see Tables 17.1 and 17.2 for an overview of experimental design) (Fig. 17.4). The treatments consisted of sowing two different grass mixtures and controls with no seeds sown. Grass mixture 1 consisted of *Dactylis glomerata* L. and *Festuca rubra* L. (50:50 by seed numbers), applied at the density of 2000 germinated seeds of both per m^2. Grass mixture 2 comprised *Festuca arundinacea* Schreber and *F. rubra* (35:65), at densities of 1000 and 2000 germinated seeds/m^2, respectively. To support the establishment of a competitive plant community, mesophytic native species (associated with sites of average soil humidity), growing well in lowland riverside meadows, were used (for descriptions, growth characteristics, palatability and relation to soil types of grass species used in the experiments, see Okonuki (1984)). The selection of species was based on several criteria: (i) the ecotypes of perennial grasses used in these experiments are genetically close to wild ecotypes of these species in Latvia (Antonijs and Rumpans, 2002; Jansone, 2002); (ii) *Festuca*

Table 17.2. The effects of sowing grass mixtures and subsequent repeated cutting of vegetation with a lawn mower (see text for details) on the density of *H. sosnowskyi* Manden and of grasses and broadleaved species. Differences in data collated at the end of Year 1 (2003) and Year 2 (2004) were tested separately by using ANOVA. *H. sosnowskyi* plants were excluded from values referring to broadleaved plant species. In the second year three replicates of each treatment had to be excluded due to patchy recruitment of grasses from the soil seed bank. Means ± SE are shown. Note that the statistical analysis needs to be interpreted with caution because of the design involving pseudoreplicates.

Treatment	No. of replicates	Year	Number of plants/m^2			
			H. sosnowskyi (adults)	*H. sosnowskyi* (seedlings)	Grass species	Broadleaved species
Cutting	10	1	6.5 ± 0.9	70.0 ± 20.9	102.0 ± 71.1	83.0 ± 14.6
	7	2	3.1 ± 1.4	62.6 ± 20.7	35.7 ± 24.9	64.3 ± 12.5
Cutting and sowing	10	1	6.7 ± 1.4	114.0 ± 37.3	583.0 ± 73.1	92.0 ± 10.7
of grass mixture 1	7	2	0.7 ± 0.4	86.4 ± 22.4	372.9 ± 50.2	47.1 ± 15.2
Cutting and sowing	10	1	7.1 ± 1.4	132.0 ± 31.1	320.0 ± 50.2	70.0 ± 24.5
of grass mixture 2	7	2	2.8 ± 0.9	88.6 ± 22.6	262.9 ± 53.6	35.7 ± 7.2
LSD$_{95}$ (*P* value)	10	1	3.7 (0.94)	88.5 (0.35)	190.2 (<0.001)	51.1 (0.68)
	7	2	4.1 (0.41)	64.8 (0.65)	132.3 (<0.001)	35.8 (0.27)

arundinacea, *F. rubra* and *D. glomerata* grow well under agroecological management normally applied in the study region, as demonstrated by field trials in Latvia (Berzins *et al.*, 2002); and (iii) these species have been demonstrated to prevent invasions of alien species. Milbau and Nijs (2004) suggested moderately productive, species-rich communities with a prevalence of species with horizontally oriented leaves, such as *F. arundinacea*.

Replicates were arranged so as to cover the range of soil types: forest sod–podzolic soil, forest sod–podzolic gley soil, floody sod–podzolic gley soil (Karklins, 1995), soil pH (7.3–8.0), organic matter content (0.6–1.7 mg/kg) and chemical properties (42–159 mg K$_2$O/kg, 39–89 mg P$_2$O$_5$/kg). For technical reasons it was not possible to avoid a design with pseudoreplications (Hurlbert, 1984). Based on observed heterogeneity of the site, the plots were arranged in blocks corresponding to the treatments. Three blocks (with five replicates) were located in areas heavily infested by *H. sosnowskyi* (8–17 plants/m^2) and three in areas with a low density of the invader (1–7 plants/m^2).

In order to reduce abundance and prevent further spread of *H. sosnowskyi*, vegetative plants were mechanically cut three times during the first year (2003) and four times in the second year (2004). After the plants of *H. sosnowskyi* were cut and their residues removed, grass mixtures were sown on 21 May 2003 and the soil manually harrowed with a rake to embed seeds to a depth of 1 cm. Plots were covered with a crop cover 'Capatex' for one month to protect the seeds from birds. Vegetation in the plots was cut using a hand lawn mower during the two following growing seasons on 21 May, 25 June, 23 July, 18 September 2003 and 18 May, 19

Fig. 17.4. Experimental area on the riverside of the Vaive, Latvia where cutting and sowing of grass mixtures was used to reduce density of *H. sosnowskyi* Manden. Photo: O. Treikale.

June, 28 July and 31 August 2004. Before cutting, plant species density was recorded by counting individuals of each species in the central 0.1 m^2 of each plot.

Results of the experimental revegetation experiment

The grass mixtures grew well on all soil types and formed a dense sward. At the end of the first growing season, the density of grasses in treated plots was higher than in control plots (Table 17.1) and the difference remained significant at the end of the second year (Table 17.2). At the end of the first year, the cover of grasses in treated plots was almost twice as high as in the control plots (Table 17.1).

In the first year the newly emerged *H. sosnowskyi* seedlings successfully competed with naturally occurring native grasses *Agropyron repens* (L.) P. Beauv. and *Poa pratensis* L., but in the second year the seedling density of *H. sosnowskyi* decreased. Similar effects were evident among the broadleaved species, with the most pronounced decrease observed in plots sown with the grass mixtures, presumably due to increased competition from newly established grass species (Table 17.2).

Differences in the number of seedlings and adult plants of *H. sosnowskyi* between treatments were not significant, although there was a tendency for reduction in seedling number in plots treated with grass mixture in the first but not the second year (Tables 17.1 and 17.2). Population renewal would have occurred from nearby stands of *H. sosnowskyi*. However, the number of adult plants, which was rather high after the first season (6.5–7.1 plants/m^2), was reduced to 0.7–3.1 plants/m^2 by the end of the second year. The numbers of *H. sosnowskyi* seedlings went down from 70.0–132.0 plants/m^2 to 62.6–88.6 plants/m^2 in successive years (Table 17.2).

Discussion and Conclusions

Though a reduction of adult plants and seedlings was observed in this experiment, sowing grass mixtures did not reduce the density of seedlings compared to areas that were only cut. This could be because of high *H. sosnowskyi* propagule pressure from the surroundings or the timing of the sowing of the grass mixtures. If test plots are small, there is a risk that they may be influenced by seeds from the surroundings.

Even though we did not find statistically significant decreases in numbers of seedlings or adult plants of *H. sosnowskyi*, there was an increase in vegetation density due to development of the grass sward of sown grasses as well as naturally occurring species. Continuous area management (cutting) over a large area will result in depletion of the hogweed seed bank in the soil (see Moravcová *et al.*, Chapter 5, this volume). Of course, permanent effort is a necessary procedure to avoid reinvasion and maintain area quality. This is especially important in riverside areas with giant *Heracleum* spp. upstream, since distribution of hogweed seeds by water is effective.

The size of the area over which this control strategy is used determines the efficacy of control and risk of reinvasion, because in larger areas the possibility of population renewal from the neighbourhood decreases. On the other hand, re-vegetation as a control strategy for giant *Heracleum* spp. on riversides would be labour intensive, and could be impractical for larger areas. The timing of the control action should be important. A strategy of waiting for the out-shading of seedlings by the older *Heracleum* plants, followed by cutting and sowing seed mixtures should be more effective. If the grass sward is established in the autumn, the germinating seedlings of *Heracleum* spp. seeds will face maximum competition the following spring.

Examples from wetlands and agricultural ecosystems (see also Nielsen *et al.*, Chapter 14, this volume) have shown encouraging results where opportunities for *H. sosnowskyi* to invade denuded areas have been reduced by use of native plant species. Sowing mixtures of native, commercially available grass species could create an intermediate stage towards restoration of the original vegetation, a stage during which the invasive weed is reduced and eventually eradicated from the area. Another advantage of grass coverage close to riverside areas is the reduced risk of erosion.

Acknowledgements

We thank colleagues in the GIANT ALIEN project for their input during the project and Petr Pyšek for critical reading of the manuscript. The study was supported by the project 'Giant Hogweed (*Heracleum mantegazzianum*) a pernicious invasive weed: developing a sustainable strategy for alien invasive plant management in Europe', funded within the 'Energy, Environment and Sustainable Development Programme' (grant no. EVK2-CT-2001-00128) of the European Union 5th Framework Programme.

References

Antonijs, A. and Rumpans, J. (2002) Productivity and botanical composition of perennial grass swards. *LLMZA, Agronomijas Vēstis* 4, 151–155. [In Latvian.]

Berzins, P., Jansone, B., Bumane, S., Sparnina, M. and Luksa, S. (2002) Results obtained in breeding of forage grasses and legumes. *LLMZA, Agronomijas Vēstis* 4, 181–191. [In Latvian.]

Booth, M.S., Caldwell, M.M. and Stark, J.M. (2003) Overlapping resource use in three Great Basin species: implications for community invasibility and vegetation dynamics. *Journal of Ecology* 91, 36–48.

DeGregorio, R., Schonbeck, M.W., Levine, J., Iranzo-Berrocal, G. and Hopkins, H. (1995) Bigflower vetch and rye vs. rye alone as a cover crop for no-till sweet corn. *Journal of Sustainable Agriculture* 5, 7–18.

Hobbs, R.J. (1991) Disturbance as a precursor to weed invasion in native vegetation. *Plant Protective Quarterly* 6, 99–104.

Hobbs, R.J. and Huenneke, L. (1992) Disturbance, diversity, and invasion: implications for conservation. *Conservation Biology* 6, 324–337.

Hobbs, R.J. and Humphries, S.E. (1995) An integrated approach to the ecology and management of plant invasions. *Conservation Biology* 9, 761–770.

Hurlbert, S.H. (1984) Pseudoreplications and the design of ecological experiments. *Ecological Monographs* 54, 187–211.

Jansone, B. (2002) Genetic resources of forage grasses and legumes in Latvia. *LLMZA, Agronomijas Vēstis* 4, 176–180. [In Latvian.]

Julien, M. and White, G. (1997) Biological control of weeds: theory and practical application. *ACIAR Monograph* 49, 1–192.

Karklins, A. (1995) *Internationally Recognized Soil Classification Systems.* Latvian University of Agriculture, Jelgava. [In Latvian.]

Laivins, M. and Gavrilova, G. (2003) *Heracleum sosnowskyi* in Latvia: sociology, ecology and distribution. *Latvijas Vegetacija* 7, 45–65. [In Latvian.]

Milbau, A. and Nijs, I. (2004) The influence of neighbourhood richness, light availability and species complementarity on the invasibility of grassland gaps. In: *Third International Conference on Biological Invasions NEOBIOTA: From Ecology to Control.* Zoological Institute, University of Bern, Switzerland, p. 14.

Nielsen, C., Ravn, H.P., Nentwig, W. and Wade, M. (eds) (2005) *The Giant Hogweed Best Practice Manual. Guidelines for the Management and Control of an Invasive Weed in Europe.* Forest and Landscape Denmark, Hørsholm, Denmark.

Okonuki, S. (1984) *World Graminous Plants.* Nippon Soda Co. Ltd, Tokyo, Japan.

Olukalns, A., Berzins, A., Lapins, D. and Lejins, A. (2005) Studies of hogweed (*Heracleum sosnowskyi*) in Latvia 2002–2004. In: *Abstracts from the Final Workshop of the Giant Alien project.* University of Giessen, p. 44.

Perry, L. and Galatowitsch, S.M. (2003) A test of two annual cover crops for controlling *Phalaris arundinacea* invasion in restored sedge meadow wetlands. *Restoration Ecology* 11, 297–307.

Pieret, N., Bedin, B. and Mahy, G. (2005) Seed demography and spatial heterogenity of life cycle stages in two populations of *Heracleum mantegazzianum*. In: *Abstracts from the Final Workshop of the Giant Alien project*. University of Giessen, p. 28.

Rasins, A. and Fatare, I. (1986) Hogweed cow parsnip – *Heracleum sosnowskyi* Manden. – dangerous weed of the Latvian flora. In: *Rare Plants and Animals*. LSSR State Planning committee, Scientific Research Institute of Scientific-Technical Information and Technical-Economical Problems. Riga, pp. 8–10. [In Latvian.]

Rejmánek, M. (1989) Invasibility of plant communities. In: Drake, J.A., Mooney, H.A., di Castri, F., Groves, R.H., Kruger, F.J., Rejmánek, M. and Williamson, M. (eds) *Biological Invasions: A Global Perspective*. Wiley, Chichester, UK, pp. 369–388.

Richardson, D.M., Allsopp, N., D'Antonio, C.M., Milton, S.J. and Rejmánek, M. (2000) Plant invasions – the role of mutualisms. *Biological Reviews* 75, 65–93.

Sheley, R.L. and Krueger-Mangold, J. (2003) Principles for restoring invasive plant-infested rangeland. *Weed Science* 51, 260–265.

Simmons, M.T. (2005) Bullying the bullies: the selective control of an exotic, invasive annual (*Rapistrum rugosum*) by oversowing with a competitive native species (*Gaillardia pulchella*). *Restoration Ecology* 13, 609–615.

Tabaka, L. and Baronina, V. (1979) Floristic structure of the reactionary zone under intensive utilization. In: *Flora and Vegetation of the Latvian SSR: Geobotanical Region Nord-Vidzeme*. Science, Riga, pp. 103–107. [In Russian.]

Turnbull, L.A., Crawley, M.J. and Rees, M. (2000) Are plant populations seed-limited? A review of seed sowing experiments. *Oikos* 88, 225–238.

Zavaleta, E., Hobbs, R.J. and Mooney, H.A. (2001) Viewing invasive species removal in a whole-ecosystem context. *Trends in Ecology and Evolution* 16, 454–459.

18 Model-assisted Evaluation of Control Strategies for *Heracleum mantegazzianum*

NANA NEHRBASS AND ECKART WINKLER

UFZ Centre for Environmental Research Leipzig-Halle, Leipzig, Germany

Strike by night! They are defenceless! They all need the sun to photosensitize their venom

(Genesis, 1971)

Introduction

Management and control of alien invasive species often remain the task of local agencies. Hence, communication and scientific documentation about management success is often insufficient. Nevertheless, as invasive species have become a global problem, so global efforts are needed and efficient management is desirable (Pullin *et al.*, 2004). The giant hogweed, *Heracleum mantegazzianum* Sommier & Levier, has been recognized as an invasive species since the 1960s (Pyšek and Prach, 1994; Tiley *et al.*, 1996). Different management techniques have been applied and each has its advocates, with varying reports of success (see Pyšek *et al.*, Chapter 7, Nielsen *et al.*, Chapter 14, Buttenschøn and Nielsen, Chapter 15, and Ravn *et al.*, Chapter 17, this volume; Nielsen *et al.*, 2005). The most important techniques have been qualitatively described in different publications (Table 18.1). Additionally, there is a wide range of 'grey literature' about control options and their efficiency available to practitioners in the form of brochures and on the Internet (for a comprehensive list, see National Invasive Species Information Center (2006)). Potential options are cutting, chopping, pulling, grazing, hoeing, digging, ploughing, increased shading and different chemical treatments, and distinctions can be made between those suitable for riparian habitats and others. As long-term strategies, biological control and even genetic manipulation have been suggested (Dodd *et al.*, 1994).

At present, there is no consistent best practice advice available. This is partly because of the very varied ecological situations that have to be addressed, but also due to a lack of systematic empirical studies. Within the framework of the GIANT ALIEN project we created an individual-based model

Table 18.1. Management options and their suitability for the control of *H. mantegazzianum.*

Control method	Suitable for management			Reference
	Yes	No	Undecided	
Chemical (other)			x^1	Tiley and Philp (1994)
	x^1			Caffrey (1994)
	x^1			Dodd *et al.* (1994)
Chemical (glyphosate)	x^1			Caffrey (1994)
	x^1			Dodd *et al.* (1994)
				Lundström and Darby (1994)
Chemical (selective)	x^2			Lundström and Darby (1994)
Cutting			x^2	Tiley and Philp (1994)
			x	Dodd *et al.* (1994)
		x		Caffrey (1994)
		x		Lundström and Darby (1994)
Chopping	x			Tiley and Philp (1994)
Pulling			x^2	Dodd *et al.* (1994)
Digging	x^2			Dodd *et al.* (1994)
Ploughing	x^3			Dodd *et al.* (1994)
Increased shading			x	Tiley and Philp (1997)
Grazing (pigs, cattle)	x			Tiley and Philp (1994)
Grazing (sheep)	x			Andersen (1994)
			x	Lundström and Darby (1994)
Biocontrol (virus)		x		Sampson (1994)
Biocontrol (herbivory)		x		Sampson (1994)
Genetic manipulation		x		Dodd *et al.* (1994)

[1] In many European countries, chemical control is not permitted on public land, especially in the vicinity of water courses. Apart from impact on native flora and fauna, drastic erosion following chemical treatment has been observed. The degree of control achieved with chemical treatments does not increase when the dosage is increased beyond the recommendations of the manufacturer (Caffrey, 1994).
[2] Only suitable for small stands or follow-up control after other treatment methods.
[3] Efficient, but not possible in most stands.

(Nehrbass, 2006) of local dynamics of *H. mantegazzianum* using recent empirical findings from Germany (Hüls, 2005) and the Czech Republic (*see* Pergl *et al.*, Chapter 6, this volume; Nehrbass *et al.*, 2006a) for model parameterization and validation. This model was used to reproduce the local dynamics of the species, an understanding of which is a necessary prerequisite for predicting any further spread over larger areas (Nehrbass *et al.*, 2006b). It is also important to pay attention to the local dynamics when trying to reduce invasions at sites where *H. mantegazzianum* is already established. In the present study we used the model to study the development of established populations under different management scenarios. These virtual experiments with different control strategies allowed us to detect potential vulnerable stages in the plant's life cycle.

H. mantegazzianum is a monocarpic perennial with a large, but short-term persistent seed bank (*see* Moravcová *et al.*, Chapter 5, this volume;

Krinke *et al.*, 2005). Thus, survival of a population strongly depends on individual plant vigour and the production of offspring. The species does not reproduce vegetatively, and population renewal depends entirely on seed production (see Perglová *et al.*, Chapter 4, this volume; Moravcová *et al.*, 2005 and Chapter 5, this volume). Hence, for model construction and parameterization we focused on plant traits related to flowering, and thus reproductive success. In management practice the flowering stage is usually the one that is easiest to identify or to handle (see Pyšek *et al.*, Chapter 7, this volume). Alternatively, control is applied to all stages of the population. The preferred control method should be effective and form the basis of an integrated weed management approach (see Ravn *et al.*, Chapter 17, this volume), a process of maximizing control effect and minimizing negative ecological, economic and social impacts. Therefore, when selecting a target stage for control, ecological aspects should be considered. In this study we aim to demonstrate how ecological modelling, combined with sensitivity analyses, can support decision making for appropriate management techniques.

The study starts with a short description of the individual-based simulation model. Then we present simulations evaluating different control options (grazing/mowing, umbel cutting, prevention of seedling establishment and reduction of suitable habitat), which were incorporated into the model and applied with different intensities and duration. Finally we discuss the consequences of the simulation results with a view to efficient control of *H. mantegazzianum*.

Heracleum mantegazzianum Individual-based Model

Model construction

A stochastic, spatially explicit, individual-based model (IBM) was developed to describe local population dynamics of *H. mantegazzianum*. The model is rule-based; life-history and dispersal rules were developed and parameterized with empirical data (Hüls, 2005).

Time: The model uses discrete time-steps. One time step represents a year, the interval between one flowering period and the next, corresponding to the censuses of the field study described by Hüls (2005). Each time step includes growth of individuals, flowering, offspring dispersal and establishment, and death.

Space: Landscape is represented by a two-dimensional grid, consisting of 2500 cells (50 × 50). Cell size (2.5 m^2) was chosen based on the size of the permanent plots monitored in the field experiment. We express habitat quality as carrying capacity K, which is incorporated as ceiling capacity. The population within a cell can grow without restriction until this capacity is reached, after which there is no establishment of new seedlings. The mean carrying capacity (K = 50 plants) was estimated from empirical data on the maximum density of adult plants in the field study.

Plants: Plants are represented as individuals. Thus, demographic stochasticity was included in the model. Each plant is characterized by three traits: age a, plant height h and number of rosette leaves l.

Growth: Initial height h_0 of new plants is assigned randomly following a truncated normal distribution with mean $m = 52$ cm and standard deviation $s = 27$ cm. Subsequent height increase (in cm/year) follows a deterministic linear relationship derived from the empirical data of Hüls (2005):

$$h(t+1) = 1.45^*h(t) + 26.56. \tag{1}$$

The maximum height that a vegetative individual can reach in the model before it stops growing is $h = 220$ cm. New plants start with a random number of leaves l_0, where l_0 obeys a truncated normal distribution with $m = 1.5$ and $s = 0.65$, rounded to the nearest integer. At each annual step vegetative plants increase or decrease their number of leaves randomly, with the increment Δl being calculated by the same rule as the initial number l_0.

Flowering: Whether a plant makes the transition to the flowering state is determined at the beginning of each annual step (i.e. before growth). For a plant to flower, it has to match or exceed a minimum size (number of leaves $l \geq 3$, height $h > 95$ cm) and a minimum age ($a \geq 2$ years). All plants matching these criteria are subject to a global flowering probability p_F, varying from year to year and drawn from a truncated normal distribution $N(0.8,0.3)$.

Seeds, seedlings and new plants: Seed production and dispersal are not modelled individually; the model focuses on the fate of offspring. Each flowering plant produces a number of potential seedlings. The number of seedlings is normally distributed ($m = 25$ and $s = 25$) close to the values found empirically ($m = 28$ and $s = 21$; $n = 2$), and independently of the size of the mother plant. New seedlings S are distributed in the cell of origin and around the home cell according to the following percentages (with subsequent truncation of numbers to the lower integer):

Home: S_H = $S^*0.72$; minimum = 1
Radius 1: S_1 = $S^*0.15/8$
Radius 2: S_2 = $S^*0.07/16$
Radius 3: S_3 = $S^*0.04/24$
Radius 4: S_4 = $S^*0.01/32$ $\qquad\qquad$ (2)

To mimic long-distance dispersal, 0.1% of all seedlings are randomly placed into cells over the whole grid with uniform probability. New plants can only become established in cells where carrying capacity K has not yet been reached. Other kinds of density regulation were not included because of negative empirical results or insufficient information. Thus, density acts only on new individuals, but not on adult, established ones (Pergl *et al.*, 2006). We assumed that there is no permanent seed bank.

Death: After seed production flowering individuals die. For all other plants there is a probability of $p = 0.5$ of dying.

The model was validated against empirical results taken from the sites where the original data for the simulation were obtained. Details were given by Nehrbass (2006) and by Nehrbass *et al.* (2006a) for a similar model.

Implementation of control measures

The aim of successful control is the local eradication of *H. mantegazzianum*. Practitioners have developed different techniques to achieve this goal. In our model, four of the most common management types are incorporated: grazing/mowing, umbel cutting, prevention of seedling establishment and reduction of suitable habitat. Each treatment was varied in its intensity, affecting 10–100% of the population in steps of 10%. The experimental control (0%) was a population without any management and thus was used as the reference data set.

We tested two different treatment scenarios. In the first scenario, management was run over 5 years, with different management intensities. In the second, 90% management was continued for 55 years. Management options are incorporated into the simulation model as follows:

Grazing/mowing/(spraying): This treatment is applied before growth or flowering takes place. It affects all cells and in each cell, the pre-set percentage of individuals is randomly selected. Controlled plants are set back to a size of $h = 0$. Regeneration takes place in the following year, according to the growth function for new plants (eq. 1).

Umbel cutting: Treatment is applied to the pre-set percentage of randomly selected reproductive plants in each cell. Individuals are 'cut' and hence do not produce seeds in the same year. However, they do not die or lose size and hence can reproduce in the following year (if they are not selected for control again).

Prevention of seedling establishment: This control is implemented as a reduction of the number of seedlings, which otherwise would have established. It is implemented to the chosen percentage of all new plants. The resulting 'gaps' are not filled by secondary recruitment in the same year and do not affect the behaviour of other seedlings.

Habitat reduction: Habitat reduction is evenly applied to all cells and reduces the carrying capacity K by the chosen percentage. It is assumed that changes will take place completely within one year and that plants can only exist in suitable habitat. Hence, 100% reduction leads to an immediate extinction of all plants in all cells.

Simulations

Each scenario was repeated 50 times. At first, a population was created mimicking a successful invasion of the species in a region: each simulation was initiated with ten new plants, which were all placed in one randomly chosen cell. Then, control was started at a time when the population had reached constant growth rate with steady increase of individual number ($t = 45$). Control ran for five consecutive years (scenario 1) or until the end of the simulation (scenario 2) and considered the entire population according to the implementation of management option and intensity. Changes in population development under controls were compared with performance in unmanaged simulation runs. Survival probability P was calculated by dividing the number of populations I surviving until year t through the number of all populations of the simulation series that had survived until the beginning of control at time t_c:

$$P = I(t)/I(t_c) \tag{3}$$

Thus, only those simulation runs were considered in which the populations did not go extinct before control was started.

Model Predictions on the Effect of Control Measures

Before the start of control, 80% of the populations starting with a low number of individuals became established and invaded the grid. Those populations failing to become sustainable usually died out in the initial phase (before $t = 10$) and were not considered in the evaluation of control measures.

Without any control (0% intensity), occupation of the complete grid was reached after approximately 85 time steps. Due to saturation of the grid, the growth rate of populations then approximated $R = 1$. At the time control started ($t = 45$) the average growth rate was $R = 1.16 \pm 0.31$.

Figure 18.1 compares different management options applied over a period of 5 years with different intensity in their effect on population growth rate R. Only when management intensity was 100% was a local eradication of

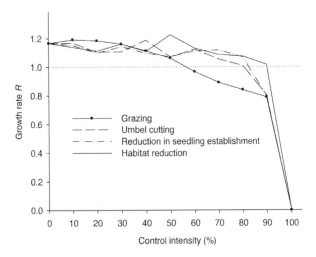

Fig. 18.1. Comparison between different management options for *H. mantegazzianum*. Management was applied to the populations over 5 years at different intensities. Growth rates R were averaged over the 5-year treatment period. Reduction of growth rate R below the threshold value $R = 1$ is a necessary condition for successful control.

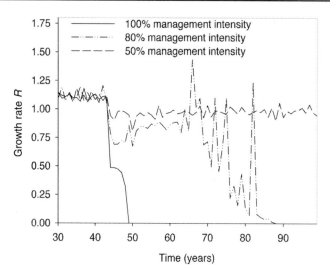

Fig. 18.2. Example results for the long-term effect of different intensities of grazing for *H. mantegazzianum*. Only 100% management led to an eradication of the population within a 5-year time period. The two other options only slightly reduced growth rates. In the 80% treatment demographic, stochasticity led to extinction after approximately 44 years of control.

the populations possible within a 5-year period. A reduction of growth rates below the threshold value $R = 1$ is a necessary, but not yet sufficient, condition for successful local eradication of the populations in the longer term. Control with only 90% intensity fulfilled this condition for all management options, except for habitat reduction. In the case of grazing, even management intensities of 80% and less reduced growth rate below $R = 1$. In all other scenarios, population densities were reduced, but growth rates remained above $R = 1$, thus enabling populations to increase steadily.

Even in cases where management intensity was high enough to reduce R to values smaller than 1, this might not lead to extinction of the population within a reasonable time. In an example comparing 'grazing' at 80% and 50% control intensity, both reduced R to values < 1 (Fig. 18.2). The former treatment reduced individual numbers markedly (not shown) and thus led to increased influence of demographic stochasticity, which resulted in large fluctuations and finally to extinction of the population, in this example after approximately 44 years. The lowest control intensity allowed the population to survive at a lower growth rate fluctuating around $R = 1$.

The reduction of population growth-rate over time clarified the different effects of control options on population growth rates (Fig. 18.3). Grazed populations showed the strongest response to control over the management period (Fig. 18.3A). The second strongest response was to umbel cutting (Fig. 18.3B), followed by seedling establishment (Fig. 18.3C). For control intensities < 60% the prevention of seedling establishment became more reliable than umbel cutting due to lower annual fluctuations in growth rate. Habitat reduction was the least effective measure in both scenarios (Fig. 18.3D).

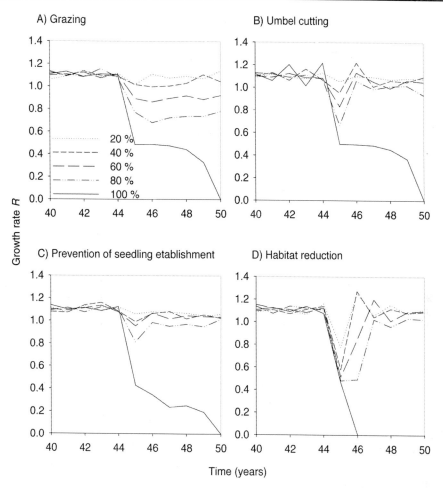

Fig. 18.3. Effect of different management options on growth rate *R* of *H. mantegazzianum* over time. The four different management strategies started at *t* = 45 years and were applied for 5 years.

The second scenario investigated the effect of duration of control. Incomplete management (90%) was applied over 55 years. The qualitative ranking of management success for the different methods was not altered by lower impact but was by prolonged treatment. Survival probability *P* for the populations was successfully reduced to *P* = 0 in three of the treatments. Only habitat reduction was not able to affect population survival rates within this time span (Fig. 18.4). Grazing again was the most successful strategy, reaching *P* = 0 after 38 years of treatment, while umbel cutting achieved *P* = 0 after 54 years of treatment. However, seedling reduction was not able to reduce *P* fully to zero in the management period, as the lowest value reached in 55 years of treatment was *P* = 0.07. Habitat reduction had the least effect and led to only one extinction (*P* = 0.98) within the simulation period.

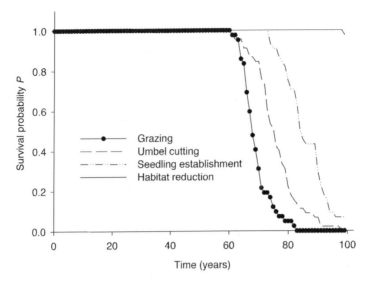

Fig. 18.4. Development of population-survival probability P of *H. mantegazzianum* under continuous treatment of 90% control, starting in year $t = 45$ of population development. Grazing and umbel cutting reached $P = 0$ for the populations most quickly, after 38 ($t = 83$) and 54 years ($t = 99$) respectively. Prevention of seedling establishment and habitat reduction did not reach $P = 0$ within a management period of 55 years.

From Modelling Results to Practical Efforts

The results presented do not encourage optimism for successful eradication of a noxious invasive plant species, even at a local scale. However, this may be realistic, as a species that is easy to control would not be considered a problem. Results of this virtual experiment demonstrate that local eradication is only possible with management effort affecting close to 100% of all plants in the area. One successful plant might produce enough offspring to recolonize an entire neighbourhood. However, if growth rates (R) can be maintained steadily at levels below one, populations will go extinct at some point.

Grazing/mowing as a control method was applied to all individuals, independent of their life-cycle stage. Technically, sheep grazing is possible, but meat or milk from these animals cannot be used because of changes in taste (Andersen and Calov, 1996). Grazing showed impact even at lower values of control intensity, and was the most successful approach in the simulation. In the model, a loss of above-ground biomass did not directly result in a higher mortality. However, together with a global annual death rate of $d = 0.5$, this management option increased the probability of a plant failing to reach the reproductive stage. With 100% application of any of the strategies, the populations would be eradicated in all cases after a 5-year treatment.

Umbel cutting is often considered a resource-saving alternative to grazing or mowing. In the simulation we assume that it is possible to implement umbel cutting in large stands, and that those plants that have been cut cannot produce seeds. Although treated plants produce some seed by means of regeneration (see Pyšek *et al.*, Chapter 7, this volume), the resulting amounts are rather small and were not considered in the model. Also, the treated plants die after the umbels are cut and do not have an opportunity to flower again. (See Pyšek *et al.*, Chapter 7 this volume.)

The prevention of seedling establishment was applied to seedlings before establishment. Under field conditions such an approach might be less efficient than in the model, since we did not consider factors such as cohort recruitment or density-dependent regulation of seedling establishment, which may lead to a higher survival probability for non-affected seedlings.

Reduction of suitable habitat (carrying capacity of the habitat for *H. mantegazzianum*) might be the most difficult task under field conditions and thus it is the most hypothetical management approach tested in the model. Hence, results have to be even more carefully interpreted than for the other management scenarios. Most projects for restoration of an ecosystem are likely to be accompanied by one of the three other management strategies, so that practical experience is more limited than for the other methods. If the reduction of suitable habitat were accomplished by another plant species it would be a slow process, if the other species were not a relatively vigorous invader itself. Also important to note is that any control intensity of less than 100% has only a small effect on the population's growth rates (Fig. 18.1). The plant itself is not diminished in its invasion potential, only in the numbers of individuals able to become established locally. This consideration weighs against the fact that a reduction of suitable habitat down to zero leads to an immediate eradication of the population. This is not a very realistic result as it is difficult to change habitat conditions completely from one year to the next, as was done in the simulation.

The results provide insight into the response of the population to different control measures. Thus, they can form the basis for provisional control recommendations. The effects of control methods shown by the model have not yet been verified in any empirical study. Nor should one forget the model assumptions and the uncertainties resulting from them. In each management scenario, there are ecological simplifications, which might influence applied management success. Such simplifications include: no effect of underground biomass (grazing); no primary or secondary regrowth of umbels immediately after umbel cutting; no seed-bank; density-dependent seedling establishment or cohort recruitment (prevention of seedling establishment); and no gradient in habitat suitability (habitat reduction). A combination of treatments was not considered, as the interactions and potential synergistic effects are largely unknown. However, modelling may elucidate the effect of such interactions and give rise to hypotheses that can be tested by observation or experiments. Additionally there could be other decisive characteristics of the species' life cycle, which have not yet been noted. Acquisition and use of resources by plants is only indirectly expressed by 'height' and 'number of leaves'. Even if the ecology of a species is well captured, we have to consider that management in real situations will never reach an intensity of 100% in its application. Nevertheless, the results give hints as to which strategy might be worth pursuing to find a sustainable management option, and they demonstrate that there is no easy way to control a noxious alien plant species, such as the giant hogweed.

This optimistic, but nevertheless cautious, conclusion is supported by a few other studies that include control measures in population-dynamic models of invasive plants: for *Acacia saligna* (Labill.) H.L. Wendl. in South Africa (Higgins *et al.*, 1997) or for *Acacia nilotica* (L.) Delile in Australia (Kriticos *et al.*, 2003). These studies made use of a large amount of varied information on life history, the interaction with other species and the impact of environmental factors, and the models were developed to be used in situations where there were conflicts of interests. However, experiences in a practical, interactive use of these models are still lacking or not yet published. Control of invasive species remains a difficult task that should be tackled by an integration of empirical and theoretical methods.

We were able to demonstrate that models can be used to support management decisions to the extent that ecological information about the species is available. Pursuing such studies of possible control measures in a virtual environment can potentially save resources and support successful management. Several management options should always be considered, especially for environmentally sensitive areas, and be compared with each other: not only with regard to the immediate success with respect to the target species but also possible side effects on other species. With the help of model simulations, such situations should become more transparent. Thus, the combination of empirical work and simulation modelling has turned out to be a promising approach to management of any species that pose or are suspected to pose problems to the environment.

Acknowledgements

We thank all participating scientists of the GIANT ALIEN project for ideas and suggestions during the course of the investigations on giant hogweed. Especially we are grateful to Wolfgang Nentwig, Charlotte Nielsen and Matthew Cock for their thoughts on potential management methods. The study was supported by the project 'Giant Hogweed (*Heracleum mantegazzianum*) a pernicious invasive weed: developing a sustainable strategy for alien invasive plant management in Europe', funded within the 'Energy, Environment and Sustainable Development Programme' (grant no. EVK2-CT-2001-00128) of the European Union 5th Framework Programme. Finally we thank Petr Pyšek, Tomáš Herben and Phil Lambdon for their critical and helpful comments on the manuscript.

References

Andersen, U.V. (1994) Sheep grazing as a method of controlling *Heracleum mantegazzianum*. In: de Waal, L., Child, L., Wade, M. and Brock, J. (eds) *Ecology and Management of Invasive Riverside Plants*. Wiley, Chichester, UK, pp. 77–92.

Andersen, U.V. and Calov, B. (1996) Long-term effects of sheep grazing on giant hogweed (*Heracleum mantegazzianum*). *Hydrobiologia* 340, 277–284.

Caffrey, J.M. (1994) Spread and management of *Heracleum mantegazzianum* (Giant Hogweed) along Irish river corridors. In: de Waal, L., Child, L., Wade, M. and Brock, J. (eds) *Ecology and Management of Invasive Riverside Plants*. Wiley, Chichester, UK, pp. 67–76.

Dodd, F.S., de Waal, L., Wade, M. and Tiley, G.E.D. (1994) Control and management of *Heracleum mantegazzianum* (Giant Hogweed). In: de Waal, L., Child, L., Wade, M. and Brock, J. (eds) *Ecology and Management of Invasive Riverside Plants*. Wiley, Chichester, UK, pp. 111–126.

Higgins, S.I., Azorin, E.J., Cowling, R.M. and Morris, M.J. (1997) A dynamic ecological-economic model as a tool for conflict resolution in an invasive-alien-plant, biological control and native-plant scenario. *Ecological Economics* 22, 141–154.

Hüls, J. (2005) Populationsbiologische Untersuchung von *Heracleum mantegazzianum* Somm. & Lev. in Subpopulationen unterschiedlicher Individuendichte. PhD Thesis, Justus-Liebig-Universität Giessen, Germany.

Krinke, L., Moravcová, L., Pyšek, P., Jarošík, V., Pergl, J. and Perglová, I. (2005) Seed bank of an invasive alien, *Heracleum mantegazzianum*, and its seasonal dynamics. *Seed Science Research* 15, 239–248.

Kriticos, D.J., Brown, J.R., Maywald, G.F., Radford, I.D., Nicholas, D.M., Sutherst, R.W. and Adkins, S.W. (2003) SPAnDX: a process-based population-dynamic model to explore management and climate change: impacts on an invasive alien plant, *Acacia nilotica*. *Ecological Modelling* 163, 187–208.

Lundström, H. and Darby, E.J. (1994) The *Heracleum mantegazzianum* (Giant Hogweed) problem in Sweden: suggestions for its management and control. In: de Waal, L., Child, L., Wade, M. and Brock, J. (eds) *Ecology and Management of Invasive Riverside Plants*. Wiley, Chichester, UK, pp. 93–100.

Moravcová, L., Perglová, I., Pyšek, P., Jarošík, V. and Pergl, J. (2005) Effects of fruit position on fruit mass and seed germination in the alien species *Heracleum mantegazzianum* (*Apiaceae*) and the implications for its invasion. *Acta Oecologica* 28, 1–10.

National Invasive Species Information Center (2006) Giant hogweed. http://www.invasivespeciesinfo.gov/aquatics/hogweed.shtml (accessed 6 June 2006).

Nehrbass, N. (2006) Using ecological models to study plant invasion: insights from the case of *Heracleum mantegazzianum* Sommier et Levier. PhD thesis, University of Marburg, Germany.

Nehrbass, N., Winkler, E., Pergl, J., Perglová, I. and Pyšek, P. (2006a) Empirical and virtual investigation of the population dynamics of an alien plant under the constraints of local carrying capacity: *Heracleum mantegazzianum* in the Czech Republic. *Perspectives in Plant Ecology, Evolution, and Systematics* 7, 253–262.

Nehrbass, N., Müllerová, J., Pergl, J., Pyšek, P. and Winkler, E. (2006b) Aerial photographs as a way of surveying an alien invasive plant *Heracleum mantegazzianum*: answers from a simulation model. *Biological Invasions* (in press).

Nielsen, Ch., Ravn, H.P., Netwig, W. and Wade, M. (eds) (2005) *The Giant Hogweed Best Practice Manual. Guidelines for the Management and Control of Invasive Weeds in Europe*. Forest and Landscape, Hørsholm, Denmark.

Pergl, J., Perglová, I., Pyšek, P. and Dietz, H. (2006) Population age structure and reproductive behavior of the monocarpic perennial *Heracleum mantegazzianum* (*Apiaceae*) in its native and invaded distribution range. *American Journal of Botany* 93, 1018–1028.

Pullin, A.S., Knight, T.M., Stone, D.A. and Charman, K. (2004) Do conservation managers use scientific evidence to support their decision-making? *Biological Conservation* 119, 245–252.

Pyšek, P. and Prach, K. (1994) How important are rivers for supporting plant invasions? In: de Waal, L., Child, L., Wade, M. and Brock, J. (eds) *Ecology and Management of Invasive Riverside Plants*. Wiley, Chichester, UK, pp. 19–26.

Sampson, C. (1994) Cost and impact of current control methods used against *Heracleum mantegazzianum* (Giant Hogweed) and the case for instigating a biological control programme. In: de Waal, L., Child, L., Wade, M. and Brock, J. (eds) *Ecology and Management of Invasive Riverside Plants*. Wiley, Chichester, UK, pp. 55–66.

Tiley, G.E.D. and Philp, B. (1994) *Heracleum mantegazzianum* (Giant Hogweed) and its control in Scotland. In: de Waal, L., Child, L., Wade, M. and Brock, J. (eds) *Ecology and Management of Invasive Riverside Plants*. Wiley, Chichester, UK, pp. 101–110.

Tiley, G.E.D. and Philp, B. (1997) Observations on flowering and seed production in *Heracleum mantegazzianum* in relation to control. In: Brock, J., Wade, M., Pyšek, P. and Green, D. (eds) *Plant Invasions: Studies from North America and Europe*. Backhuys, Leiden, The Netherlands, pp. 123–138.

Tiley, G.E.D., Dodd, F.S. and Wade, P.M. (1996) Biological flora of British Isles. *Heracleum mantegazzianum* Sommier & Levier. *Journal of Ecology* 84, 297–319.

19 Master of All Traits: Can We Successfully Fight Giant Hogweed?

PETR PYŠEK,[1,2] MATTHEW J.W. COCK[3], WOLFGANG NENTWIG[4] AND HANS PETER RAVN[5]

[1]Institute of Botany of the Academy of Sciences of the Czech Republic, Průhonice, Czech Republic; [2]Charles University, Praha, Czech Republic; [3]CABI Switzerland Centre, Delémont, Switzerland; [4]University of Bern, Bern, Switzerland; [5]Royal Veterinary and Agricultural University, Hørsholm, Denmark

> *Giant Hogweed lives*
>
> (Genesis, 1971)

Introduction

The Caucasian *Apiaceae* giant hogweed, *Heracleum mantegazzianum* Sommier & Levier, has been with us in Europe for almost two centuries (see Jahodová *et al.*, Chapter 1, this volume), but as is the case with most invasive species, it has received serious attention only in recent decades. A need for practical solutions to the problems posed by ecological and economic impacts, and scientific appeal associated with one of the most spectacular plant invasions in Europe, were the main reasons for giant hogweed becoming the subject of the international GIANT ALIEN project, which included participants from various parts of Europe with a wide range of expertise (Nielsen *et al.*, 2005).

It is not surprising that during the work on the GIANT ALIEN project some myths were unmasked: the plant is almost never truly biennial in the field (Pergl *et al.*, Chapter 6, this volume); it does not produce over 100,000 fruits (Perglová *et al.*, Chapter 4, this volume); the seeds do not survive for 15 years in the soil (Moravcová *et al.*, Chapter 6, this volume); and it is not polycarpic (Pyšek *et al.*, Chapter 7, this volume). Surprisingly, the reputation of giant hogweed is even worse than it deserves – because of its attractiveness to the public it became a tabloid archetypal plant invader. It even attracted the attention of artists – long before scientists in Europe began to recognize the problem of this species, *H. mantegazzianum* was the subject of the song 'The

Return of Giant Hogweed' by the rock group Genesis. The lyrics reproduced as mottos for chapters in this book indicate that the group made a good job in covering various aspects of the giant hogweed invasion.

The present chapter documents, by comparing with other plant invaders worldwide, why *H. mantegazzianum* is considered an aggressive invasive species and summarizes how the knowledge on the biology, ecology and distribution of *H. mantegazzianum* achieved during the GIANT ALIEN project can be interpreted in terms of the invasion potential of this species. From plant ecological and zoological perspectives, it highlights the features that allow *H. mantegazzianum* to be a successful invader, and what implications these circumstances have for its management and potential control.

What are the Attributes of an Aggressive Invader and How Does *Heracleum mantegazzianum* Compare with Other Invasive Species?

Extensive stands with a high cover

For any species, there are several assumptions that need to be met if it is to be viewed as an aggressive invader. The first assumption is the capability of creating large and dominant stands. Although in Central Europe two-thirds of *H. mantegazzianum* records relate to plants scattered in invaded vegetation without being dominant (see Thiele *et al.*, Chapter 8, this volume), it often forms dominant stands with a high cover (Fig. 19.1). Further, once suitable environmental conditions are met and the species starts to dominate the invaded vegetation, the stands can be very extensive. In the Slavkovský les study area, Czech Republic (see Perglová *et al.*, Chapter 4, this volume), on the basis of aerial photographs it is estimated that in ten 60-ha landscape sections, populations of *H. mantegazzianum* at present completely cover 41.9 ha (see Pyšek *et al.*, 2005; Chapter 3, this volume). This represents 7% of the total land in the region. These figures can be related to biomass, using the data from Tiley *et al.* (1996), who report a yield of 57 t/ha dry mass (including roots), based on measurements in the west of Scotland.

High rate of spread

Extensive populations can only result from a high rate of spread in the past, but these two measures, although closely related, do not necessarily indicate the same pattern of invasion. A species can be locally abundant and form extensive stands in a geographically limited region, yet it may not be widespread over a large region. Easy dispersal and rapid spread in the past also indicate a potential for the invasion to continue in the future, because they allow a species to spread over considerable distances from source populations and reach new areas. Quantitative information on the rate of *H. mantegazzianum* spread is available for the second half of the 20th century from the

Fig. 19.1. In the Slavkovský les study area, Czech Republic, populations of *H. mantegazzianum* cover approximately 7% of the landscape. Photo: P. Pyšek.

Czech Republic for both local and regional scales (Fig. 19.2). Interestingly, the pattern and rate of spread is similar at both scales; if the increase in the invaded area at local scale and in the number of squares at country scale is compared statistically, there is no difference in the slopes of both plots (for details, see Pyšek *et al.*, Chapter 3, this volume). This indicates that *H. mantegazzianum* spreads at the national scale at the same rate as locally in the region of its introduction to the country, and that the constraints imposed to its spread by landscape features and availability of suitable habitats are similar at both scales. That the invasion at the country scale is of a similarly high rate as that in the very suitable region of the Slavkovský les (Pyšek *et al.*, 2005; Chapter 3, this volume) indicates that the species seems to be little constrained by environmental settings. When suitable habitats are available, it spreads at a high and constant rate.

Comparing the rate of *Heracleum mantegazzianum* invasion with that of other species

Heracleum mantegazzianum regularly appears on global lists of the most invasive species (e.g. Cronk and Fuller, 1995; Weber, 2003) and is listed in the Global Invasive Species Database (Lowe *et al.*, 2000) among the 100 worst invasive alien species. What is its position among plant invaders of global significance expressed in quantitative terms? Exact data are unfortunately

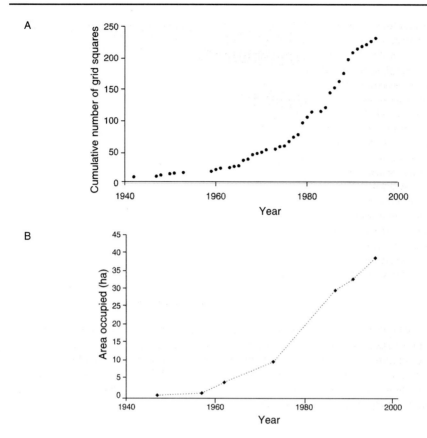

Fig. 19.2. *Heracleum mantegazzianum* spread was very fast at both national and local scales in the Czech Republic in the second half of the 20th century. (A) The pattern of increase in the cumulative number of squares ca. 11 × 12 km from which the species has been recorded up to the given time is similar to that in the actual population size in the Slavkovský les study area. (B) The population size is given in hectares, estimated as pooled value from ten sites of 60 ha each, monitored by using aerial photographs (Müllerová *et al.*, 2005; Pyšek *et al.*, Chapter 3, this volume). Thus, for example, 30 ha corresponds to 5% of the total landscape occupied.

limited, but recent reviews make it possible to compare the rate of invasion in the past.

The maximum rate of spread during the exponential phase of invasion was compared with that of other invasive neophytes in the Czech Republic (see Pyšek *et al.*, Chapter 3, this volume; Williamson *et al.*, 2005). On a regional scale and with global focus, there are two records of how fast the geographical distribution of *H. mantegazzianum* was increasing (Fig. 19.3). Using a simple measure of the number of grid squares divided by the period between initial and final mapping shows that the geographical range of major invasive species increases by hundreds to thousands of square kilometres a year (Pyšek and Hulme, 2005). The invasion of *H. mantegazzianum* in

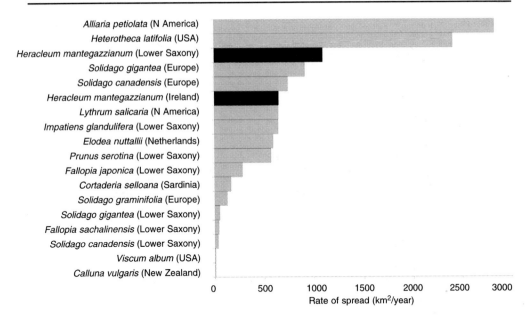

Fig. 19.3. The rate of increase in distribution of *H. mantegazzianum* at a regional scale compared to other species. The rate of spread in various regions is expressed as the number of grid squares divided by the length of the period over which the distribution was compared. For example, over the 16 years covered by the study of Schepker and Kowarik (1998), the distribution range in Lower Saxony, Germany, was increasing by 1089 km²/year. Based on data from Pyšek and Hulme (2005).

Europe is comparable with such spectacular invasions as that of *Alliaria petiolata* (Bieb.) Cavara & Grande (Weber, 1998) or *Heterotheca latifolia* Buckley (Plummer and Keever, 1963) in North America (Fig. 19.3).

In terms of the rate of increase in population size on a local scale, *H. mantegazzianum* spread by 1261 m²/year (see Müllerová *et al.*, 2005; Pyšek *et al.*, Chapter 3, this volume), which compares it to such prominent invasions as those of *Rhododendron ponticum* L. in the UK (Fuller and Boorman, 1977) or *Caulerpa taxifolia* (Vahl) C. Agandh in the Mediterranean Sea (Meyer *et al.*, 1998) (Table 19.1).

Rapid spread alone may not necessarily result in a dramatic invasion if the spread is not associated with the ability of invading populations to persist in invaded locations. Wade *et al.* (1997) surveyed *H. mantegazzianum* occurrence in Ireland and found that of the 96 historical sites, the species was still present at 43 in 1993. This represents 45% persistence and illustrates a remarkable ability of the species to thrive in the places once invaded. This value is lower than that reported for prominent clonally spreading invaders, such as *Fallopia* spp. (Pyšek *et al.*, 2001), but it needs to be noted that *H. mantegazzianum* was a target of control efforts in recent decades. It may be hypothesized that without locally successful eradications, the persistence would be higher.

Table 19.1. Comparison of the rate of invasion by *H. mantegazzianum* at a local scale with that of other species. The area covered by stands of the invading species was monitored over time or reconstructed from aerial photographs (see Pyšek *et al.*, Chapter 3). Data from Pyšek and Hulme (2005) – see references to primary sources therein.

Species	Life form	Region	Habitat	Span of study (years)	Rate of spread (m^2/year)
Caulerpa taxifolia	Alga	France	Sea bottom	5	2000
Heracleum mantegazzianum	Herb	Czech Republic	Pastures, disturbed	45	1260
Rhododendron ponticum	Shrub	UK	Forest, dunes	20	1100
Spartina anglica C.E. Hubb	Grass	New Zealand	Seashore	9	22
Spartina anglica	Grass	New Zealand	Seashore	4	13
Spartina anglica	Grass	New Zealand	Seashore	41	1

Impact of *Heracleum mantegazzianum* on resident vegetation

Although impact is difficult to measure and includes various aspects (Williamson, 1998), that of *H. mantegazzianum* on resident vegetation is manifested through changes in vegetation structure, cover and species composition. It depends on the type of invaded plant communities and their successional status, being especially marked in ruderal grasslands and ruderal pioneer vegetation (see Thiele and Otte, Chapter 9, this volume). *H. mantegazzianum* excludes resident species or reduces their abundance, hence decreases the local species richness of invaded communities (Pyšek and Pyšek, 1995). Nevertheless, conflicts with nature conservation are unlikely as *H. mantegazzianum* was not reported to invade habitats of conservation concern, nor did the regional populations of common native species seem to be endangered, as documented for Germany (see Thiele and Otte, Chapter 9, this volume). As pointed out in Chapter 9, the impact of *H. mantegazzianum* on resident vegetation is driven by human-induced disturbances and some native species have similar effects on species richness of plant communities if they prevail in the course of succession.

The detailed analysis carried out in Chapter 9 seems to imply that the impact of giant hogweed is overestimated in the literature. However, even if the above-mentioned effects on resident vegetation are subtle, the fact that the species is able to cover up to 10% of the landscape with dense populations must be taken as clear evidence of a serious impact. Moreover, unpublished data are available from paired invaded and uninvaded plots in the Czech Republic to demonstrate that *H. mantegazzianum* exerts a rather severe effect on species diversity of invaded sites, reducing their species richness to about half of the state prior to invasion, which is comparable to the effect of *Fallopia* spp., but stronger than that of *Lupinus polyphyllus* Lindl. and *Impatiens glandulifera* Arn. (M. Hejda, unpublished data).

To summarize, *H. mantegazzianum* has all the attributes of a successful invasive species (Fig. 19.4). Extensive and often dominant stands associated

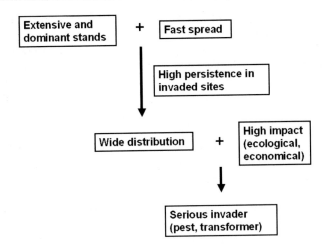

Fig. 19.4. A flow chart summarizing attributes of a successful plant invader. *Heracleum mantegazzianum* exhibits the attributes associated with a serious invader (see text for details).

with fast spread result in a wide distribution; this is possible because the persistence in localities once colonized is high. Consequently, any alien species with a wide distribution and marked impact is potentially a serious invader.

What Makes *Heracleum mantegazzianum* So Special: a Botanical Perspective

The GIANT ALIEN project produced detailed information on various aspects of the biology and ecology of *H. mantegazzianum* from both botanical and zoological perspectives. This makes it possible to highlight those species traits that contributed to such a successful and dramatic invasion. Before dealing with these traits, it needs to be noted that the size and attractiveness of this species as an ornamental led to its use in horticulture, and its early rapid dissemination between the countries of Europe via garden centres and botanical gardens. From this it follows that it was present in most countries before anyone was seriously concerned about its impact and spread as a naturalized, and later on invasive, species.

In the temperate conditions of Europe, *germination in early spring* provides the plants of *H. mantegazzianum* with the advantage of developing populations well ahead of the resident vegetation. Mean seedling density is usually around 500–700 seedlings/m^2, but extreme values can reach up to 3700. Relative growth rate of seedlings is higher than that of most native temperate herbs. Once seedlings get established after the period of self-thinning and reach the stage of juvenile plants and rosettes, their mortality is fairly low (see Pergl *et al.*, Chapter 6, this volume). Rapid growth of rosettes results in the *rapid formation of a dense cover*; placing the leaves above other species provides the plants of *H. mantegazzianum* with advantage in the competition for light.

Later in the phenological development, it is important to note that there are *no constraints to flowering*. Many aliens introduced to the temperate zone of Europe fail due to constraints to flowering; the vegetation period is too short to complete the life cycle and bear fruits (Pyšek *et al.*, 2003). This limitation, however, concerns species introduced to warmer regions. *H. mantegazzianum* occurs naturally at higher altitudes of the Greater Caucasus where the flowering period is short. Therefore, an average plant completed flowering within 36 days at the study sites in the Czech Republic and bore ripe fruits ready to release approximately 2 months after the onset of flowering (see Perglová *et al.*, 2006 and Chapter 4, this volume). That is, the populations of *H. mantegazzianum* are not constrained by climate in most of Europe as far as flowering is concerned. This was confirmed by a historical study of invasion in the Slavkovský les area, which did not detect any year-to-year variation in the flowering intensity; the proportion of plants that flowered was very stable over 45 years of invasion (Müllerová *et al.*, 2005). Moreover, plants of *H. mantegazzianum* are able to postpone flowering for some years until sufficient reserves are stored, for example if they grow under stress from grazing or unfavourable site conditions (see Pergl *et al.*, 2006; Chapter 6, this volume).

Frequent overlaps between male and female flowering phases, both between- and within-umbels, allow for *self-pollination* and self-pollinated flowers give rise to viable seeds, a high percentage of which germinate. This has important implications for invasion: if a single plant or a few individuals arrive in a new locality or region, potential invasion is not constrained by the absence of a mating partner as would be the case in a strictly outcrossing species (see Perglová *et al.*, 2006; Chapter 4, this volume).

The *extremely high fecundity* of *H. mantegazzianum* has long been recognized and the ability of this species to produce enormous number of fruits is often exaggerated in the literature (see Perglová *et al.*, Chapter 4, this volume). Although reliable estimates based on a large number of plants and accounting for geographical variation yielded lower numbers, the average 20,500 fruits per plant still provides the species with enormous reproductive potential (see Perglová *et al.*, 2006; Chapter 4, this volume). The complex information on various phases of the life cycle, obtained during the GIANT ALIEN projects, allowed us to estimate the number of *H. mantegazzianum* fruits that enter the landscape. Taking the average fruit number and the mean density of flowering plants, which is very similar in study plots in the Czech Republic and Germany (0.7 and $0.8/m^2$, respectively), the largest population in the Slavkovský les region (see Pyšek *et al.*, Chapter 3: Fig 3.4, this volume; Moravcová *et al.*, Chapter 5: Fig. 5.1, this volume), covering ca. 99,000 m^2 at the locality Žitný, is likely to produce about 1.4 billion seeds each year. To translate this figure to the whole landscape, it needs to be noted that 99,000 m^2 of *H. mantegazzianum* stands were recorded in a 60 ha section covered by aerial photographs (Müllerová *et al.*, 2005); the average fruit load would be ca. 2370 fruits of giant hogweed per m^2. Such extrapolations are prone to estimation bias, but the value obtained clearly indicates that the fruit load is enormous.

This high fecundity is associated with very *high germination rates* once dormancy is broken; the percentage of seed that germinated reached 91% in laboratory conditions and germination does not depend on the position of the fruit on the plant (Moravcová *et al.*, 2005). These high percentages are probably not realized in the field and between-year fluctuations in seed quality are likely to occur (L. Moravcová *et al.*, unpublished), yet there is no doubt that the vast majority of seeds produced are viable, with potential to contribute to the population dynamics and further spread to new locations.

The majority of fruits do not disperse far from adult plants and form a *large seed bank*. Some macro-ecological implications can be drawn using the detailed information on seed bank obtained during the project. From the size of population in the Žitný locality and average seed-bank density (see Moravcová *et al.*, Chapter 5, this volume), it can be estimated that there are 380 million non-dormant seeds in the seed bank in spring. With germination percentage exceeding 90%, the number of seeds that can potentially produce new plants is only slightly lower: $3550/m^2$, with a total of over 350 million estimated for the whole population.

The seed bank of *H. mantegazzianum* and its longevity became a kind of self-perpetuating myth; statements unsupported by data were taken from one source to another (for an account of this myth, see Moravcová *et al.*, Chapter 5, this volume). It is evident that reports on seeds surviving in the soil for 15 years, often cited in association with control strategies, are far from reality. An ongoing study monitoring seeds buried at different localities in the Czech Republic, selected in order to represent a range of climates and soil types, showed that on average 8.8% survived 1 year, 2.7% survived 2 years and 1.2% remained viable and dormant after 3 years of burial (see Moravcová *et al.*, Chapter 5, this volume). Relating these values to the seed-bank density in the autumn after the fruits are released (Krinke *et al.*, 2005), there are on average about 80 dormant seed/m^2 present in the soil after 3 years. The question of how long viable seeds are able to survive in the soil can only be answered conclusively when this experiment is finished (and possibly repeated in different climatic settings in other parts of Europe), but it seems unlikely that the longevity will be markedly longer than 3 years. Nevertheless, *H. mantegazzianum* has a *short-term persistent* seed bank which makes it possible to extend the germination period over time and respond to actual climate conditions in the site of introduction (Krinke *et al.*, 2005; Moravcová *et al.*, Chapter 5, this volume).

Last but not least, a *high regeneration ability*, especially in terms of fruit production (see Pyšek *et al.*, Chapter 7, this volume), is likely to contribute to the invasion of *H. mantegazzianum*. The species is one of the most prominent invaders in Europe, not only in terms of its invasion potential but also because of the attention it is paid by landscape managers and the public. Therefore, it is often targeted for control efforts (Nielsen *et al.*, 2005; Chapter 14, this volume; Buttenschøn and Nielsen, Chapter 15, this volume). If control is mechanical, the ability to regenerate is crucial. The high regeneration ability has been long recognized in a number of studies (for an overview, see Pyšek *et al.*, Chapter 7, this volume) and even umbels cut at late flowering or early fruiting are able to produce viable seeds (Pyšek *et al.*, 2007). It is important from the viewpoint of control strategies that, although regenerating plants produce much fewer fruits, the seeds are viable. Given the enormous fecundity, even if drastic reduction in fruit numbers is achieved, the resulting fruits are likely to ensure the renewal of targeted populations. A review of available studies conducted in different parts of Europe indicates that if plants are cut at ground level at peak flowering or beginning of fruit formation, regeneration ranges between 2.8% and 4.3% of the number of fruits produced by the control (see Pyšek *et al.*, Chapter 7: Table 7.2 and Fig. 7.2, this volume). Given the above-mentioned average fecundity exceeding 20,000 seeds per plant, the implications are obvious.

Landscape Features Set the Scene for Invasion

The role of plant traits needs to be assessed in the context of invaded communities and ecosystems. The above traits, however supportive in term of invasion success, would be hardly helpful if not realized in an environment vulnerable to invasion. Landscape characteristics are the scene upon which any invasion occurs and their importance has been repeatedly demonstrated in plant invasion literature (e.g. Campbell *et al.*, 2002; Wania *et al.*, 2006). In Europe, many invasions were shown to have accelerated in the second half of the 20th century, apparently because of landscape changes. Williamson *et al.* (2005), on the basis of the analysis of historical patterns of invasion of more than 60 species, suggested that the character of the landscape is more important than species' biological and ecological traits in determining the outcome of invasion.

For *H. mantegazzianum*, the suite of traits seems to be the main reason for invasiveness, but interactions with landscape features are also crucial. Data indicate that the species is less constrained by environmental settings than is the case in the majority of invasive alien plants (Müllerová *et al.*, 2005); it occurs under a wide spectrum of environmental conditions and successfully invades less-disturbed or semi-natural habitats. Nevertheless, habitats that suffer from massive invasion are almost always characterized by a rich resource supply, some degree of disturbance and the lack of regular management. Environmental factors constraining invasion of *H. mantegazzianum* are regular land use, shading by trees, low soil nutrient status and/or wetness (see Thiele *et al.*, Chapter 8, this volume). As for most invasions, it is also true for giant hogweed that the invasion is usually associated with poorly managed landscapes. This view is supported by the fact that in the Caucasus the species expands in disturbed habitats outside its natural ecological range (see Otte *et al.*, Chapter 2, this volume).

Herbivores and Pathogens: a Zoological Perspective

A large number of insect herbivores occur on *H. mantegazzianum* plants, but most of them occur only sporadically. Among common herbivores, specialization usually occurs mainly at family level (*Apiaceae*), sometimes on the level of a group of genera. Specialization on genus level is rare and species exclusively feeding on *H. mantegazzianum* do not seem to exist (see Hansen *et al.*, Chapter 11, this volume). During the rapid evolution of the large hogweed species, herbivores obviously did not follow with their own radiation (see Cock and Seier, Chapter 16, this volume). This results in more specialized herbivores being found beyond the native range of *H. mantegazzianum* because other *Heracleum* species occur there. A typical example is the aphid *Paramyzus heraclei* Börner, restricted to the genus *Heracleum*, occurring in both native and invaded areas. Comparably, most herbivores found in the native range of *H. mantegazzianum* were also present in the invaded part of Europe (see Hansen *et al.*, Chapter 11, this volume). A few species could not be evaluated sufficiently because they are rare or were only recently described as new species, thus the species-specific knowledge available is still poor. As a result of this, *H. mantegazzianum* has very similar herbivore populations in both native and invaded area, but the overall herbivore damage is slight. The *considerable resistance to herbivory* is probably due to very effective defence mechanisms (see Hattendorf *et al.*, Chapter 13, this volume), which permits *H. mantegazzianum* to establish the well-known tall plants.

Because *H. mantegazzianum* can store resources over a period of years before flowering, when it does flower, these resources are suddenly available for rapid utilization to grow a large, productive flowering structure. None of the associated herbivores and pathogens have sufficient impact at the population densities observed to significantly affect this flowering process. Thus, although it is possible that invertebrate herbivory may contribute to delaying the onset of flowering, once the flowering process starts, they are no longer significant. In this way, *H. mantegazzianum* also escapes its invertebrate herbivores.

Biological Control Perspective

In the GIANT ALIEN project, we specifically explored the scope for classical biological control, that is the introduction of host-specific natural enemies from the area of origin of *H. mantegazzianum* in the north-western Greater Caucasus. At the densities observed in the Greater Caucasus, none of the associated natural enemies caused sufficient damage to have a significant impact on *H. mantegazzianum* (see Seier and Evans, Chapter 12, this volume; Hattendorf *et al.*, Chapter 13, this volume). Furthermore, contrary to our expectation, we were unable to demonstrate that any of the insect or fungal natural enemies found in the Caucasus were specific to *H. mantegazzianum*, or safe enough to be introduced into Europe, all causing damage to other *Heracleum* spp. and most causing damage to parsnip, *Pastinaca sativa* L. (see Cock and Seier, Chapter

16, this volume). This result was unexpected, and in Chapter 16 we considered possible reasons for the lack of species-specific natural enemies. We hypothesized that this is because the co-evolution of *Heracleum* species and their natural enemies has been dynamic in the Caucasus region during the Quaternary. Periodic retreat into two small refugia during glacial periods, followed by expansion and mixing of populations during inter-glacial periods, has led to diversity and hybridization of the plant hosts, while containment of very small populations in the glacial refugia would not have selected for species-specificity on the part of associated natural enemies of *Heracleum* species. This hypothesis merits further examination and evaluation in the future.

Thus, the preferred biological control option by introducing exotic biological control agents is not possible. However, the alternative option of using indigenous natural enemies has not been adequately evaluated – in particular, the use of a laboratory-produced indigenous fungus as a mycoherbicide deserves study. Preliminary trials with a fungus known to attack many species in Europe have been made, but a more thorough programme, considering other species, particularly those more damaging to *H. mantegazzianum*, should be considered (see Seier and Evans, Chapter 12, this volume).

It also worth reiterating that current EU legislation would not facilitate the introduction of a weed biological control agent, and in the case of an exotic fungus, this option is more or less excluded by EU Directive 91/414 (1991) (see Cock and Seier, Chapter 16, this volume; Seier, 2005). It is important that the new EU Concerted Action Project, REBECA, starting in 2006, should address this issue, to facilitate weed biological control in Europe in the future. However, for *H. mantegazzianum*, we conclude that the introduction of exotic biological control agents is unlikely to be the solution, although it should be considered for other alien invasive plants in Europe (Sheppard *et al.*, 2006).

Conclusions: Master of all Traits

Plant ecological determinants of giant hogweed's invasiveness are simple and easy to determine. The species does not seem to possess any special characteristic/mechanism; extremely high fecundity, rapid growth, capability of self-pollination, extended germination period by means of short-term persistent seed bank, high germination, negligible impact of natural enemies – all these characteristics can be found in other plant invaders (Pyšek and Richardson, 2007).

A recent review of species traits associated with invasiveness identified, among others, fecundity, rapid growth and associated physiological measures, height, resistance to herbivory, early germination and flowering and persistent seed bank as important (Pyšek and Richardson, 2007). This volume illustrates that *H. mantegazzianum* has many such attributes (Fig. 19.5) and some of them compensate for the lack of some others typical of invasive species. For example, vigorous vegetative spatial growth is consistently recognized as an attribute of invasiveness (Callaway and Josselyn, 1992; Vila and D'Antonio, 1998; Larson, 2000; Morris *et al.*, 2002). However, the main advantage provided by this ability is efficient and rapid space pre-emption, which in

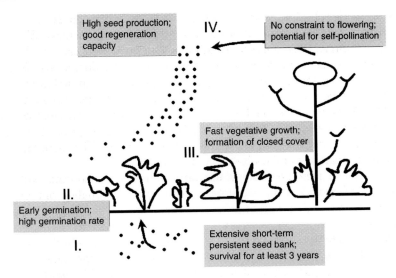

Fig. 19.5. Traits identified as contributing to the successful invasion by *H. mantegazzianum*, plotted on the scheme of the life cycle of a monocarpic plant: I – soil seed bank; II – stage of seedling recruitment; III – growth phase; IV – the terminal phase of seed production (stages according to Harper, 1977). Drawing: J. Pergl.

H. mantegazzianum is achieved by rapid growth of large leaves above the surrounding vegetation and the production of large numbers of vigorously growing offspring in the vicinity of fruiting plants.

Invasions by some plant species are facilitated by special mechanisms resulting from their specific biochemical features, e.g. substances resulting in a high flammability that changes the fire regime in invaded regions (D'Antonio and Vitousek, 1992). *H. mantegazzianum* has two different defence mechanisms: glandular trichomes and a phototoxic sap containing furanocoumarins that protect the plants against vertebrates, invertebrates, fungi, bacteria and viruses (Hattendorf *et al.*, Chapter 13, this volume). However, the role both mechanisms may play in facilitating the invasion by *H. mantegazzianum* in Europe is unclear. The resource-costly defence by glandular trichomes is less pronounced in the invaded area, thus releasing the plant from some defence costs. This would enable the invading plant populations to invest more in growth. The second defence mechanism by less costly qualitative toxic compounds did not change as suggested by the hypothesis of Blossey and Nötzold (1995). *H. mantegazzianum* showed higher levels in the invaded range of Europe even though no damaging specialized herbivores occur there. The plant appears to invest more in biochemical defence than necessary and the reason for this is not clear. However, we know that the levels and types of furanocoumarins in this species are variable, and this may reflect the original stock from which the introductions were made.

Therefore, it is a combination of superior traits associated with a single species and acting at different stages of the life cycle (Fig. 19.5) that provides

H. mantegazzianum with remarkable invasion potential and makes it a 'master-of-all-traits' of plant invasions. This has practical implications, as the species does not seem to have a weak link in its life cycle, on which the control measures could be most efficiently targeted. Appropriately conducted long-term mechanical and chemical control (Nielsen *et al.*, Chapter 14 and Pyšek *et al.*, Chapter 7, this volume) associated with suitable landscape management and revegetation schemes (Ravn *et al.*, Chapter 17, this volume) can be used, with reasonable success, to reduce the extent of invasion in heavily affected areas and prevent the species from further spread. It is unlikely, however, that the species can be completely eradicated by classical means of control. So the answer to the question raised in the title of this chapter is – yes, we can fight giant hogweed with some success but, for now … giant hogweed lives.

Acknowledgements

The study was supported by the project 'Giant Hogweed (*Heracleum mantegazzianum*) a pernicious invasive weed: developing a sustainable strategy for alien invasive plant management in Europe', funded within the 'Energy, Environment and Sustainable Development Programme' (grant no. EVK2-CT-2001-00128) of the European Union 5th Framework Programme. The authors thank all participants of the project and Irena Perglová and Jan Pergl for the comments on the manuscript and technical assistance. P.P. was supported by institutional long-term research plan no. AV0Z60050516 of the Academy of Sciences of the Czech Republic, and no. 0021620828 of the Ministry of Education of the Czech Republic, and by the grant of the Biodiversity Research Center no. LC06073.

References

Blossey, B. and Nötzold, R. (1995) Evolution of increased competitive ability in invasive non-indigenous plants: a hypothesis. *Journal of Ecology* 83, 887–889.

Callaway, J.C. and Josselyn, M.N. (1992) The introduction and spread of smooth cordgrass (*Spartina alterniflora*) in South San Francisco Bay. *Estuaries* 15, 218–225.

Campbell, G.S., Blackwell, P.G. and Woodward, F.I. (2002) Can landscape-scale characteristics be used to predict plant invasions along rivers? *Journal of Biogeography* 29, 535–543.

Cronk, Q.C.B. and Fuller, J.L. (1995) *Plant Invaders: The Threat to Natural Ecosystems*. Chapman and Hall, London.

D'Antonio, C.M. and Vitousek, P.M. (1992) Biological invasions by exotic grasses, the grass fire cycle, and global change. *Annual Review of Ecology and Systematics* 23, 63–87.

Fuller, R.M. and Boorman, L.A. (1977) The spread and development of *Rhododendron ponticum* L. on dunes at Winterton, Norfolk, in comparison with invasion by *Hippophaë rhamnoides* L. at Saltfleetby, Lincolnshire. *Biological Conservation* 12, 83–94.

Harper, J.L. (1977) *Population Biology of Plants*. Academic Press, London.

Krinke, L., Moravcová, L., Pyšek, P., Jarošík, V., Pergl, J. and Perglová, I. (2005) Seed bank of an invasive alien, *Heracleum mantegazzianum*, and its seasonal dynamics. *Seed Science Research* 15, 239–248.

Larson, K.C. (2000) Circumnutation behavior of an exotic honeysuckle vine and its native congener: influence on clonal mobility. *American Journal of Botany* 87, 533–538.

Lowe, S., Browne, M., Boudjelas, S. and De Poorter, M. (2000) *100 of the World's Worst Invasive Species: A Selection from the Global Invasive Species Database*. World Conservation Union, Auckland, New Zealand.

Meyer, U., Meinesz, A. and de Vaugelas, J. (1998) Invasion of the accidentally introduced tropical alga *Caulerpa taxifolia* in the Mediterranean Sea. In: Starfinger, U., Edwards, K., Kowarik, I. and Williamson, M. (eds) *Plant Invasions: Ecological Mechanisms and Human Responses*. Backhuys, Leiden, The Netherlands, pp. 225–234.

Moravcová, L., Perglová, I., Pyšek, P., Jarošík, V. and Pergl, J. (2005) Effects of fruit position on fruit mass and seed germination in the alien species *Heracleum mantegazzianum* (*Apiaceae*) and the implications for its invasion. *Acta Oecologica* 28, 1–10.

Morris, L.L., Walck, J.L. and Hidayati, S.N. (2002) Growth and reproduction of the invasive *Ligustrum sinense* and native *Forestiera ligustrina* (*Oleaceae*): implications for the invasion and persistence of a nonnative shrub. *International Journal of Plant Sciences* 163, 1001–1010.

Müllerová, J., Pyšek, P., Jarošík, V. and Pergl, J. (2005) Aerial photographs as a tool for assessing the regional dynamics of the invasive plant species *Heracleum mantegazzianum*. *Journal of Applied Ecology* 42, 1042–1053.

Nielsen, C., Ravn, H.P., Nentwig, W. and Wade, M. (eds) (2005) *The Giant Hogweed Best Practice Manual. Guidelines for the Management and Control of an Invasive Alien Weed in Europe*. Forest and Landscape Denmark, Hørsholm, Denmark.

Pergl, J., Perglová, I., Pyšek, P. and Dietz, H. (2006) Population age structure and reproductive behavior of the monocarpic perennial *Heracleum mantegazzianum* (*Apiaceae*) in its native and invaded distribution range. *American Journal of Botany* 93, 1018–1028.

Perglová, I., Pergl, J. and Pyšek, P. (2006) Flowering phenology and reproductive effort of the invasive alien plant *Heracleum mantegazzianum*. *Preslia* 78, 265–285.

Plummer, G.L. and Keever, C. (1963) Autumnal daylight weather and camphor-weed dispersal in the Georgia piedmont region. *Botanical Gazette* 124, 283–289.

Pyšek, P. and Hulme, P.E. (2005) Spatio-temporal dynamics of plant invasions: linking pattern to process. *Ecoscience* 12, 302–315.

Pyšek, P. and Pyšek, A. (1995) Invasion by *Heracleum mantegazzianum* in different habitats in the Czech Republic. *Journal of Vegetation Science* 6, 711–718.

Pyšek, P. and Richardson, D.M. (2007) Traits associated with invasiveness in alien plants: where do we stand? In: Nentwig, W. (ed.) *Biological Invasions*. Springer, Berlin, pp. 97–125.

Pyšek, P., Mandák, B., Francírková, T. and Prach, K. (2001) Persistence of stout clonal herbs as invaders in the landscape: a field test of historical records. In: Brundu, G., Brock, J., Camarda, I., Child, L. and Wade, M. (eds) *Plant Invasions: Species Ecology and Ecosystem Management*. Backhuys, Leiden, pp. 235–244.

Pyšek, P., Sádlo, J., Mandák, B. and Jarošík, V. (2003) Czech alien flora and a historical pattern of its formation: what came first to Central Europe? *Oecologia* 135, 122–130.

Pyšek, P., Krinke, L., Jarošík, V., Perglová, I., Pergl, J. and Moravcová, L. (2007) Timing and extent of tissue removal affect reproduction characteristics of an invasive species *Heracleum mantegazzianum*. *Biological Invasions* (in press doi 10.107/s10530-006-90380).

Schepker, H. and Kowarik, I. (1998) Invasive North American blueberry hybrids (*Vaccinium corymbosum × angustifolium*) in northern Germany. In: Starfinger, U., Edwards, K., Kowarik, I. and Williamson, M. (eds) *Plant Invasions: Ecological Mechanisms and Human Responses*. Backhuys, Leiden, The Netherlands, pp. 253–260.

Seier, M.K. (2005) Fungal pathogens as classical biological control agents for invasive alien weeds – are they a viable concept for Europe? In: Nentwig, W., Bacher, S., Cock, M.J.W.,

Hansörg, D., Gigon, A. and Wittenberg, R. (eds) *Biological Invasions – from Ecology to Control. Neobiota* 6. Institute of Ecology of the TU Berlin, Berlin, Germany, pp. 165–176.

Sheppard, A.W., Shaw, R.H. and Sforza, R. (2006) Top 20 environmental weeds for classical biological control in Europe: a review of opportunities, regulations and other barriers to adoption. *Weed Research* 46, 93–117.

Tiley, G.E.D., Dodd, F.S. and Wade, P.M. (1996) Biological flora of the British Isles. 190. *Heracleum mantegazzianum* Sommier et Levier. *Journal of Ecology* 84, 297–319.

Vila, M. and D'Antonio, C.M. (1998) Fruit choice and seed dispersal of invasive vs. noninvasive *Carpobrotus* (*Aizoaceae*) in coastal California. *Ecology* 79, 1053–1060.

Wade, M., Darby, E.J., Courtney, A.D. and Caffrey, J.M. (1997) *Heracleum mantegazzianum*: a problem for river managers in the Republic of Ireland and the United Kingdom. In: Brock, J.H., Wade, M., Pyšek, P. and Green, D. (eds) *Plant Invasions: Studies from North America and Europe*. Backhuys, Leiden, The Netherlands, pp. 139–151.

Wania, A., Kühn, I. and Klotz, S. (2006) Plant richness patterns in agricultural and urban landscapes in Central Germany – spatial gradients of species richness. *Landscape and Urban Planning* 75, 97–110.

Weber, E. (1998) The dynamics of plant invasions: a case study of three exotic goldenrod species (*Solidago* L.) in Europe. *Journal of Biogeography* 25, 147–154.

Weber, E. (2003) *Invasive Plant Species of the World: A Reference Guide to Environmental Weeds*. CABI Publishing, Wallingford, UK.

Williamson, M. (1998) Measuring the impact of plant invaders in Britain. In: Starfinger, U., Edwards, K., Kowarik, I. and Williamson, M. (eds) *Plant Invasions: Ecological Mechanisms and Human Responses*. Backhuys, Leiden, The Netherlands, pp. 57–68.

Williamson, M., Pyšek, P., Jarošík, V. and Prach, K. (2005) On the rates and patterns of spread of alien plants in the Czech Republic, Britain and Ireland. *Ecoscience* 12, 424–433.

Index